AF342434

Springer Series in Materials Science

Volume 260

Series Editors

Robert Hull, Center for Materials, Devices, and Integrated Systems, Rensselaer Polytechnic Institute, Troy, NY, USA

Chennupati Jagadish, Research School of Physics and Engineering, Australian National University, Canberra, ACT, Australia

Yoshiyuki Kawazoe, Center for Computational Materials, Tohoku University, Sendai, Japan

Jamie Kruzic, School of Mechanical & Manufacturing Engineering, UNSW Sydney, Sydney, NSW, Australia

Richard M. Osgood, Department of Electrical Engineering, Columbia University, New York, USA

Jürgen Parisi, Universität Oldenburg, Oldenburg, Germany

Udo W. Pohl, Institute of Solid State Physics, Technical University of Berlin, Berlin, Germany

Tae-Yeon Seong, Department of Materials Science & Engineering, Korea University, Seoul, Korea (Republic of)

Shin-ichi Uchida, Electronics and Manufacturing, National Institute of Advanced Industrial Science and Technology, Tsukuba, Ibaraki, Japan

Zhiming M. Wang, Institute of Fundamental and Frontier Sciences - Electronic, University of Electronic Science and Technology of China, Chengdu, China

The Springer Series in Materials Science covers the complete spectrum of materials research and technology, including fundamental principles, physical properties, materials theory and design. Recognizing the increasing importance of materials science in future device technologies, the book titles in this series reflect the state-of-the-art in understanding and controlling the structure and properties of all important classes of materials.

More information about this series at https://link.springer.com/bookseries/856

Tian-You Fan · Wenge Yang · Hui Cheng ·
Xiao-Hong Sun

Generalized Dynamics of Soft-Matter Quasicrystals

Mathematical Models, Solutions and Applications

Second Edition

 Springer

Tian-You Fan
School of Physics
Beijing Institute of Technology
Beijing, China

Hui Cheng
Hebei University of Engineering
Handan, China

Wenge Yang
Center for High Pressure Science
and Technology Advanced Research
Shanghai, China

Xiao-Hong Sun
Zhengzhou University
Zhengzhou, China

ISSN 0933-033X   ISSN 2196-2812   (electronic)
Springer Series in Materials Science
ISBN 978-981-16-6627-8   ISBN 978-981-16-6628-5   (eBook)
https://doi.org/10.1007/978-981-16-6628-5

This Springer imprint is published by the registered company Springer Nature Singapore Pte Ltd.
The registered company address is: 152 Beach Road, #21-01/04 Gateway East, Singapore 189721,
Singapore

Preface to the Second Edition

Due to limited experimental data reported since the publication of the first edition, the study on generalized dynamics of soft-matter quasicrystals has not achieved great progress. In this second edition, we still mainly focus on the mathematical model but provide some applications additional to the first edition.

Nevertheless, there are still some steady progress in the experimental study in the field. The observation of the 10-fold symmetry soft-matter quasicrystal is an example,[1] the importance of which will be beyond that of the first discovered 12-fold one, although the details have not been openly discussed yet. This finding extended the presence of soft-matter quasicrystals in a broader range and would bring fruitful expectations in both experimental and theoretical studies.

Another progress is the experimental and theoretical study on the correlations between the Frank-Kasper phase and soft-matter quasicrystals in giant molecules. This contribution was made by the group of Prof. Stephen Z. D. Cheng and will be introduced in Chap. 2 in the second edition.

In theoretical aspects, we will also introduce the recent progress.

In this circumstance, we strengthen the discussions on possible applications of the theory and the method in new areas, for example, thermodynamic stability, three-dimensional problems, device physics, liquid crystals, general soft matter, etc., which will be introduced in Chaps. 13–16. Of course, the applications are in the preliminary phase.

Thanks to the suggestions from the readers, we paid more attention to supplement recent computational models and simulation results in the second edition to visualize the theoretical formulas. In addition, although errors and mistakes in the first edition have been corrected, there might still be errors and mistakes. The authors are grateful to any criticism from the readers so that we can continuously improve the book.

---

[1] The 10-fold symmetry quasicrystals in soft matter will be reported by Expanding Quasiperiodicity in Soft Matter: Decagonal Quasicrystals by Hierarchical Packing Frustrations. Authors: Yuchu Liu, Tong Liu, Xiaoyun Yan, Qing-Yun Guo, Huanyu Lei, Zongwu Huang, Rui Zhang, Yu Wang, Jing Wang, Feng Liu, Feng-Gang Bian, E. W. Meijer, Takuzo Aida, Mingjun Huang, Stephen Z. D. Cheng, Proc. Natl. Acad. Sci., in press, 2021.

The first author thanks the National Natural Science Foundation of China and the Alexander von Humboldt Foundation of Germany for their support over the last decades. The work is also partially supported by the School of Physics, Beijing Institute of Technology.

We thank Professors T. C. Lubensky from the University of Pennsylvania, the USA, Stephen Z. D. Cheng from the University of Akron, the USA and South China University of Technology, Yu-Gui Yao from Beijing Institute of Technology, Wei-Qiu Chen from Zhejiang University, Xian-Fang Li from Central South University, Ming-Jun Huang from South China University of Technology, China and C. Peng from the University of Memphis, the USA for their kind advice and helpful discussions.

The authors are grateful to researcher in metallic materials Prof. Zi-Tong Li, former student Fang Wang from Beijing Institute of Technology for their hard work during the preparation of this second edition.

Due to their important contributions to the second edition, Wenge Yang, Hui Cheng, and Xiao-Hong Sun are invited as co-authors to take charge of the book with me.

Beijing, China   Tian-You Fan
December 2021

Preface to the First Edition

Since 2004, quasicrystals have been discovered in various kinds of soft matters, including liquid crystals, colloids, polymers, and nanoparticles. In particular, 18-fold symmetry quasicrystals in colloids were observed in 2011. More recently the quasicrystals with 12-fold symmetry were also found in surfactants. The formation mechanisms of these kinds of quasicrystals are associated closely with the self-assembly of spherical building blocks by supramolecules, compounds, and block copolymers, and so on, which is quite different from that of the metallic alloy quasicrystals. They can be identified as soft-matter quasicrystals, exhibiting natures of quasicrystals with soft-matter characters. Soft-matter behavior is between solid and simple fluid, while the quasicrystals form in highly ordered structures with crystalline-forbidden symmetry. These features are very complex yet extremely interesting and attractive. Since 2004, soft-matter quasicrystal studies have attracted attention in mathematics, physics, chemistry, and materials science.

So far all observed soft-matter quasicrystals are two-dimensional quasicrystals. It is well known that two-dimensional quasicrystals consist of only two distinct types, one presents 5-, 8-, 10-, and 12-fold symmetries, the other 7-, 9-, 14-, and 18-fold ones according to the group symmetry theory. Therefore, two terminological phrases can be defined such as the first and second kinds of two-dimensional quasicrystals respectively. The two-dimensional solid quasicrystals observed so far only belong to the first kind, while soft-matter quasicrystals discovered up to now belong to both kinds. It is likely other soft-matter quasicrystals beyond 12- and 18-fold symmetries could be found. Therefore, it is quite important to develop a suite of general principles to describe elasticity and fluid dynamics.

It is quite difficult to study these new phases due to the complexity of their structures and limited experimental data including the basic physical constants. It is true for theoretical studies as well. For example, the symmetry groups of soft-matter quasicrystals have yet been thoroughly investigated, and only some preliminary work has been reported so far (the details are not included in the book). In conjunction with this issue, the study on constitutive laws for phason and phonon-phason coupling is still difficult.

Despite only very limited experimental and theoretical studies reported so far, there is great potential to push these directions forward. For example, the soft-matter quasicrystals as a new ordered phase are connected with symmetry breaking, like those discussed in solid quasicrystals. Thus, elementary excitations such as phonon and phason are important issues in the study of quasicrystals based on the Landau phenomenological theory. For soft-matter quasicrystals, furthermore, another elementary excitation—fluid phonon will be considered besides phonon and phason. According to the Landau school, the liquid acoustic wave is a fluid phonon (refer to Lifshitz E M and Pitaevskii L P, Statistical Physics, Part 2, Pergamon, Oxford, 1980). This is suitable for describing the liquid effect of soft matter, which can be seen as complex liquids or structured liquids. The elementary excitations—phonon, phason, and fluid phonon and their coupling terms constitute the main feature of these new phases. They will be discussed as a major issue through this book. The fluid phonon was first introduced in the quasicrystals study. Correspondingly, the equation of state should also be introduced in soft-matter quasicrystal study. Combining these two key concepts with the hydrodynamics principle established in solid quasicrystals, the dynamics of soft-matter quasicrystals can be constructed. For solid quasicrystals, there has been tremendous progress over last decades, for example, Lubensky TC, Symmetry, elasticity, and hydrodynamics in quasiperiodic structures, in Introduction to Quasicrystals, ed by Jaric MV, Boston: Academic Press, 190-289,1988; Hu ZC et al, Symmetry groups, physical property tensors, elasticity and dislocations in quasicrystals, Rep. Prog. Phys., 63(1), 1-39, 2000; Fan T Y, Mathematical Theory of Elasticity of Quasicrystals and Its Applications, Beijing: Science Press/Heidelberg, Springer-Verlag, 1st edition, 2010, 2nd edition 2016. Based on the development of the theory of solid quasicrystals, we will extend the quantitative analysis into the rich phenomena of soft-matter quasicrystals.

Some applications will be discussed on the distribution, deformation, and motion of soft-matter quasicrystals. The mathematical principles and applications required are briefly reviewed in the first six chapters of this book (for more details, refer to Chaikin J and Lubensky TC, Principles of Condensed Matter Physics, New York: Cambridge University Press, 1995). The computational applications on soft-matter quasicrystals are quite preliminary, but they verified partially the mathematical model and explored the distinguished dynamic behavior from solid quasicrystals. In addition, specimens and flow modes adopted in the computation modeling might be intuitive, observable, and can be verified easily.

The author thanks the National Natural Science Foundation of China and Alexander von Humboldt Foundation of Germany for their financial support over the years and Professors U. Messerschmidt from Max-Planck Institut fur Mikrostruktur-physik in Halle, H.-R. Trebin from Stuttgard Universitaet in Germany, T. C. Lubensky from the University of Pennsylvania, Stephen Z. D. Cheng from the University of

Akron in the USA, H. H. Wensink from Utrecht University in Netherland, Xian-Fang Li from Central South University in China, and Wei-Qiu Chen from Zhejiang University in China for their cordial encouragement and helpful discussions.

Beijing, China
December 2016

Tian-You Fan

Contents

Notations

$\mathbf{r}$	Radius vector
D	Domain
S	Boundary of domain
S_u	Boundary part where the displacements are given
S_t	Boundary part where the tractions are given (or S_σ where the applied stresses are given)
ρ	Mass density (g/cm^3)
p	Fluid pressure (Pa = N/m^2)
$\mathbf{u}$	Phonon type displacement field (cm)
$\mathbf{w}$	Phason type displacement field (or second phason displacement field for second kind quasicrystals with 7-, 9-, 14-, 18-fold symmetry) (cm)
$\mathbf{V}$	Fluid velocity field (or fluid phonon field) (cm/s)
$\varepsilon_{ij} = \frac{1}{2}\left(\frac{\partial u_i}{\partial x_j} + \frac{\partial u_j}{\partial x_i}\right)$	Phonon strain tensor
$w_{ij} = \frac{\partial w_i}{\partial x_j}$	Phason strain tensor (or second phason strain tensor for second kind quasicrystals with 7-, 9-, 14-, 18-fold symmetry)
$\dot{\xi}_{ij} = \frac{1}{2}\left(\frac{\partial V_i}{\partial x_j} + \frac{\partial V_j}{\partial x_i}\right)$	Fluid phonon deformation tensor (1/s)
σ_{ij}	Phonon stress tensor (Pa)
H_{ij}	Phason stress tensor (or second phason stress tensor for second kind quasicrystals with 7-, 9-, 14-, 18-fold symmetry) (Pa)
σ'_{ij}	Viscous stress tensor (Pa)
$p_{ij} = -p\delta_{ij} + \sigma'_{ij}$	Fluid phonon stress tensor (Pa)
C_{ijkl}	Phonon elastic coefficient tensor (Pa)
K_{ijkl}	Phason elastic coefficient tensor (or second kind phason elastic coefficient tensor for quasicrystals with 7-, 9-, 14-, 18-fold symmetry) (Pa)
R_{ijkl}	Phonon-phason coupling elastic coefficient tensor (u–w coupling elastic coefficient tensor) (Pa)

η	First viscosity coefficient of fluid ($0.1\,\mathrm{Pa}\cdot\mathrm{s} = \mathrm{Poise}$)
η/ρ	First kinetic viscosity coefficient of fluid (cm^2/s)
ζ	Second viscosity coefficient of fluid ($0.1\,\mathrm{Pa}\cdot\mathrm{s} = \mathrm{Poise}$)
ζ/ρ	Second kinetic viscosity coefficient of fluid (cm^2/s)
Γ_u	Phonon dissipation coefficient ($\mathrm{m}^3\cdot\mathrm{s/kg}$)
Γ_w	Phason dissipation coefficient (or second kind phason dissipation coefficient tensor for quasicrystals with 7-, 9-, 14-, 18-fold symmetry) ($\mathrm{m}^3\cdot\mathrm{s/kg}$)
$\mathbf{v}$	First phason type displacement field (only for second kind quasicrystals) (cm)
$v_{ij} = \dfrac{\partial v_i}{\partial x_j}$	First phason strain tensor (only for second kind quasicrystals)
τ_{ij}	First phason stress tensor (only for second kind quasicrystals) (Pa)
r_{ijkl}	Phonon-first phason coupling elastic coefficient tensor (or u–v coupling elastic coefficient tensor only for second kind quasicrystals) (Pa)
Γ_v	The first kind phason dissipation coefficient of quasicrystals ($\mathrm{m}^3\cdot\mathrm{s/kg}$)

Chapter 1
Introduction to Soft Matter

Soft-matter quasicrystals are observed in liquid crystals, colloids, polymers, and surfactants, etc., which brings new family members to soft matter with crystallographic forbidden symmetry. Soft matter is a type of common material, introduced by de Gennes [1] in 1991, including liquid crystals, colloids, polymers, foams, emulsions, surfactants, biomacromolecules, etc. They are neither ideal solid nor simple fluid, but presents characteristics of both solid and fluid, and belongs to an intermediate phase between isotropic fluid and ideal solid macroscopically. Sometimes one calls them anisotropic fluids, structured fluids, or complex fluids [2–5], more exactly speaking, as anisotropic liquids, structured liquids, or complex liquids.

As pointed out by Guo [6], if every atom of a molecule possesses the thermal energy $k_B T$ in an ideal solid, e.g., solid crystal, here k_B being the Boltzmann constant, T the absolute temperature, the thermal energy per unit volume $k_B T / l_0^3$, may characterize an entropy state of the crystal, here $l_0 \sim 0.1$ nm the typical lattice size or interatomic distance. For soft-matter systems, the structure and dynamic properties are related to mesoscopic size $l \sim 10 - 100$ nm (e.g., the size of the long-chain of polymers, or the size of self-assembly structures, etc.). Fluctuation, thermal motion, and self-organization or self-assembly are often induced by entropy with thermal energy per unit volume $k_B T / l^3$. Apparently, at room temperature, the thermal energy per unit volume of soft matter is lower by 6–9 orders of magnitude than that of the ideal crystals. This may explain the softness of soft matter from the perspective of the intra-structure of materials. In contrast, the ideal solid presents very high stiffness. The distinction between soft matter and ideal solid is significant. The thermal energy per unit volume concept may provide a basis by some analogies between soft matter and ideal solid. The other differences between soft matter and conventional materials will be discussed in the following description, but won't be elaborated in detail or in depth here.

For simplicity, we here only consider hydrodynamics, or generalized dynamics, of soft-matter quasicrystals. More strictly speaking, only the fluidity or the flow effect from the perspective of fluid is considered apart from elasticity and interaction between fluidity and elasticity of the matter. The fluidity, elasticity, and their

T.-Y. Fan et al., *Generalized Dynamics of Soft-Matter Quasicrystals*, Springer Series in Materials Science 260, https://doi.org/10.1007/978-981-16-6628-5_1

interaction are the only portions of the behavior of soft matter, which would help us to understand the distribution, deformation, and motion of soft-matter quasicrystals in a macroscope. In this case, the micro-scale structures of the matter have not been concerned. Although the mesoscale structures are important for soft matter, it has not been considered in general in our presentation apart from some special exceptions. In this sense, the modeling on hydrodynamics or generalized dynamics of soft matter and soft-matter quasicrystals is a macro-and continuum-medium-study, with low-frequency and long-wavelength characteristics, which have been discussed in solid quasicrystals, and will be extended to the soft-matter case.

Among various kinds of soft-matter systems, liquid crystals are typical and relatively well studied, and their material constants have been reported more in detail. From a macroscopic and continuum point of view, a similar process from liquid crystals can be extended to the study of soft-matter quasicrystals. For example, the generalized Newton's fluid law can approximately be used in some cases, and the generalized Hooke's elasticity law can also be applied, but the deformation in soft-matter quasicrystals is more complex. The deformation of liquid crystals consists of bulk deformation and local curvature variation. For the bulk deformation the conventional generalized Hooke's law still holds, but the deformation induced by curvature variation needs additional terms to be included, which are beyond the discussion in this chapter, and we will return to this issue in Chap. 16 where we discuss the curvature of smectic A liquid crystals. As an intermediate phase between simple fluid and ideal solid, the soft matter presents many behavior differences from those of isotropic liquids and ideal crystals. For example, in ordinary liquid and nematic liquid crystals, there is only one acoustic wave, i.e., longitudinal sound wave. In solid crystals and amorphous solids, there are three acoustic wave speeds under the linear deformation, i.e., $c_1 = \sqrt{\frac{\lambda+2\mu}{\rho}}$ or $\left(or \quad c_1 = \sqrt{\frac{L+2M}{\rho}}\right)$, $c_2 = c_3 = \sqrt{\frac{\mu}{\rho}}\left(or \quad c_2 = c_3 = \sqrt{\frac{M}{\rho}}\right)$, as discussed in Chaps. 7–11 of this book. Smectic A liquid crystal has only one nonzero displacement component, the longitudinal shear state in the elastic deformation, so it is often categorized as a one-dimensional crystal. For a pure solid, the acoustic wave speed is $\sim \sqrt{\frac{E}{\rho}}$ where E is the elastic modulus, ρ the mass density; For a pure fluid, the acoustic speed is $\sim \sqrt{\frac{\partial p}{\partial \rho}}$, p the fluid pressure. For smectic A liquid crystals, there are both acoustic wave speeds $\sim \sqrt{\frac{E}{\rho}}$ and $\sim \sqrt{\frac{\partial p}{\partial \rho}}$, where the first speed often depends on the angle between wave vector $\mathbf{k}$ and the normal vector $\mathbf{n}$ of the layer of the smectic A liquid crystal. In general, soft matter behaves differently from simple fluid and ideal solid. Due to complicated nonlinear behavior, the spectra and dispersion relations of soft matter cannot be easily determined, so as the wave speeds. Often one introduces $\sqrt{\frac{\lambda+2\mu}{\rho}}, \sqrt{\frac{\mu}{\rho}}$ and $\sqrt{\frac{\partial p}{\partial \rho}}$ (in some cases we denote $\sqrt{\frac{\partial p}{\partial \rho}} = c_4$ for simplicity) to describe wave propagating speeds in the soft matter as a coarse approximation, and the realistic wave speeds in the matter present quite different in magnitude and nature, and so far the relevant mechanism is not yet clear. The successful introduction to the computation can partly reveal these questions.

For liquid crystals, the dynamic viscosity coefficient η (in the unit 1 Poise $=$ 0.1 Pa s) is introduced to describe the fluid effect. Sometimes, the kinetic viscosity coefficient η/ρ also is used, note that the unit is cm^2/s rather than Poise. The elastic modulus E (in the unit 10^8 erg/cm$^3 = 10^7$ Pa $= 10$ MPa) is used to describe the bulk deformation. The Poisson ratio ν may be negative unlike that for solid. These basic material constants are fundamental and useful for us to deal with the mechanical problems of soft-matter quasicrystals in the late chapters. The viscosity of liquid crystals is quite large (about 100 times that of water), and liquid crystals present a certain degree of elasticity behavior. In general, people do not consider elasticity in simple fluid, and viscosity in ideal solid (at least they are not so important). In the following chapters, we carry out the analysis and computation on distribution, deformation, and motion of soft-matter quasicrystals following the successful experiences from the study of the liquid crystal. Apart from these, some physical constants, e.g., the phonon dissipation coefficient Γ_u and phason dissipation coefficient Γ_w for soft-matter quasicrystals are not available from experimental reports, so we just take relevant values from solid quasicrystals [7] as references.

Another important feature of motion in soft matter is its small Reynolds number Re. According to the definition, Re $= \frac{\rho U a}{\eta}$, where a represents the characteristic size of the matter or flow field. Because the characteristic velocity U is small and the viscosity coefficient η is large, in general, the Reynolds number is small, i.e., Re $= 10^{-4} \sim 1$. In this case, the force due to viscosity is much larger due to inertia. Omitting the inertia terms, we can take the Stokes assumption in the equation of motion sometimes, like in the classical fluid dynamics. This simplifies the governing equations which are still very complicated. Not like in classical fluid dynamics where one has obtained quite a lot of approximate analytic solutions, the analytic solution or even approximate analytic solution cannot be derived. Although the governing equations in classical fluid dynamics are complex, they are much simpler than those in generalized dynamics of soft matter. It should be pointed out that the Stokes approximation in a two-dimensional case leads to the famous Stokes paradox— there was no solution. Oseen [8] analyzed the Stokes paradox physically in depth. To overcome the paradox, the Stokes approximation equations must be modified. They should be replaced by Oseen approximation equations and yield reasonable solutions in the two-dimensional case. Further discussion on this issue can be found in Sommerfeld [9], Sleozkin [10], and Kochin et al. [11]. When we discuss the soft matter dynamics, especially the two-dimensional problems, we will get in touch with similar problems, and the Oseen theory provides a useful guideline. Note that the Ref. [10] points out Oseen approximation holds for the cases Re $<$ 10, which is helpful for the study of soft matter.

In addition, the above introduction regarding the soft matter is very limited and preliminary, which only provides the most elementary knowledge for presentation and application in the current chapter. Readers are suggested to refer to monographs [2–5] for a broader understanding of soft matter and related research findings. The generalized dynamics and possible generalized dynamics will be introduced in the subsequent chapters.

References

1. de Gennes, P.D.: Soft matter. Mod. Phys. Rev. **64**,544–548; Angw Chem, **31**, 842–845 (1992)
2. Witten, T.A., Pincus, P.A.: Structured Fluids: Polymers Surfactants. Oxford University Press, New York, Colloids (2004)
3. Kleman, M.: Soft Matter Physics: An Introduction. Springer, Berlin (2003)
4. Motiv, M.: Sensitive Matter: Foams Liquid Crystals and Other Materials. Harvard University Press, New York, Gels (2010)
5. Israelachvili, N.J.: Intermolecular and Surface Forces. Academic Press, New York (2010)
6. Guo, H.X.: Coarse graining model of polymers. In: Computer Simulation of Soft Matter and Theoretical Methods, Chemical Industry Press, Beijing, in Chinese (2010)
7. Fan, T.Y.: Mathematical Theory of Elasticity of Quasicrystals and Its Applications. Science Press, Beijing, Springer-Verlag, Heidelberg, 1st edn (2010), 2nd edn (2016)
8. Oseen, C.W.: Ueber die Stokes'sche Formel und ueber eine verwandte Aufgabe in der Hydrodynamik, Ark Math Astronom Fys **6**(29), (1910); Oseen, C.W.: Neuere Methoden und Ergibnisse in der Hydrodynamik, AkademischeVerlagsgesellschaft, Leipzig (1927)
9. Sommerfeld, A.: Vorlesungen ueber theoretische Physik, Band II, Mechanik der deformierbaren Medien. Verlag Harri Deutsch, Thun. Frankfurt/M. (1992)
10. Sleozkin, N.A.: Incompressible Viscous Fluid Dynamics. Gostehizdat Press, Moscow, in Russian (1959)
11. Kochin, N.E., Kibel'iI, A., Roze, N.V.: Theoretical Hydrodynamics. Government Press of Phys-Math Literature, Moscow, in Russian (1953)

Chapter 2
Discovery of Soft-Matter Quasicrystals and Their Properties

2.1 Experimental Observation of Quasicrystalline Phases in Soft Matter

Quasicrystals have long-range orientational order but no translational symmetry. As a consequence, sharp diffraction spots can occur but are unable to be described by 230 crystallographic space groups in both real and reciprocal spaces. There are three types of quasicrystals: one-, two- and three-dimensional quasicrystals. In one-dimensional quasicrystals, the quasiperiodic arrangement of atoms is along one direction, while the plane perpendicular to which has a regular two-dimensional periodic arrangement. There are several sub-classes of one-dimensional quasicrystals. One typical arrangement along the unique quasiperiodic direction follows a Fibonacci sequence. In two-dimensional quasicrystals, there is a quasiperiodic two-dimensional plane with a periodic arrangement perpendicular to this plane, resulting in a true layer structure within which no transitional symmetry exists. Typical two-dimensional quasicrystals include pentagonal, octagonal, decagonal, and dodecagonal quasicrystals with 5-, 8-, 10- and 12-fold symmetry respectively. In the three-dimensional quasicrystals, the atomic arrangement is quasiperiodic in all three directions, in which the icosahedral quasicrystal is the typical one. Three independent vectors used in traditional crystallography are not enough to index the diffraction peaks in quasicrystals, instead, at least four linearly independent vectors are needed [1]. The necessary n vectors span independently in n $(n > 3)$ dimensional space. In other words, the quasicrystals in three-dimensional space can be constructed from a periodic crystal in a higher n dimensional space. The real structure of quasicrystal in three-dimensional physical space can be obtained by appropriate projection/section technique preserving the symmetries from n dimensional space [2]. Five miller indices are needed for describing a two-dimensional polygonal quasicrystal and six miller indices for three-dimensional icosahedral quasicrystal.

Before the discovery of alloy quasicrystals, Roger Penrose created a set of prototiles to tile a plane quasi-periodically with a strict marching rule for preserving the fivefold rotational symmetry. Later on, this famous Penrose tiling motivated much

T.-Y. Fan et al., *Generalized Dynamics of Soft-Matter Quasicrystals*, Springer Series in Materials Science 260, https://doi.org/10.1007/978-981-16-6628-5_2

work on the theory of quasicrystals. The third type of Penrose tiling (P3) is composed of two types of rhombi tiles with different angles. An equivalent Penrose tiling can also be constructed in three-dimensional space, resulting in icosahedral symmetry. The fivefold Penrose tiling is a geometry expression of pentagonal quasicrystals similar to other 2d symmetries including 8-, 10-, and 12-fold, as illustrated in Fig. 2.1a–c for examples [3]. The corresponding lattice vectors in reciprocal space can be obtained through projection from higher dimensional spaces as shown in the middle panel of Fig. 2.1. Four linear independent planar vectors are needed to describe the related rotational symmetry. The simulated diffraction patterns exhibit 8-, 10-, and 12-fold symmetries as shown in the bottom panel of Fig. 2.1. The fifth vector is along with the plan's normal direction.

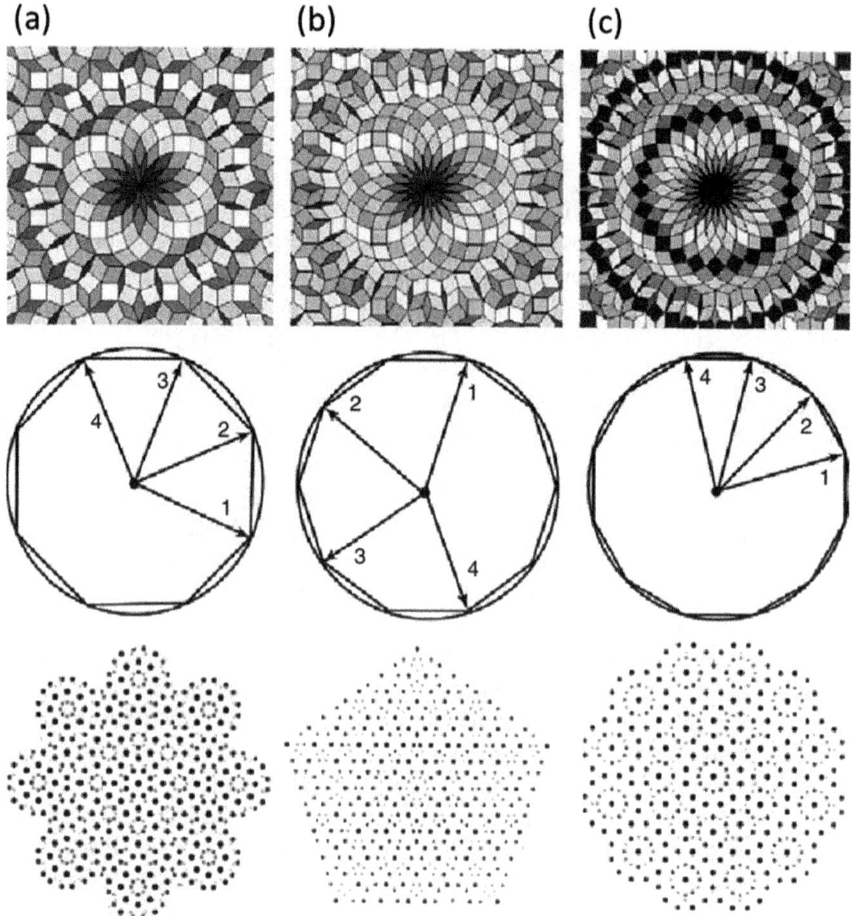

Fig. 2.1 Two-dimensional tilings (top), projection of the unit vectors of the polygonal reciprocal lattices in four-dimensional space (middle) and simulated diffraction patterns (bottom) in the quasiperiodic plane for **a** octagonal, **b** decagonal, and **c** dodecagonal quasicrystals [3]

After the first report of soft-matter quasicrystals by Zeng et al. in 2004 [4], there is rapid development in this field. Especially, the dodecagonal quasicrystalline phase (DDQC) with 12-fold rotational symmetry has been frequently observed in diverse soft-matter systems, such as dendrimers [4], block copolymers [5], colloidal particles [6], and giant molecules [7, 8]. The DDQC phase belongs to two-dimensional quasicrystals and has a strong tie to periodic Frank-Kasper phases. Formation of the Frank-Kasper and quasicrystalline phases results from the subtle competitions between thermodynamic driving forces (both enthalpic and entropic origins) and molecular shapes [9], which might open up a new route to design materials with tailored properties.

At this moment, the physical model of two-dimensional quasicrystals is considered as the combination of framework and decoration atoms [10]. The framework is the tiling pattern on a two-dimensional plane which has been discussed in Fig. 2.1. The dodecagonal tiling pattern proposed in Fig. 2.1c is mainly composed of a rhombus with the same edge length. The rhombus structure is relatively rare in real quasicrystals, and triangles and squares occupy the most space. The ideally quasiperiodic DDQC structure is still not fully determined yet, but many methods have been developed to create a tiling pattern with 12-fold symmetry using only triangles and squares. The deterministic DDQC with triangle-square tiling is presented in Fig. 2.2a generated by the projection method [11]. The 12-fold symmetry of the tiles results from the long-range orientational order of the edges restricted in 12 directions with intervals of $30°$ and equal frequency. Three various tiling methods of $3^2 \cdot 4 \cdot 3 \cdot 4$, $3^3 \cdot 4^2$, and 3^6 can be identified, among which the 3^6 has two different orientations rotated by $30°$ and causes also two orientations for dodecagons consisting of 12 triangles and 6 squares. According to the self-similarity principle, the second generation of dodecagons on a larger scale can be constructed by connecting centers of the smallest dodecagons, illustrated by red lines in Fig. 2.2a. They have $2 + \sqrt{3}$ times larger edge longer than that of original ones. Continuing this process can obtain higher generations. Considering the corresponding areas of equilateral triangle and square, it is amazing to note, squares and triangles each occupy exactly half of the total area [12].

Besides the deterministic quasicrystal tiling, various random tilings can be constructed using the same number of triangles and squares, as an example shown in Fig. 2.2b. In detail, a square-triangle tiling contains exclusively equilateral squares and triangles. The random tiling belongs to an assemble of tiles that satisfy certain boundary conditions and requires long-range bond orientational order with 12-fold diffraction symmetry. These assembles thus have many probabilities from random tiles [12]. A proposed mechanism of random tiling formation is the special local atom site relocation, phason flips, which are related to phase strain (Fig. 2.2c). A pair of rhombi shapes can be created accompanied by the change of energy states in the local sites. Each rhombus then follows A-type, B-type, or bounce phason flips through the square-triangle tiling, in a way analogous to the operation of a zip fastener. This pair of two rhombi would sooner or later annihilate back into a square and triangle, acting the role as a zipper of squares and triangles in their whole life and creating randomness. Why do the quasicrystals always have random tilings? A reasonable answer is minimizing the free energy in random tiling caused by the maximized density of

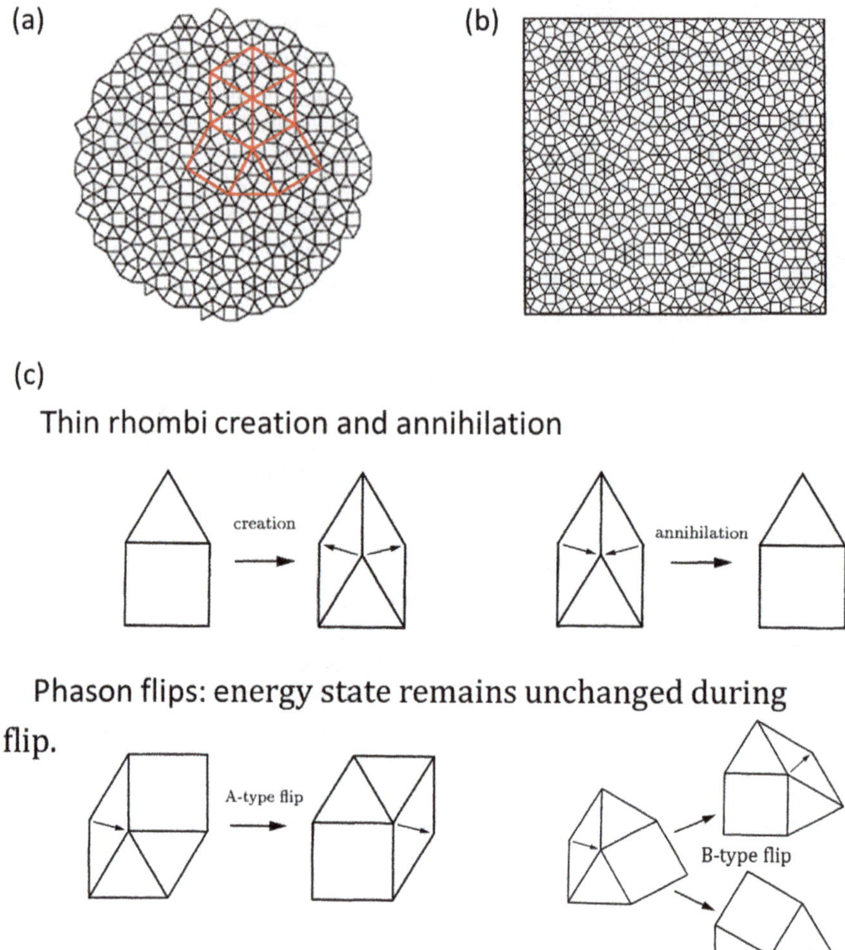

Fig. 2.2 **a** Deterministic dodecagonal quasilattice generated by the projection method [11]. **b** DDQC with random tiling of triangles and squares. **c** Scheme of phason flips in random tiling DQC [12]

entropy. This entropy arises from coordinated atomic relocations, where each such relocation is equivalent to a rearrangement of tiles. In the condition of either infinite temperature or zero Hamiltonian, the maximally random tiling would occur and the phason strain diminishes [13].

In real space DDQC is built from the combination of the random tiling framework and the decoration units, resulting in a layered structure with 12-fold symmetry. To get a better understanding of the real space structure of DDQC, the readers can follow the reference about the binary nanoparticle blends [6]. A quasicrystalline superlattice forms in a binary nanocrystal blend of 6.8 nm $CoFe_2O_4$ and 12.0 nm Fe_3O_4. A three-dimensional structure model was constructed with electron tomography and direct imaging of surface topography. Although the nanocrystals are inorganic, the DDQC superlattice share very similar structural features as those DDQC phases observed in soft matters.

In 2004, Zeng et al. reported the first DDQC structure identified in the soft matter [4]. The single-domain dendrimer (Fig. 2.3a) reveals the distinctive 12-fold symmetry SAXS pattern (Fig. 2.3c). When the sample is oriented with x-ray direction along the 12-fold axis, a different diffraction pattern can be obtained with one periodic direction and this feature can be repeated every 30° along the 12-fold axis (Fig. 2.3c). The full quasicrystal structure is constructed by decoration with three basic tiling units, the square and two triangles (Fig. 2.3b). The simulated diffraction pattern of DQC marked by open circles matched rather well with the experimental one. Note that this quasiperiodic structure exists in the scaled-up micellar phases, representing an unprecedented mode of organization in soft matter.

DDQC phase was later revealed by Bates et al. in diblock or tetrablock copolymers [5], thanks to the almost unlimited structural flexibility of block copolymers

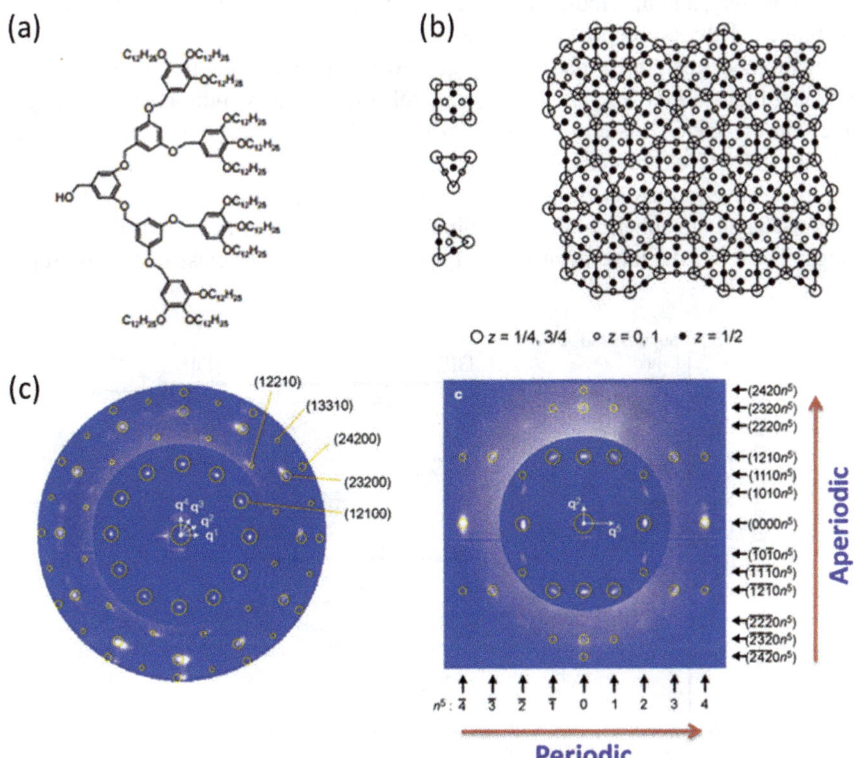

Fig. 2.3 **a** Chemical structure of the dendrimer resulting DQC structure. **b** Three basic decorating tiles for generating the model of the DDQC. **c** Single-domain diffraction pattern along the 12-fold axis (on the left) and diffraction pattern perpendicular to the 12-fold axis [4]

by tuning chemical compositions, molecular mass, and molecular topology. In a sphere (micelle) forming poly (isoprene-*b*-lactide) (IL) diblock copolymer melt, the supercooled disordered state evolves into a metastable DDQC that transforms with time to the σ phase, where the lifetime of the DDQC phase is strongly temperature-dependent. Besides, the DDQC phase can evolve only from the supercooled disordered state. Direct cooling from the high-temperature BCC phase would lead to the Frank-Kasper σ phase, bypassing the DDQC phase [5]. This systematic work strongly suggests that the DDQC structure is a metastable intermediate state before the formation of the equilibrium Frank-Kasper σ phase in many soft-matter systems. Exceptionally, as in metals or alloys, the structure evolution is mediated by charge exchange, the transition in diblock copolymers is mediated by mass exchange via redistribution of diblock copolymer molecules among deformed micelles (Fig. 2.4).

More recently, Cheng et al. observed the DDQC phases in the giant molecular systems, including giant surfactants [7], and giant shape amphiphiles [8]. For example, in a giant surfactant series that has one molecular nanoparticle head and four polystyrene tails, a unique sequence of HEX $\rightarrow$ A15 $\rightarrow$ σ $\rightarrow$ DDQC $\rightarrow$ BCC phase was determined with increasing v_f^{PS} [7]. The characteristic structural features of DDQC were observed in both reciprocal and real spaces.

Interestingly, a nonperiodic random tiling DDQC pattern was discovered in ABC star-branched three-component polymer copolymer systems [13, 14]. The strong segregation between three distinct polymers would drive the junction points aligned into one-dimensional length, resulting in self-organized two-dimensional tilings. Note that the DDQC pattern constructed from ABC type star polymers is distinct from the DDQC we discussed dendrimers, giant surfactants, diblock copolymers, and other soft systems since it's purely two-dimensional tiling by columnar phase domains. Thus, no spherical building blocks are involved in these two-dimensional tilings. These results have been similarly reported in another class of thermotropic

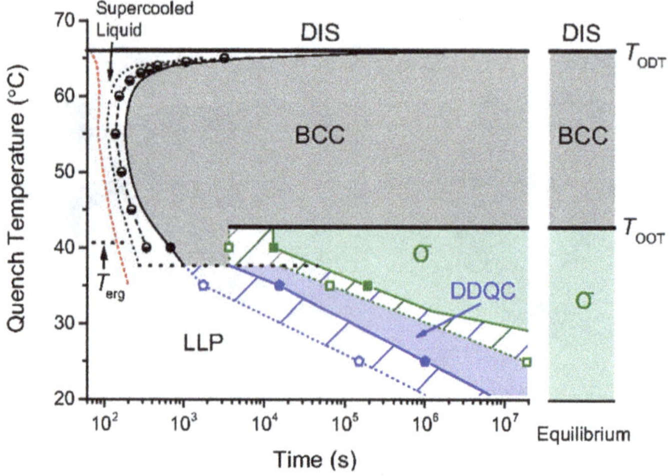

Fig. 2.4 Time–temperature–transformation diagram of poly(1,4-isoprene-*b*-DL-lactide) diblock copolymer (IL 5.1) [5]

columnar liquid crystal phases, which often have the T-shaped bolaamphiphiles and facial amphiphiles [9, 16]. Generally, the aromatic rod-like mesogen cores lie perpendicular to the column axis, constructing the cells of columns that are composed of flexible side groups. The aromatic rod walls take a polygonal tiling in two-dimension.

It's still unknown why the quasicrystal with 12-fold is so dominated in soft matters. Seeking quasicrystal with other symmetries including tenfold or 18-fold would be very intriguing and meaningful to soft-matter study. Here it is worthy to mention that Chen et al. discovered that large-area tenfold quasicrystalline superlattice could self-organize from truncated tetrahedral quantum dots with anisotropic patchiness [17]. Moreover, Foerster et al. reported the quasicrystal phase of PI_{30}-b-PEO_{120} micelles self-assembled in concentrated aqueous solution [18]. In the concentration region between 13 and 18%, an intriguing phase sequence occurred with decreasing temperature: Q18 < 15 °C < Q12 < 25 °C < FCC. Here the Q18 and Q12 represent the quasicrystal phase with 18-fold and 12-fold symmetries. The corresponding SAXS patterns are presented in Fig. 2.5. Note that quasicrystal phases are spontaneously formed from single-component block copolymer micelles via self-assembly in concentrated solutions. This lyotropic block copolymer colloidal system provides the only example of quasicrystal with 18-fold symmetry.

Note that in all the experimental results of soft matters or even in most simulations quasicrystals mostly occur as aperiodic only in two dimensions but periodic along

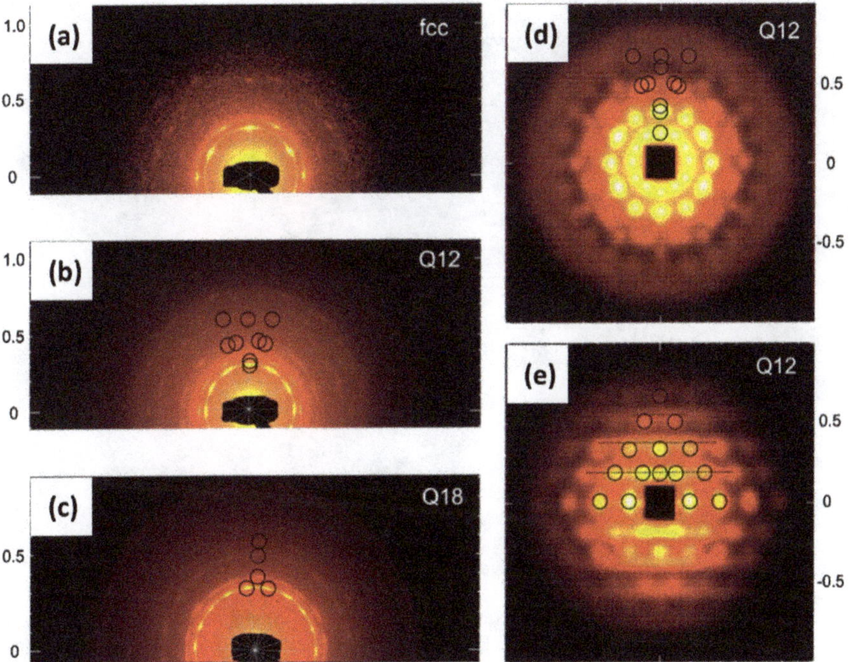

Fig. 2.5 Synchrotron SAXS patterns of FCC phase (**a**) and the quasicrystalline phases Q12 (**b**) and Q18 (**c**). The SANS patterns have also been recorded for the Q12 phase, parallel (**d**) and normal (**e**) to the 12-fold rotation axis [18]

with the 12-fold symmetry or occasionally other symmetry. The three-dimensional icosahedral quasicrystals (IQC) so far have only been observed in complicated metal alloys via the cooperation of several different atomic elements. A very recent intriguing molecular dynamics simulation result by Glotzer et al. shows that an IQC can be obtained from a one-component system of uniform particles interacting via a tunable, isotropic pair (Fig. 2.6a–d) [19]. Almost at the same time, other groups reported that entropy and spherical confinement suffice for the formation of icosahedral clusters consisting of up to 100,000 particles [20]. Specifically, tens of thousands of nanoparticles (below 10 nm size) compressed under spherical confinement spontaneously crystallize into icosahedral clusters with obvious icosahedral symmetry morphology (Fig. 2.6e–g). This forming dynamics and structure assembly process

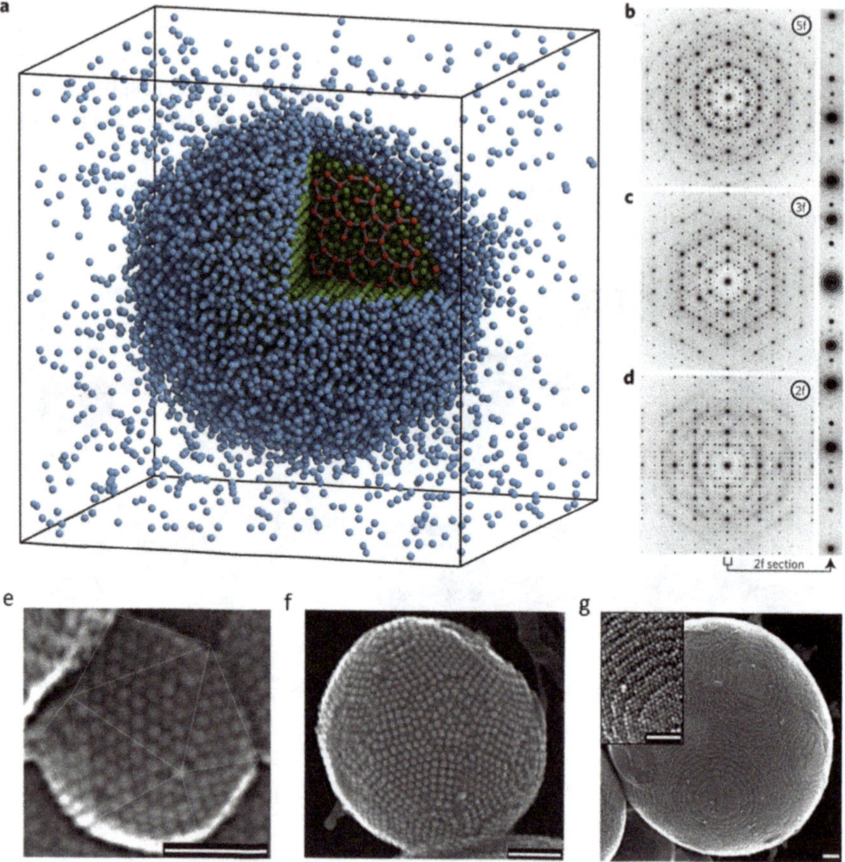

Fig. 2.6 Three-dimensional icosahedral quasicrystal phase (**a**) Computational self-assembly of a one-component icosahedral quasicrystal. Diffraction patterns along five-fold (**b**), three-fold (**c**), and two-fold (**d**) axes exhibit Bragg peaks and weak diffuse scattering [19]. (e–g) Entropy-driven experimental formation of large icosahedral colloidal clusters by spherical confinement [20]

of IQC in this finding may help on the right design in soft matters or nanoparticle systems to form three-dimensional IQC structures.

The discovery of quasicrystals in soft matter opens up the scope of the quasicrystal study. A distinct feature of quasicrystal in soft matter is that the constructing spherical motifs are assembled from multiple, identical molecular constituents. These spherical motifs scaling from several to tens of nanometers spontaneously form volume partition asymmetry during self-assembly. The thermodynamic environments of the quasicrystal state in the soft matter could be different from the metallic alloys, regarding the enthalpic and entropic factors of constitutional molecules and supramolecular motifs. More importantly, the underlying dynamics of quasicrystal state in the soft matter would be different and thus need systematic study in both theoretical calculation and experimental investigation.

The formation of quasicrystal from soft matter constitutes a new source of quasicrystals, but how can one distinguish the matter from soft-matter quasicrystals? If the quasicrystals present the two characters by de Gennes, i.e., the fluidity and internal structure, then they belong to the soft-matter quasicrystals, otherwise not. Some quasicrystals are formed from soft matter, but there is almost no fluidity, we still label them as solid quasicrystals or close to solid quasicrystals. For example, the quasicrystals generated from the giant surfactants do not belong to soft matter, as the fluid viscosity is too stiff ($\sim 10^5$ Poise).

The 18-fold symmetry quasicrystals in soft-matter were reported in Ref. [18], whose diffraction pattern and Penrose tiling are shown by Figs. 2.7 and 2.8 respectively.

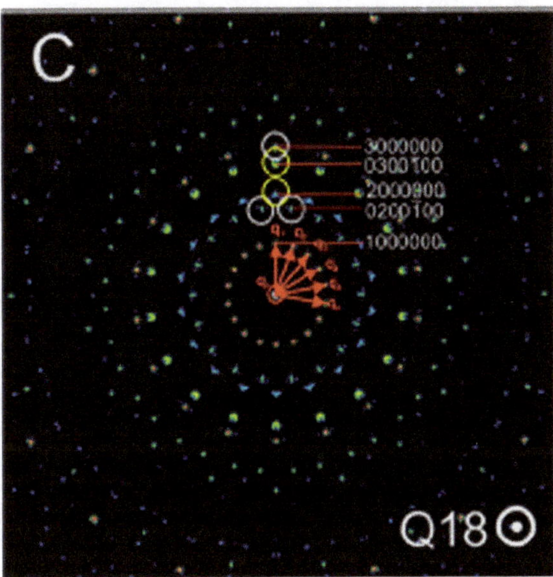

Fig. 2.7 Diffraction pattern of 18-fold symmetry quasicrystal in colloid

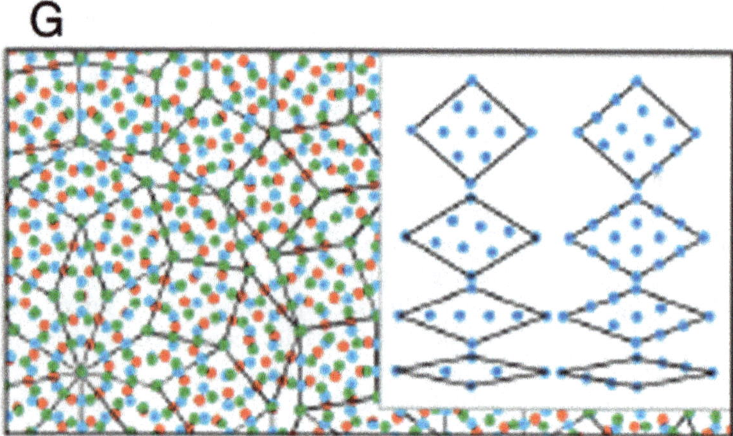

Fig. 2.8 Penrose tiling of 18-fold symmetry quasicrystal in soft matter

2.2 Characters of Soft-Matter Quasicrystals

Based on the experimental results the soft-matter quasicrystals observed in different kinds of soft matter, their forms and structures are quite different from each other. This book is unable specially and in detail to study soft matter. Our object is only to study soft-matter quasicrystals, and for this purpose, we have to understand a preliminary and necessary knowledge on soft matter. The nature of soft matter is an intermediate phase between ideal solid and simple fluid, or call it is as a complex fluid or structured fluid, more exactly a complex liquid or structured liquid, which is one of soft condensed matter.

All soft-matter quasicrystals observed so far are two-dimensional. During the process of their formation, it is accomplished via chemical processes, such that crystal-quasicrystal transition, liquid crystal-quasicrystal transition, etc. In the formation process of quasicrystals from colloids, static electricity plays some rules as the particles in colloids usually have charges. These complex physical–chemical effects will not be discussed in our presentation on soft-matter quasicrystals in detail. The main attention hereafter focuses on mechanical behavior and the continuous theory of soft-matter quasicrystals.

The thermodynamics of soft-matter quasicrystals is briefed here. In Refs. [21] and [22], Lifshitz et al. studied thermodynamics and considered the stability of the new phases, which we will elaborate on in detail in Chap. 13. For studying generalized dynamics of soft-matter quasicrystals, the equation of state is necessary. Theoretical work from Fan and co-workers [23, 24] will be introduced in Chap. 4, however, the model needs experimental verification. For simplicity, in some cases, we take an incompressible complex fluid model on soft matter, for which one does not need to use the equation of state in the analysis.

Due to the lack of experimental data, numerical analysis can help us to obtain some results on the mechanical and physical behavior of the matter. For example, after our computation, Cheng et al. [25] finds the compressibility of soft-matter quasicrystals is quite large, e.g., $\delta\rho/\rho_0 = 10^{-7} \sim 10^{-5}$, $\delta\rho = \rho - \rho_0$, while for solid quasicrystals, $\delta\rho/\rho_0 = 10^{-13}$ refer to Cheng et al. [26]. In addition, in some cases the ratio $\left|p_{yy}/\sigma_{yy}\right|$ for soft-matter quasicrystals is considerable, i.e. the orders of magnitude difference between fluid stress and elastic stress in soft-matter quasicrystals can be present, while the ratio between viscous stress and elastic stress is about $\left|\sigma'_{yy}/\sigma_{yy}\right| \sim 10^{-15}$ for solid quasicrystals, where σ_{yy} denotes the elastic stress, σ'_{yy} the solid viscous stress, p_{yy} the fluid stress. The gigantic distinctions in the hydrodynamic behavior between soft-matter quasicrystals and solid quasicrystals also reveal the great differences in nature between these two kinds of matters. The accuracy of these computational results largely depends on the experimental measurements of these physical parameters.

In addition, some general characters of soft matter hold for soft-matter quasicrystals too, e.g., the motion of soft-matter quasicrystals is in small Reynolds number, etc. similar to that of general soft matter.

2.3 Some Concepts Concerning Possible Generalized Dynamics on Soft-Matter Quasicrystals

Solid quasicrystals are formed with metallic alloys, while soft-matter quasicrystals are formed in liquid crystals, colloids, polymers, and surfactants, etc., latter belong to some kinds of soft matter. These soft matters can exist for quite a long period, and they belong to nontraditional materials, which we are not very familiar with them yet. Soft matter is the common title, introduced by Gennes [27] in 1991 for liquid crystals, colloids, polymers, foams, emulsions, surfactants, biomacromolecules, etc. They are neither ideal solid nor simple fluid, but present characters of both solid and fluid, belong to an intermediate phase between isotropic fluid and ideal solid macroscopically. Sometimes one calls them anisotropic fluid or structured fluid [28–31]. In Chap. 1 we have mentioned in brief these.

In the book, we will give some descriptions of the distribution, deformation, and motion of soft-matter quasicrystals by considering their behavior on behalf of soft matter as well as their quasiperiodic symmetry characteristics from dynamics point of view. Based on the knowledge and data accumulated in soft matter science and solid quasicrystal study, we will constitute a generalized dynamics of soft-matter quasicrystals, starting from the Landau symmetry breaking and elementary excitation principle. We introduce three kinds of elementary excitations or quasiparticles: phonons, phasors, and fluid phonon, in which the first two come from solid quasicrystal theory, and the fluid phonon is originated from the Landau school but not introduced in solid quasicrystals [32]. These quasiparticles and their interactions follow partly the conservation laws and symmetry breaking rule, which leads to some nonlinear partial differential equations. Solving relevant initial and boundary conditions of these equations can determine the distribution, deformation, and motion of the matter. The verification of the possible dynamics can be realized through comparison between the solutions mentioned above with experimental results, or/and with solutions of solid quasicrystal or/and with solutions of classical fluid dynamics. The relevant discussions will be introduced in the following chapters.

2.4 First and Second Kinds of Two-Dimensional Quasicrystals

Up to now, soft-matter quasicrystals discovered so far, 12-, 18-, and unreported tenfold symmetry quasicrystals are two-dimensional quasicrystals, but they are categorized into two kinds from a symmetry point of view. The 12-fold symmetry quasicrystals are similar to those of 5-, 8-, and tenfold symmetry quasicrystals, so they may be classified as the first kind of two-dimensional quasicrystals according to quasiperiodic structure [33], while 18- and possible 7-, 9-, and 14-fold symmetry quasicrystals belong to the second kind of two-dimensional quasicrystals [34]. Based on the analysis, the first kind of two-dimensional quasicrystals has one kind of phason elementary excitations, while the second kind of two-dimensional quasicrystals has the first and second kinds of phason elementary excitations. The concept of the second phasons was suggested by Hu et al. [35], in which they developed a hypothesis on six-dimensional embedding space, the discussion was based on group representation theory. In Chaps. 7, 8, and 9 we will discuss the first kind of two-dimensional soft-matter quasicrystals, and in Chaps. 10 and 11, we will discuss the second kind of ones. Compared with the first kind of two-dimensional quasicrystals, the theory on the second one is developing, so there are very limited results.

The symmetry of the first kind of soft-matter quasicrystals can be drawn from the analysis of solid quasicrystals. The point groups are listed in Table 2.1.

Table 2.1 Systems and point groups of the first kind of two-dimensional quasicrystal

Systems	Names (for solid)	Point groups
fivefold symmetry	Pentagonal quasicrystals	$5,\ \bar{5}$
		$5m,\ 52,\ \bar{5}m$
tenfold symmetry	Decagonal quasicrystal	$10,\ \overline{10},\ 10/m$
		$10mm,\ 1022,\ \overline{10}m2,\ 10/mmm$
eightfold symmetry	Octagonal quasicrystals	$8,\ \bar{8},\ 8/m$
		$8mm,\ 822,\ \bar{8}m2,\ 8/mmm$
12-fold symmetry	Dodecagonal quasicrystals	$12,\ \overline{12},\ 12/m$
		$12mm,\ 1222,\ \overline{12}m2,\ 12/mmm$

Table 2.2 Point groups of the second kind of two-dimensional quasicrystals

Axis	Plus m_h mapping m_h	Plus m_v	Plus 2_h	Plus multi-operators
7	$\overline{14}$	$7m$	72	$\overline{14}m2$
$\overline{7}$		$\overline{7}m$	$\overline{7}m$	
14	$14/m$	$14mm$	1422	$14/mmm$
$\overline{14}$		$\overline{14}m2$	$\overline{14}m2$	
9	$\overline{18}$	$9m$	92	$\overline{18}m2$
$\overline{9}$		$\overline{9}m$	$\overline{9}m$	
18	$18/m$	$18mm$	1822	$18/mmm$
$\overline{18}$		$\overline{18}m2$	$\overline{18}m2$	

The symmetries on the second kind of soft-matter quasicrystals have not well been studied. Tang and Fan [36] put forward the point group classification and the group representation on the structure.

Based on the Schoenflies method, the point groups of 7-, 14-, 9-, and 18-fold symmetry quasicrystals are listed in Table 2.2.

The group representation theory including the character tables of the second kind of two-dimensional quasicrystals given in Ref. [36] is quite complicated. The derivation will not be conducted here. The key results useful here is the determination of quadratic invariants of strain tensors of phonons, first and second phasons and their couplings (i.e., the determination of all independent nonzero components of physical modulus tensors of the material), and the constitutive equations, which will be discussed in Chaps. 10 and 11, respectively.

2.5 Motivation of Our Discussion in the Book

The formation mechanism of soft-matter quasicrystals is very interesting. We gave a preliminary introduction in Sect. 2.1.

Soft-matter quasicrystals present some applications and potential applications associated with their structures and properties. Here we aim to discuss only macroscopic dynamics concerning the distribution, deformation, and motion of the material, or their generalized dynamics. With limited experimental data, the discussion is mainly on calculation with the assistant of mathematical physics and computational physics. A few preliminary case studies may help the reader to understand some macroscopic behavior of soft-matter quasicrystals. Although we limit to the macroscopic discussion and not in-depth formation mechanism, the equation of state will be used in the study on the mesoscopescale. In this new edition of this book, some new contents are included, for example, the possible applications in photonic band-gap introduced in Chap. 14, which go beyond soft-matter quasicrystals.

References

1. Abe, E., Yan, Y., Pennycook, S.J.: Quasicrystals as cluster aggregates. Nat. Mater. **3**, 759–767 (2004)
2. Yamamoto, A.: Crystallography of quasiperiodic crystals. Acta Crystallogr. A **52**, 509–560 (1996)
3. Graef, M.D., Mchenry, M.E.: Structure of Materials: an Introduction to Crystallography, Diffraction and Symmetry, 2nd edn. Cambridge University Press, Cambridge (2012)
4. Zeng, X., Ungar, G., Liu, Y., Percec, V., Dulcey As, E., Hobbs, J.K.: Supramolecular dendritic liquid quasicrystals. Nature **428**, 157–160 (2004)
5. Gillard, T.M., Lee, S., Bates, F.S.: Dodecagonal quasicrystalline order in a diblock copolymer melt. Proc. Natl. Acad. Sci. **113**, 5167–5172 (2016)
6. Ye, X., Chen, J., Eric Irrgang, M., Engel, M., Dong, A., Glotzer, S.C., Murray, C.B.: Quasicrystalline nanocrystal superlattice with partial matching rules. Nat. Mater. **16**, 214–219 (2017)
7. Yue, K., Huang, M., Marson, R.L., He, J., Huang, J., Zhou, Z., Wang, J., Liu, C., Yan, X., Wu, K., Guo, Z., Liu, H., Zhang, W., Ni, P., Wesdemiotis, C., Zhang, W.B., Glotzer, S.C., Cheng, S.Z.D.: Geometry induced sequence of nanoscale Frank-Kasper and quasicrystal mesophases in giant surfactants. Proc. Natl. Acad. Sci. **113**, 14195–14200 (2016)
8. Feng, X., Liu, G., Guo, D., Lang, K., Zhang, R., Huang, J., Su, Z., Li, Y., Huang, M., Li, T., Cheng, S.Z.D.: Transition kinetics of self-assembled supramolecular dodecagonal quasicrystal and Frank-Kasper σ Phases in ABn Dendron-like giant molecules. ACS Macro Lett. **8**, 875–881 (2019)
9. Ungar, G., Zeng, X.: Frank-Kasper, quasicrystalline and related phases in liquid crystals. Soft Matter **1**, 95–106 (2005)
10. Ishimasa, T.: Dodecagonal quasicrystals still in progress. Isr. J. Chem. **51**, 1216–1225 (2011)
11. Baake, M., Klitzing, R., Schlottman, M.: Fractally shaped acceptance domains of quasiperiodic square-triangle tilings with dodecagonal symmetry. Physica A **191**, 554–558 (1992)
12. Oxborrow, M., Henley, C.: Random square-triangle tilings: a model for twelvefold-symmetric quasicrystals. Phys. Rev. B **48**, 6966–6998 (1993)
13. Leung, P., Henley, C., Chester, G.: Dodecagonal order in a two-dimensional Lennard-Jones system. Phys. Rev. B **39**, 446–458 (1989)
14. Asai, Y., Takano, A., Matsushita, Y.: Creation of cylindrical morphologies with extremely large oblong unit lattices from ABC block terpolymer blends. Macromolecules **48**, 1538–1542 (2015)
15. Hayashida, K., Dotera, T., Takano, A., Matsushita, Y.: polymeric quasicrystal: mesoscopic quasicrystalline tiling in star polymers. Phys. Rev. Lett. **98**, 195502 (2007)
16. Zeng, X., Kieffer, R., Glettner, B., Nurnberger, C., Liu, F., Pelz, K., Prehm, M., Baumeister, U., Hahn, H., Lang, H., Gehring, G.A., Weber, C.H.M., Hobbs, J.K., Tschierske, C., Ungar, G.: Complex multicolor tilings and critical phenomena in tetraphilic liquid crystals. Science **331**, 1302–1306 (2011)
17. Nagaoka, Y., Zhu, H., Eggert, D., Chen, O.: Single-component quasicrystalline nanocrystal superlattices through flexible polygon tiling rule. Science **362**, 1396–1400 (2018)
18. Fischer, S., Exner, A., Zielske, K., Perlich, J., Deloudi, S., Steurer, W., Lindner, P., Foerster, S.: Colloidal quasicrystals with 12-fold and 18-fold diffraction symmetry. Proc. Nat. Acad. Sci. **108**, 1810–1814 (2011)
19. Engel, M., DamascenoP, F., Phillips, C.L., Glotzer, S.C.: Computational self-assembly of a one-component icosahedral quasicrystal. Nat. Mater. **14**, 109–116 (2014)
20. de Nijs, B., Dussi, S., Smallenburg, F., Meeldijk, J.D., Groenendijk, D.J., Filion, L., Imhof, A., van Blaaderen, A., Dijkstra, M.: Entropy-driven formation of large icosahedral colloidal clusters by spherical confinement. Nat. Mater. **14**, 56–60 (2015)
21. Lifshitz, R., Diamant, H.: Soft quasicrystals-Why are they stable? Phil. Mag. **87**, 3021–3030 (2007)

22. Barkan, K., Diamant, H., Lifshitz, R.: Stability of quasicrystals composed of soft isotropic particles, Phys. Rev. B **83**, 172201 (2011)
23. Fan, T.Y., Sun, J.J.: Four phonon model for studying thermodynamics of soft-matter quasicrystals. Phil. Mag. Lett. **94**, 112–117 (2014)
24. Fan, T.Y.: Equation system of generalized hydrodynamics of soft-matter quasicrystals. Appl. Math Mech. **37**, 331–347, in Chinese (2016); arXiv:1908.06425[cond-mat.soft]. 15 Oct 2019
25. Cheng, H., Fan, T.Y.: Dynamics of decagonal soft-matter quasicrystals. Sci. China-Phys. Mech. Astron. submitted (2021)
26. Cheng, H., Fan, T.Y., Wei, H.: Solutions of hydrodynamics of quasicrystals with 5- and 10-fold symmetry. Appl. Math. Mech. **37**, 1393–1404 (2016)
27. de Gennes, P.D.: Soft matter. Mod. Phys. Rev. **64**,544–548 (1992); Angw. Chem. **31**, 842–845 (1992)
28. Witten, T.A., Pincus, P.A.: Structured Fluids: Polymers Colloids, Surfactants. Oxford University Press, New York (2004)
29. Kleman, M.: Soft Matter Physics: An Introduction. Springer, Berlin (2003)
30. Motiv, M.: Sensitive Matter: Foams Gels, Liquid Crystals and Other Materials. Harvard University Press, New York (2010)
31. Israelachvili, N.J.: Intermolecular and Surface Forces. Academic Press, New York (2010)
32. Lifshitz, E.M., Pitaevskii, L.: Statistical Physics, Part 2. Pergamon, Oxford (1980)
33. Fan, T.Y., Tang, Z.Y.: Three-dimensional generalized dynamics of soft-matter quasicrystals. Appl. Math. Mech. **38**, 1195–1207 (2017), in Chinese; Adv. Mat. Sci. Eng. **2020**, Article 1D 4875854 (2020)
34. Fan, T.Y.: Generalized hydrodynamics of soft-matter second kind of two-dimensional quasicrystals. Appl. Math. Mech. **38**, 189–199, in Chinese (2017); arXiv: 1908.06430[cond-mat.soft]. 15 Oct 2019
35. Hu, C.Z., Ding, D.H., Yang, W.G., Wang, R.H.: Possible two-dimensional quasicrystals structures with six-dimensional embedding space. Phys. Rev. B **49**, 9423–9427 (1994)
36. Tang, Z.Y., Fan, T.Y.: Point groups and group representation theory of second kind of two-dimensional quasicrystals, unpublished work (2017)

Chapter 3
Introduction on Elasticity and Hydrodynamics of Solid Quasicrystals

Elasticity and hydrodynamics of solid quasicrystals are the basis of the dynamics of soft-matter quasicrystals. A brief review of these topics is given in this chapter, which may be beneficial for understanding the dynamics of soft-matter quasicrystals.

3.1 Physical Basis of Elasticity of Quasicrystals, Phonons, and Phasons

Immediately after the discovery of quasicrystals, Bak [1] published the principle framework of the elasticity of quasicrystals, in which he extended the crystalline description into higher dimensions. The core is the Landau theory on symmetry breaking and elementary excitation for condensed matter. Bak [1, 2] pointed out that ideally, one would like to explain the structure by first-principles calculations that take into account of the actual electronic configuration of constituent atoms. However, such calculations are practically impossible to date as a too-large system is involved. Instead, he suggested that the Landau phenomenological theory [3] could be used to describe the structural transition. In this theory, the condensed phase is described with a symmetry-breaking order parameter which transforms as an irreducible representation of the symmetry group of a liquid with full translational and rotational symmetry. According to Landau theory, the order parameter of quasicrystals is the density wave in reciprocal lattice. For the density of the ordered, low temperature, d-dimensional quasicrystal can be expressed as a Fourier series by the extended formula defined by Anderson [4] (the expansion exists due to the periodicity in lattice or reciprocal lattice of higher dimensional space).

The original version of this chapter was revised: The equation in this chapter has been amended. The correction to this chapter is available at
https://doi.org/10.1007/978-981-16-6628-5_18

T.-Y. Fan et al., *Generalized Dynamics of Soft-Matter Quasicrystals*, Springer Series in Materials Science 260, https://doi.org/10.1007/978-981-16-6628-5_3

$$\rho(\mathbf{r}) = \sum_{G \in L_R} \rho_{\mathbf{G}} \exp\{i\mathbf{G} \cdot \mathbf{r}\} = \sum_{G \in L_R} |\rho_G| \exp\{-i\Phi_G + i\mathbf{G} \cdot \mathbf{r}\}, \qquad (3.1.1)$$

where $\mathbf{G}$ is a reciprocal vector, and L_R denotes the reciprocal lattice (for the definitions on the reciprocal vector and reciprocal lattice, refer to Chap. 1 of [5]), $\rho_{\mathbf{G}}$ is a complex number represented by

$$\rho_{\mathbf{G}} = |\rho_{\mathbf{G}}| e^{-i\Phi_{\mathbf{G}}} \qquad (3.1.2)$$

with an amplitude of $|\rho_{\mathbf{G}}|$ and the phase angle $\Phi_{\mathbf{G}}$. Note that $\rho(r)$ is a real value, therefore $|\rho_{\mathbf{G}}| = |\rho_{-\mathbf{G}}|$, and $\Phi_{\mathbf{G}} = -\Phi_{-\mathbf{G}}$.

There exists a set of N base vectors, $\{\mathbf{G}_n\}$, such that each $\mathbf{G} \in L_R$ can be written as $\sum m_n \mathbf{G}_n$ for integers m_n. Furthermore, $N = kd$, where k is the mutual number of the incommensurate vectors in the d-dimensional quasicrystal. For the first kind of quasicrystals, $k = 2$. A convenient parametrization of the phase angle is given by

$$\Phi_n = \mathbf{G}_n^{\parallel} \cdot \mathbf{u} + \mathbf{G}_n^{\perp} \cdot \mathbf{w} \qquad (3.1.3)$$

in which $\mathbf{u}$ is analogous to the phonon as in conventional crystals, while $\mathbf{w}$ can be understood as the phason degrees of freedom in quasicrystals, which describe the local rearrangement of the unit cell based on the Penrose tiling. Both are functions of the position vector in the physical space only. $\mathbf{G}_n^{\parallel}$ is the reciprocal vector in the physical space $E_{\parallel}^3$ and $\mathbf{G}_n^{\perp}$ is the conjugate vector in the perpendicular space $E_{\perp}^3$. It can be seen that the above-mentioned Bak's hypothesis is a natural development of Anderson's theory [4].

Almost at the same time, Levine et al. [6], Lubensky et al. [7–9], Kalugin et al. [10], Torian and Mermin [11], Jaric [12], Duneau and Katz [13], Socolar et al. [14], Gahler and Rhyner [15] carried out the study on the elasticity of quasicrystals. Although the researchers studied the elasticity from different descriptions, e.g., the unit-cell description based on the Penrose tiling is adopted too, the density wave description based on the Landau phenomenological theory on symmetry breaking of condensed matter has played the central role and been widely acknowledged. This means there are two elementary excitations of low-energy, i.e., the phonon field $\mathbf{u}$ and the phason field $\mathbf{w}$ for quasicrystals, with the vector $\mathbf{u}$ in the parallel space $E_{\parallel}^3$ and the vector $\mathbf{w}$ in the perpendicular space $E_{\perp}^3$, respectively. So the total displacement field for quasicrystals is

$$\bar{\mathbf{u}} = \mathbf{u}^{\parallel} \oplus \mathbf{u}^{\perp} = \mathbf{u} \oplus \mathbf{w}, \qquad (3.1.4)$$

where $\oplus$ represents the direct sum.

According to the argument of Bak et al.

$$\mathbf{u} = \mathbf{u}(\mathbf{r}^{\|}), \, \mathbf{w} = \mathbf{w}(\mathbf{r}^{\|}), \tag{3.1.5}$$

i.e., the displacement vector $\mathbf{u}$ and $\mathbf{w}$ are a function of the position vector $\mathbf{r}^{\|}$ in parallel space $E_{\|}^3$ only, which leads to Eq. (3.1.2). For simplicity, the superscript $\mathbf{r}^{\|}$ is removed in the subsequent discussions.

With the basic formulas (3.1.3) and (3.1.4) and other fundamental conservation laws well-known in physics, the macroscopic basis of the continuous medium model of elasticity of solid quasicrystals can be set up. To some extent, the discussion herein is an extension to that in the elasticity of crystals.

3.2 Deformation Tensors

In the theory of crystal elasticity, it is introduced that the deformation of the phonon field lies in the relative displacement (i.e., the rigid translation and rotation do not result in deformation), which can be expressed by

$$d\mathbf{u} = \mathbf{u}' - \mathbf{u}.$$

If we set up an orthogonal coordinate system (x_1, x_2, x_3) or (x, y, z), then we have

$$\mathbf{u} = (u_x, u_y, u_z) = (u_1, u_2, u_3) \quad \text{and} \quad du_i = \frac{\partial u_i}{\partial x_j} dx_j, \tag{3.2.1}$$

where $\partial u_i / \partial x_j$ has the meaning of the gradient of the vector $\mathbf{u}$. In some publications, one denotes

$$\nabla \mathbf{u} = \frac{\partial u_i}{\partial x_j} = \begin{bmatrix} \frac{\partial u_x}{\partial x} & \frac{\partial u_x}{\partial y} & \frac{\partial u_x}{\partial z} \\ \frac{\partial u_y}{\partial x} & \frac{\partial u_y}{\partial y} & \frac{\partial u_y}{\partial z} \\ \frac{\partial u_z}{\partial x} & \frac{\partial u_y}{\partial y} & \frac{\partial u_y}{\partial z} \end{bmatrix} \tag{3.2.2}$$

and

$$\begin{bmatrix} \frac{\partial u_x}{\partial x} & \frac{\partial u_x}{\partial y} & \frac{\partial u_x}{\partial z} \\ \frac{\partial u_y}{\partial x} & \frac{\partial u_y}{\partial y} & \frac{\partial u_y}{\partial z} \\ \frac{\partial u_z}{\partial x} & \frac{\partial u_z}{\partial y} & \frac{\partial u_z}{\partial z} \end{bmatrix} = \begin{bmatrix} \frac{\partial u_x}{\partial x} & \frac{1}{2}\left(\frac{\partial u_x}{\partial y} + \frac{\partial u_y}{\partial x}\right) & \frac{1}{2}\left(\frac{\partial u_x}{\partial z} + \frac{\partial u_z}{\partial x}\right) \\ \frac{1}{2}\left(\frac{\partial u_x}{\partial y} + \frac{\partial u_y}{\partial x}\right) & \frac{\partial u_y}{\partial y} & \frac{1}{2}\left(\frac{\partial u_y}{\partial z} + \frac{\partial u_z}{\partial y}\right) \\ \frac{1}{2}\left(\frac{\partial u_x}{\partial z} + \frac{\partial u_z}{\partial x}\right) & \frac{1}{2}\left(\frac{\partial u_y}{\partial z} + \frac{\partial u_z}{\partial y}\right) & \frac{\partial u_z}{\partial z} \end{bmatrix}$$

$$+ \begin{bmatrix} 0 & -\frac{1}{2}\left(\frac{\partial u_y}{\partial x} - \frac{\partial u_x}{\partial y}\right) & -\frac{1}{2}\left(\frac{\partial u_z}{\partial x} - \frac{\partial u_x}{\partial z}\right) \\ -\frac{1}{2}\left(\frac{\partial u_x}{\partial y} - \frac{\partial u_y}{\partial x}\right) & 0 & -\frac{1}{2}\left(\frac{\partial u_z}{\partial y} - \frac{\partial u_y}{\partial z}\right) \\ -\frac{1}{2}\left(\frac{\partial u_x}{\partial z} - \frac{\partial u_z}{\partial x}\right) & -\frac{1}{2}\left(\frac{\partial u_y}{\partial z} - \frac{\partial u_z}{\partial y}\right) & 0 \end{bmatrix}$$

$$= \frac{1}{2}\left(\frac{\partial u_i}{\partial x_j} + \frac{\partial u_j}{\partial x_i}\right) - \frac{1}{2}\left(\frac{\partial u_j}{\partial x_i} - \frac{\partial u_i}{\partial x_j}\right) = \varepsilon_{ij} + \omega_{ij}$$

$$\varepsilon_{ij} = \frac{1}{2}\left(\frac{\partial u_i}{\partial x_j} + \frac{\partial u_j}{\partial x_i}\right) \tag{3.2.3}$$

$$\omega_{ij} = \frac{1}{2}\left(\frac{\partial u_i}{\partial x_j} - \frac{\partial u_j}{\partial x_i}\right) \tag{3.2.4}$$

This expression means that the gradient of the phonon vector $\mathbf{u}$ can be decomposed into two parts, ε_{ij} and ω_{ij}, ε_{ij} contributes to the deformation energy, and ω_{ij} represents a kind of rigid rotations. We consider only ε_{ij}, which is the phonon deformation tensor, called strain tensor, and is a symmetric tensor: $\varepsilon_{ij} = \varepsilon_{ji}$.

Similarly, for the phason field, we have

$$dw_i = \frac{\partial w_i}{\partial x_j} dx_j \tag{3.2.5}$$

and

$$\nabla \mathbf{w} = \frac{\partial w_i}{\partial x_j} = \begin{bmatrix} \frac{\partial w_x}{\partial x} & \frac{\partial w_x}{\partial y} & \frac{\partial w_x}{\partial z} \\ \frac{\partial w_y}{\partial x} & \frac{\partial w_y}{\partial y} & \frac{\partial w_y}{\partial z} \\ \frac{\partial w_z}{\partial x} & \frac{\partial w_z}{\partial y} & \frac{\partial w_z}{\partial z} \end{bmatrix} \tag{3.2.6}$$

Although it can be decomposed into symmetric and asymmetric parts, all components $\frac{\partial w_i}{\partial x_j}$ contribute to the deformation of quasicrystals, the phason deformation tensor, or the phason strain tensor, which is defined by

$$w_{ij} = \frac{\partial w_i}{\partial x_j}. \tag{3.2.7}$$

It describes the local rearrangement of atoms in a cell and is an asymmetric tensor, namely $w_{ij} \neq w_{ji}$.

The difference between ε_{ij} and w_{ij} given by Eqs. (3.2.3) and (3.2.7) originates from the physical properties of phonon modes and phason modes. This difference in behavior can also be explained by the group theory, i.e., they follow different irreducible representations for some symmetry transformations for all quasicrystal systems with crystallographic forbidden symmetry. But for crystallographic symmetry quasicrystal, like the cubic quasicrystals, the phason modes can be under the same irreducible representation, which we will discuss separately in Chap. 9 of the book authored by Fan [5].

3.3 Stress Tensors and Equations of Motion

The gradient of the displacement field $\mathbf{w}$ manifests the local rearrangement of atoms in a cell in quasicrystals. It needs external forces to drive the atoms through barriers when they make the local rearrangement in a cell (like the flip between the different kits in Penrose tiling), suggesting that there are other kinds of body forces and tractions apart from the conventional body forces $\mathbf{f}$ and tractions $\mathbf{T}$ for deformed quasicrystals, which are named the generalized body forces (density) $\mathbf{g}$ and generalized tractions (the generalized area forces density) $\mathbf{h}$.

First, we study the static case.

Let's name the stress tensor σ_{ij} corresponding to ε_{ij} as the phonon stress tensor, and that H_{ij} to w_{ij} as the phason stress tensor, then we have the following equilibrium equations

$$\left.\begin{array}{l} \frac{\partial \sigma_{ij}}{\partial x_j} + f_i = 0 \\ \frac{\partial H_{ij}}{\partial x_j} + g_i = 0 \end{array}\right\} \quad (x, y, z) \in \Omega \tag{3.3.1}$$

based on the momentum conservation law.

Applying the angular momentum conservation law to the phonon field

$$\frac{\mathrm{d}}{\mathrm{d}t} \int_{\Omega} \mathbf{r}^{\parallel} \times \rho \dot{\mathbf{u}} d\Omega = \int_{\Omega} \mathbf{r}^{\parallel} \times \mathbf{f} d\Omega + \int_{\Omega} \mathbf{r}^{\parallel} \times \mathbf{T} d\Gamma \tag{3.3.2}$$

and by using the Gauss theorem, it follows that

$$\sigma_{ij} = \sigma_{ji}, \tag{3.3.3}$$

which indicates that the phonon stress tensor is symmetric.

Since $\mathbf{r}^{\parallel}$ and $\mathbf{w}(\mathbf{g}, \mathbf{h})$ transform under different representations of the point groups, more precisely, the former transforms like a vector, but the latter does not, the product representation: $\mathbf{r}^{\parallel} \times \mathbf{w}$, $\mathbf{r}^{\parallel} \times \mathbf{g}$ and $\mathbf{r}^{\parallel} \times \mathbf{h}$ do not produce any vector representations. For the phason field, there is no equation analogous to (3.3.2), and it follows that, generally

$$H_{ij} \neq H_{ji}. \tag{3.3.4}$$

This result holds for all quasicrystal systems except the case for three-dimensional cubic quasicrystals.

In dynamic cases, the deformation process is quite complicated and there are different arguments. Levine et al. [6] claimed that phonon modes and phason modes are different based on their role in six-dimensional hydrodynamics. Phonons are wave propagation while phasons are diffusive with very large diffusive time. Physically the phason modes represent a relative motion of the constituent density waves. Dolinsek et al. [16, 17] further developed the point of view of Levine et al. and argued about the atom flip or atom hopping concept for the phason dynamics. But according to Bak [1, 2], the phason describes particular structural disorders or structure fluctuations in quasicrystals, and it can be formulated based on a six-dimensional space description. Since there are six continuous symmetries, there exist six hydro-dynamic vibration modes. In the following text, we give a brief introduction on elastodynamics based on Bak's argument as well as the argument of Levine et al.

Ding et al. [18] derived that

$$\left.\begin{aligned} \frac{\partial \sigma_{ij}}{\partial x_j} + f_i &= \rho \frac{\partial^2 u_i}{\partial t^2} \\ \frac{\partial H_{ij}}{\partial x_j} + g_i &= \rho \frac{\partial^2 w_i}{\partial t^2} \end{aligned}\right\} \quad (x, y, z) \in \Omega,\ t > 0 \qquad (3.3.5)$$

based on the momentum conservation law. This derivation is carried out from Bak's argument, in which ρ is the mass density of quasicrystals.

The above result cannot be obtained following the argument of Lubensky et al. [19], instead, it becomes

$$\left.\begin{aligned} \frac{\partial \sigma_{ij}}{\partial x_j} + f_i &= \rho \frac{\partial^2 u_i}{\partial t^2} \\ \frac{\partial H_{ij}}{\partial x_j} + g_i &= \kappa \frac{\partial w_i}{\partial t} \end{aligned}\right\} \quad (x, y, z) \in \Omega,\ t > 0 \qquad (3.3.6)$$

in which $\kappa = 1/\Gamma_w$, and Γ_w is the kinetic coefficient of the phason field. Fan et al. [5], and Rochal and Norman [20] proposed the same equations based on Levine et al. [6] for linear cases and by omitting the fluid velocity field. The hydrodynamics formulation (3.3.6) by Lubensky et al. [19] based on the Landau symmetry-breaking principle can be treated as the elasto-/hydro-dynamic equation of quasicrystals. In particular, the second equation of (3.3.6) presents the dissipation feature of motions of phason degrees in dynamic processes, and it is irreversible thermodynamically.

For a detailed discussion on the great challenge on dynamics of quasicrystals, please refer to Chaps. 10 and 16 in the text and Appendix III of the book [5].

3.4 Free Energy Density and Elastic Constants

For the free energy density or the strain energy density of a quasicrystal $F\left(\varepsilon_{ij}, w_{ij}\right)$, it is difficult to obtain the general expression, so we take a Taylor expansion in the neighborhood of $\varepsilon_{ij} = 0$ and $w_{ij} = 0$, and retain up to the second-order term as follows,

$$
\begin{aligned}
F\left(\varepsilon_{ij}, w_{ij}\right) &= \frac{1}{2}\left[\frac{\partial^2 F}{\partial \varepsilon_{ij} \partial \varepsilon_{kl}}\right]_0 \varepsilon_{ij}\varepsilon_{kl} + \frac{1}{2}\left[\frac{\partial^2 F}{\partial \varepsilon_{ij} \partial w_{kl}}\right]_0 \varepsilon_{ij}w_{kl} + \frac{1}{2}\left[\frac{\partial^2 F}{\partial w_{ij} \partial w_{kl}}\right]_0 w_{ij}w_{kl} \\
&\quad + \frac{1}{2}\left[\frac{\partial^2 F}{\partial w_{ij} \partial \varepsilon_{kl}}\right]_0 w_{ij}\varepsilon_{kl} \\
&= \frac{1}{2}C_{ijkl}\varepsilon_{ij}\varepsilon_{kl} + \frac{1}{2}R_{ijkl}\varepsilon_{ij}w_{kl} + \frac{1}{2}K_{ijkl}w_{ij}w_{kl} + \frac{1}{2}R'_{ijkl}w_{ij}\varepsilon_{kl} \\
&= F_u + F_w + F_{uw}
\end{aligned}
\tag{3.4.1}
$$

where F_u, F_w, and F_{uw} denote the contributions by a phonon, phason, and phonon-phason coupling respectively.

$$
C_{ijkl} = \left[\frac{\partial^2 F}{\partial \varepsilon_{ij} \partial \varepsilon_{kl}}\right]_0
\tag{3.4.2}
$$

is the phonon elastic constant tensor, which has been discussed in Chap. 2 already, and

$$
C_{ijkl} = C_{klij} = C_{jikl} = C_{ijlk}.
\tag{3.4.3}
$$

The tensor can be expressed by a symmetric matrix.

$$
[C]_{9\times9}.
$$

In Eq. (3.4.1), another elastic constant tensor is defined as

$$
K_{ijkl} = \left[\frac{\partial^2 F}{\partial w_{ij} \partial w_{kl}}\right]_0,
\tag{3.4.4}
$$

where the suffixes j, l belong to space $E_{\parallel}^3$, i, k belong to space $E_{\perp}^3$, and

$$
K_{ijkl} = K_{klij}.
\tag{3.4.5}
$$

All components K_{ijkl} can also be expressed by a symmetric matrix.
$[K]_{9\times9}.$

In addition,

$$R_{ijkl} = \left[\frac{\partial^2 F}{\partial \varepsilon_{ij} \partial w_{kl}} \right]_0 \tag{3.4.6}$$

and

$$R'_{ijkl} = \left[\frac{\partial^2 F}{\partial w_{ij} \partial \varepsilon_{kl}} \right]_0 \tag{3.4.7}$$

are the elastic constants of phonon-phason coupling. It should be noted that the suffixes i, j, l belong to space $E_\parallel^3$, k belongs to space $E_\perp^3$, and

$$R_{ijkl} = R_{jikl}, \ R'_{ijkl} = R_{klij}, \ R'_{klij} = R_{ijkl}, \tag{3.4.8}$$

but

$$R_{ijkl} \neq R_{klij}, \ R'_{ijkl} \neq R'_{klij}, \tag{3.4.9}$$

all the components of which can be expressed in symmetric matrixes as

$$[R]_{9\times9} \ \text{and} \ \left[R'\right]_{9\times9},$$

and

$$[R]^T = \left[R'\right], \tag{3.4.10}$$

where T denotes the transpose operator. The composition of four matrixes, $[C], [K], [R]$ and $\left[R'\right]$ forms a matrix of dimension 18×18

$$[C, K, R] = \begin{bmatrix} [C] & [R] \\ [R'] & [K] \end{bmatrix} = \begin{bmatrix} [C] & [R] \\ [R]^T & [K] \end{bmatrix}. \tag{3.4.11}$$

If the strain tensor is expressed by a row vector with 18 elements, i.e.,

$$\left[\varepsilon_{ij}, w_{ij}\right] = \begin{bmatrix} \varepsilon_{11}, \varepsilon_{22}, \varepsilon_{33}, \varepsilon_{23}, \varepsilon_{31}, \varepsilon_{12}, \varepsilon_{32}, \varepsilon_{13}, \varepsilon_{21}, \\ w_{11}, w_{22}, w_{33}, w_{23}, w_{31}, w_{12}, w_{32}, w_{13}, w_{21} \end{bmatrix}, \tag{3.4.12}$$

Its transpose denotes the array vector, and the free energy density (or strain energy density) may be expressed by

$$F = \frac{1}{2}\left[\varepsilon_{ij}, w_{ij}\right]\begin{bmatrix} [C] & [R] \\ [R]^T & [K] \end{bmatrix}\left[\varepsilon_{ij}, w_{ij}\right]^T, \qquad (3.4.13)$$

which is identical to that given by Eq. (3.4.1).

3.5 Generalized Hooke's Law

For the application of the theory of elasticity of quasicrystals to any science or
engineering problem, one must determine the displacement field and the stress field.
A relationship between strain and stress needs to be set up. This relationship is
called the generalized Hooke's law of quasicrystalline material. From the free energy
density (3.4.1) or (3.4.13) we have

$$\sigma_{ij} = \frac{\partial F}{\partial \varepsilon_{ij}} = C_{ijkl}\varepsilon_{kl} + R_{ijkl}w_{kl}$$

$$H_{ij} = \frac{\partial F}{\partial w_{ij}} = K_{ijkl}w_{kl} + R_{klij}\varepsilon_{kl} \qquad (3.5.1)$$

or in the form of matrixes

$$\begin{bmatrix} \sigma_{ij} \\ H_{ij} \end{bmatrix} = \begin{bmatrix} [C] & [R] \\ [R]^T & [K] \end{bmatrix}\begin{bmatrix} \varepsilon_{ij} \\ w_{ij} \end{bmatrix}, \qquad (3.5.2)$$

where

$$\begin{bmatrix} \sigma_{ij} \\ H_{ij} \end{bmatrix} = \left[\sigma_{ij}, H_{ij}\right]^T.$$

$$\begin{bmatrix} \varepsilon_{ij} \\ w_{ij} \end{bmatrix} = \left[\varepsilon_{ij}, w_{ij}\right]^T \qquad (3.5.3)$$

3.6 Boundary Conditions and Initial Conditions

The above general formulas describe the basic law of elasticity of quasicrystals and
provide the key to solving those problems in the application for academic research and
engineering practice. The formulas hold in any interior of the body, i.e., $(x, y, z) \in \Omega$,
where (x, y, z) denote the coordinates of any point of the interior, and Ω the body.
The formulas consist of some partial differential equations. To solve these equations,

it is necessary to know the situation of the field variables at the boundary S of Ω. Without appropriate information at the boundary, the solution has no physical meaning. According to practical cases, the boundary S consists of two parts, S_t and S_u, i.e., $S = S_t + S_u$. At S_t the tractions are given and at S_u the displacements are prescribed. For the former case,

$$\left.\begin{array}{l} \sigma_{ij}n_j = T_i \\ H_{ij}n_j = h_i \end{array}\right\} \quad (x, y, z) \in S_t, \tag{3.6.1}$$

where n_j represents the unit outward normal vector at any point at S, and T_i, and h_i are the traction and generalized traction vectors respectively, which are given functions at the boundary. Formula (3.6.1) is called the stress boundary conditions. For the latter case,

$$\left.\begin{array}{l} u_i = \overline{u}_i \\ w_i = \overline{w}_i \end{array}\right\} \quad (x, y, z) \in S_u, \tag{3.6.2}$$

where $\overline{u}_i$ and $\overline{w}_i$ are known functions at the boundary. Formula (3.6.2) is named as the displacement boundary conditions.

If $S = S_t$ (i.e., $S_u = 0$), the problem for solving Eqs. (3.2.3), (3.2.7), (3.3.1), and (3.5.1) under boundary conditions (3.6.1) is called the stress boundary value problem. While $S = S_u$ (i.e., $S_t = 0$), the problem for solving Eqs. (3.2.3), (3.2.7), (3.3.1), and (3.5.1) under boundary conditions (4.6.2) is called the displacement boundary value problem.

If $S = S_u + S_t$, and both $S_t \neq 0$, $S_u \neq 0$, the problem for solving Eqs. (3.2.3), (3.2.7), (3.3.1), and (3.5.1) under boundary conditions (3.6.1) and (4.6.2) is called the mixed boundary value problem.

For dynamic problems, if wave Eqs. (3.3.5) are taken together with (3.2.3), (3.2.7), and (3.5.1), besides boundary conditions (3.6.1) and (3.6.2), we must provide relevant initial conditions, i.e.,

$$\left.\begin{array}{l} u_i(x, y, z, 0) = u_{i0}(x, y, z), \ \dot{u}_i(x, y, z, 0) = \dot{u}_{i0}(x, y, z) \\ w_i(x, y, z, 0) = w_{i0}(x, y, z), \ \dot{w}_i(x, y, z, 0) = \dot{w}_{i0}(x, y, z) \end{array}\right\} (x, y, z) \in \Omega \tag{3.6.3}$$

in which $u_{i0}(x, y, z, 0)$, $\dot{u}_{i0}(x, y, z, 0)$, $w_{i0}(x, y, z, 0)$, and $\dot{w}_{i0}(x, y, z, 0)$ are known functions, etc. In this case, the problem is called as initial-boundary value problem.

When the wave equations are coupled with diffusion Eqs. (3.3.6), together with (3.2.3) and (3.5.1), the initial value conditions will be

$$\left.\begin{array}{l} u_i(x, y, z, 0) = u_{i0}(x, y, z), \ \dot{u}_i(x, y, z, 0) = \dot{u}_{i0}(x, y, z) \\ w_i(x, y, z, 0) = w_{i0}(x, y, z) \end{array}\right\} (x, y, z) \in \Omega. \tag{3.6.4}$$

This is also called an initial-boundary value problem, but different from the previous one.

3.7 Solutions of Elasticity

Based on the formulation given above, systematical solutions of elasticity for different systems of solid quasicrystals can be obtained. Among those solutions, there are many analytic exact ones (refer to Fan [5]).

3.8 Hydrodynamics of Solid Quasicrystals

Hydrodynamics is one of the most important branches of the study of solid quasicrystals, refer to Lubenskyet al [19]. Although the discussion on phonon-phason dynamics, suggested by Rochal and Norman [20] and Fan et al. [21], is concerned somewhat with hydrodynamics, the description there is much simplified, and some natures of the hydrodynamics of quasicrystals have not been considered. We herein intend to give a detailed introduction on the hydrodynamics of solid quasicrystals following Lubensky et al. The theory is concerned with many aspects of physics and mathematics, which are listed in Chap. 5 of this book.

Before the discovery of quasicrystals, hydrodynamics of solid (crystals) have been developed, see, e.g., the work of Martin et al. [22], Fleming and Cohen [23], which are related to the viscosity of solid. Considering the viscosity, the numbers of field variables and field equations are extended. Nature is connected to symmetry breaking. At first, we introduce some fundamental concepts of viscosity of solid, which will be beneficial to the understanding of the hydrodynamics of quasicrystals.

3.8.1 Viscosity of Solid

The elasticity discussed in Sects. 3.1, 3.2, 3.3, 3.4, 3.5, and 3.6 is reversible, while viscosity is an irreversible deformation. The irreversibility of viscosity lies in the existence of dissipation. To study the viscosity of solid, one can use a method similar to that in fluid dynamics. By introducing the velocity of mass point $\mathbf{V} = \left(V_x, V_y, V_z \right)$ and tensor of deformation velocity

$$\dot{\xi}_{ij} = \frac{1}{2}\left(\frac{\partial V_i}{\partial x_j} + \frac{\partial V_j}{\partial x_i} \right), \tag{3.8.1}$$

the viscosity stress tensor is defined by

$$\sigma'_{ij} = 2\eta_L\left(\dot{\xi}_{ij} - \frac{1}{3}\dot{\xi}_{kk}\delta_{ij}\right) + \eta_T\dot{\xi}_{kk}\delta_{ij}, \tag{3.8.2}$$

where only the isotropic viscosity is considered. η_L is the longitudinal viscosity constant, and η_T is the transverse one. Equation (3.8.2) represents the constitutive law of viscosity of isotropic solid.

The general constitutive law of viscosity of solid is

$$\sigma'_{ij} = \eta_{ijkl}\dot{\xi}_{kl}, \tag{3.8.3}$$

where η_{ijkl} represents the viscosity coefficient tensor of anisotropic viscosity of solid.

A description of the viscosity of solid can also be provided by introducing a dissipation function R such as

$$R = \frac{1}{2}\eta_{ijkl}\dot{\xi}_{ij}\dot{\xi}_{kl}, \tag{3.8.4}$$

so that we have

$$\sigma'_{ij} = \frac{\partial R}{\partial \dot{\xi}_{ij}}, \tag{3.8.5}$$

which is similar to the strain energy in elasticity in form.

3.8.2 *Hydrodynamics of Solid Quasicrystals*

Considering both elasticity and viscosity of solid quasicrystals allows one to establish hydrodynamics (Lubensky et al. [19]). The governing equations of hydrodynamics can be described with the Poisson bracket method. There are four sets of equations: mass conservation equation, momentum conservation equations, and equations of motions of phonons and phasons due to symmetry breaking if the energy conservation equation is not taken into account. The mass conservation equation is:

$$\frac{\partial \rho(r, t)}{\partial t} = -\nabla_i(r)(\rho V_i) \tag{3.8.6}$$

and the momentum conservation law is:

$$\frac{\partial g_i(r, t)}{\partial t} = -\nabla_k(r)(V_k g_i) + \nabla_j(r)\left(\eta_{ijkl}\nabla_k(r)V_l\right)$$

$$-\left(\delta_{ij} - \nabla_i(r)u_j\right)\frac{\delta H}{\delta u_i} - \rho\nabla_i(r)\frac{\delta H}{\delta\rho},$$

$$g_j = \rho V_j \tag{3.8.7}$$

In addition, there are equations of motions of phonons due to symmetry breaking

$$\frac{\partial u_i(r,t)}{\partial t} = -V_j\nabla_j(r)u_i - \Gamma_u\frac{\delta H}{\delta u_i(r,t)} + V_i \tag{3.8.8}$$

and phason dissipation equations

$$\frac{\partial w_i(r,t)}{\partial t} = -V_j\nabla_j(r)w_i - \Gamma_w\frac{\delta H}{\delta w_i(r,t)} \tag{3.8.9}$$

where the Hamiltonian H is defined by:

$$\begin{cases} H = \int\left[\dfrac{g^2}{2\rho} + \dfrac{1}{2}A\left(\dfrac{\delta\rho}{\rho_0}\right)^2 + B\left(\dfrac{\delta\rho}{\rho_0}\right)\nabla\cdot u\right]d^d r + F_u + F_w + F_{uw}, \\ g = \rho V \end{cases} \tag{3.8.10}$$

The integral in (3.8.10) describes the contributions of momentum and variation of mass density. The last three terms of (3.8.10) denote the contributions of phonon, phason, and phonon-phason coupling. A, B are new constants of materials describing effects from a variety of mass densities, which are named Lubensky-Ramaswamy-Toner (LRT) constants due to their contributions. In Eq. 3.8.10, the superscript of the volume element of the integral represents a dimension. Equations (3.8.6)–(3.8.9) are the equations of motion of hydrodynamics for solid quasicrystals. The field variables include mass density ρ, velocity V_i (or momentum ρV_i), phonon displacement u_i, and phason displacement w_i. To write the Hamiltonian H in Eq. (3.8.10), one must give the constitutive law of quasicrystals, in which the elastic constitutive equations are discussed in detail in the first 16 chapters of the second edition of the book [5].

The equations listed above were first derived by Lubensky et al. in 1985, but with a lack of details of the derivation. We will elaborate on this in Chap. 5.

3.9 Solution of the Hydrodynamics of Solid Quasicrystals

Despite linearization in Sects. 3.1, 3.2, 3.3, 3.4, 3.5, and 3.6 and solutions of Eqs. (3.8.1)–(3.8.4) in the Fourier transform domain, there has been no solution in the time-spatial domain thus far. Cheng et al. [24] first calculate the variation of mass density and viscosity normal stress as a function of time in detail, as illustrated in Figs. 3.1 and 3.2, respectively.

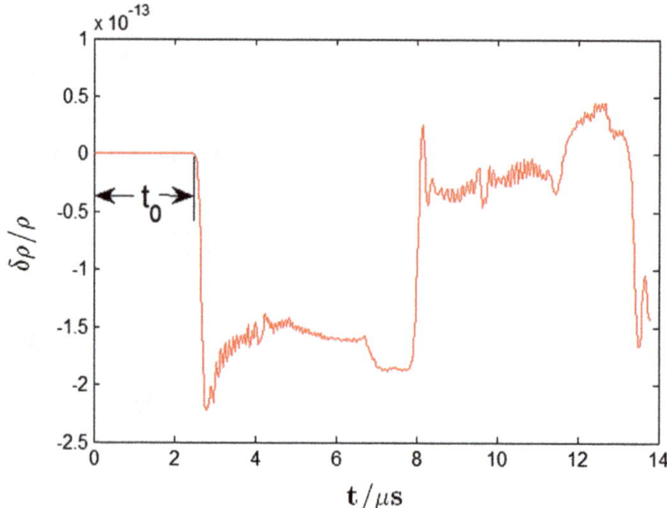

Fig. 3.1 Variation of the mass density of the computational point A_1 (or A_2) of specimen versus time

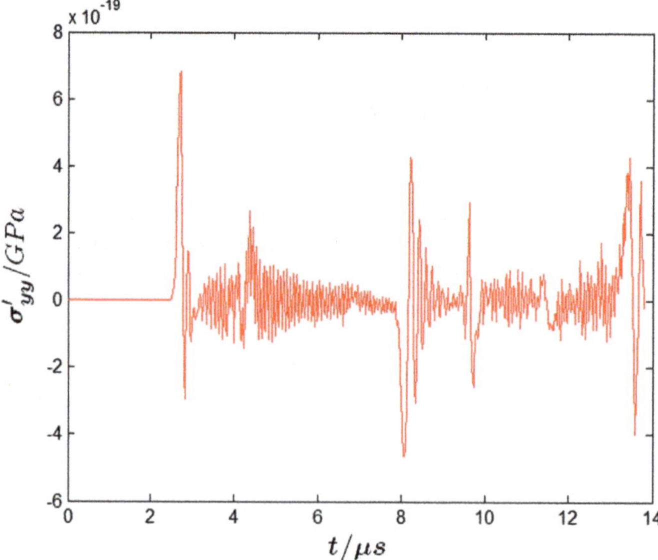

Fig. 3.2 Viscosity normal stress at the computational point A_1 (or A_2) of specimen versus time

3.10 Conclusion and Discussion

This chapter provides an introduction to the elasticity and hydrodynamics of solid quasicrystals. The elasticity of quasicrystals is familiar to many researchers in this field, and the hydrodynamics was introduced based on the work of Lubensky et al., which is a developed one from the crystal cased from Martin et al. [22] and Fleming and Cohen [23]. There are few numerical results of the hydrodynamics. Cheng et al. [24] partially solved the variation of mass density and viscosity normal stress at one location of the specimen as a function of time, in which case they have shown the difficulties of the study and further work needs to be developed.

References

1. Bak, P.: Phenomenological theory of icosahedral incommensurate ("quasiperiodic") order in Mn-Al alloys. Phys. Rev. Lett. **54**, 1517–1519 (1985)
2. Bak, P.: Symmetry, stability and elastic properties of icosahedral incommensurate crystals. Phys. Rev. B **32**, 5764–5772 (1985)
3. Landau, L.D., Lifshitz, E.M.: Theoretical Physics V: Statistical Physics, Part 1, 3rd edn. Pregamen Press, New York (1980)
4. Anderson, P. W.: Basic Notations of Condensed Matter Physics. Benjamin-Cummings, Menlo-Park (1984)
5. Fan, T.Y.: Mathematical Theory of Elasticity of Quasicrystals and Its Applications. 1st edn. Science Press, Beijing (2010); 2nd edn. Springer, Heidelberg (2016)
6. Levine, D., Lubensky, T.C., Ostlund, S., Ramaswamy, S., Steinhardt, P.J., Toner, J.: Elasticity and dislocations in pentagonal and icosahedral quasicrystals. Phys. Rev. Lett. **54**, 1520–1523 (1985)
7. Lubensky, T.C., Ramaswamy, S., Toner, J.: Dislocation motion in quasicrystals and implications for macroscopic properties. Phys. Rev. B **33**, 7715–7719 (1986)
8. Lubensky, T.C., Socolar, J.E.S., Steinhardt, P.J., Bancel, P.A., Heiney, P.A.: Distortion and peak broadening in quasicrystal diffraction patterns. Phys. Rev. Lett. **57**, 1440–1443 (1986)
9. Lubensky, T.C.: Symmetry, elasticity and hydrodynamics in quasiperiodic structures. In: Jaric, M.V. (ed.) Introduction to Quasicrystals, pp. 199–289. Academic Press, Boston (1988)
10. Kalugin, P.A., Kitaev, A., Levitov, L.S.: 6-Dimensional properties of $Al_{0.86}Mn_{0.14}$ alloy. J. Phys. Lett. **46**, 601–607 (1985)
11. Torian, S.M., Mermin, D.: Mean-field theory of quasicrystalline order. Phys. Rev. Lett. **54**, 1524–1527 (1985)
12. Jaric, M.V.: Long-range icosahedral orientational order and quasicrystals. Phys. Rev. Lett. **55**, 607–610 (1985)
13. Duneau, M., Katz, A.: Quasiperiodic patterns. Phys. Rev. Lett. **54**, 2688–2691 (1985)
14. Socolar, J.E.S., Lubensky, T.C., Steinhardt, P.J.: Phonons, phasons, and dislocations in quasicrystals. Phys. Rev. B **34**, 3345–3360 (1986)
15. Gahler, F., Rhyner, J.: Equivalence of the generalised grid and projection methods for the construction of quasiperiodic tilings. J. Phys. A: Math. Gen. **19**, 267–277 (1986)
16. Dolinsek, J., Ambrosini, B., Vonlanthen, P., et al: Atomic motion in quasicrystalline $Al_{70}Re_{8.6}Pd_{21.4}$: a two-dimensional exchange NMR study. Phys. Rev. Lett. **81**, 3671–3674 (1998)
17. Dolinsek, J., Apih, T., Simsic, M., et al.: Self-diffusion in icosahedral $Al_{72.4}Pd_{20.5}Mn_{7.1}$ and phason percolation at low temperatures studied by ^{27}Al NMR. Phys. Rev. Lett. **82**, 572–575(1992)

18. Ding, D.H., Yang, W.G., Wang, R.H., Hu, C.Z.: Generalized elasticity theory of quasicrystals. Phys. Rev. B **48**, 7003–7010 (1993)
19. Lubensky, T.C., Ramaswamy, S., Toner, J.: Hydrodynamics of icosahedral quasicrystals. Phys. Rev. B **32**, 7444–7452 (1985)
20. Rochal, S.B., Norman, V.L.: Minimal model of the phonon-phason dynamics of quasicrystals. Phys. Rev. B **66**, 144204 (2002)
21. Fan, T.Y., Wang, X.F., Li, W., Zhu, A.Y.: Elasto-hydrodynamics of quasicrystals. Phil. Mag. **89**, 501–512 (2009)
22. Martin, P.C., Parodi, O., Pershan, P.S.: Unified hydrodynamic theory for crystals, liquid crystals and normal fluids. Phys. Rev. A **6**, 2401–2420 (1972)
23. Fleming, P.D., Cohen, C.: Hydrodynamics of solids. Phys. Rev. B **13**, 500–516 (1976)
24. Cheng, H.: Fan T Yand Wei H, Solutions for hydrodynamics of 5- and 10-fold symmetry quasicrystals. Appl. Math. Mech. **37**, 1393–1404 (2016)

Chapter 4
Case Study of Equation of State in Several Structured Fluids

Equation of state, i.e., the equation connecting pressure and mass density referred here, is one of the fundamental properties for all condensed matter. Certainly, it is very important for describing the motion of conventional fluid and other substantive systems described in this book [1]. As Landau and Lifshitz [2] pointed out that in the general theory of relativity, there are equations of the gravitational field but a lack of equation of state, so the matter distribution and motion are not completely determined yet.

To that end, for solving the equation of state, one must analyze fluids from a microscopic scale. Qian [3] has paid an effort to study the equation of state of liquids from statistical physics and quantum mechanics and obtained some useful results. So far the study on the equation of state is often based on a phenomenological model, and the results are in a high degree of approximation.

Here we focus only on a special problem related to the soft-matter quasicrystals. There is a longstanding puzzle concerning the question of the equation of state in the hydrodynamics of soft matter. This chapter focuses only on several special case studies of the equation of state for a class of soft matter first, to extend the study to soft-matter quasicrystals in the later chapters. We will only list some results and not do in-depth derivation.

4.1 Introduction of Equation of State in Some Fluids

There are some successful examples of thermodynamics study in complex fluids, among them the superfluid liquid [4] He has been studied extensively, whose equation of state—the relation between pressure and mass density—is expressed by a simple form [4]

$$p = a\rho^2 + b\rho^3 + c\rho^4 \tag{4.1.1}$$

where a, b, and c are constants, determined by experimental methods. The theoretical simulation matches the experimental results very well [5, 6]. Although this is the case for the dense system and extremely low temperature, the experience especially the concept of fluid phonon developed by the Landau school [7] generated significant meaning for other complex liquids. The form of the equation can also be referred to study other complex liquids including soft matter. The constants for individual complex liquids need extensive research to be determined by experiments or/and numerical simulation. So far there are no general results on a, b, and c for all soft matter in Eq. (4.1.1), we will not directly use this formula for the analysis in this chapter.

In the studies on thermodynamics of liquid crystals, for example in refs. [8, 9], a normalized equation of state for columnar liquid crystals in the one-dimensional case is given by Wensink [8]

$$p = 3\frac{k_B T}{L}\frac{\rho}{1-\rho} \tag{4.1.2}$$

where L is the thickness of hard disks. The expression is very simple. The formula is only for the special structure of liquid crystals (it requires that $\rho < 1$, in the dimensionless unit, i.e., $\rho < \rho_0$ in dimensional unit), and the result is hardly used in general cases of soft matter in computation.

In explosion physics, [10] an equation of state

$$p = \frac{\rho_0 c_0^2}{n}\left[\left(\frac{\rho}{\rho_0}\right)^n - 1\right] \tag{4.1.3}$$

has been adopted in macroscope, and can be extended for some solutions. Here parameters n and c_0 are determined empirically. It is difficult to apply to general soft matter.

In the case of a general theory of relativity, as pointed out by Landau and Lifshitz [2], apart from the equations of the gravitational field, a corresponding equation of state is necessary, especially to the relativistic hydrodynamics. The equation of state $p = p(\rho_0, \varepsilon)$ is necessary to form a complete set of the evolution equations, in which ρ_0 denotes the rest-mass density (in sense of relativity), ε the specific internal energy. Alcubierre [11] suggested

$$p = K\rho_0^{\gamma} \tag{4.1.4}$$

$$\varepsilon = \frac{K}{\gamma - 1}\rho_0^{\gamma - 1} \tag{4.1.5}$$

here K is a constant, ρ_0 the rest-mass density (in sense of relativity), and

$$\gamma = \frac{c_p}{c_V} \tag{4.1.6}$$

in which c_p and c_V are the specific heat of the matter at constant pressure and constant volume, respectively. It is worth noting that (4.1.4) is similar to that in gas dynamics of conventional fluids.

Readers who are interested in the deep discussion can refer to the corresponding monograph, and we here do not discuss anymore in detail.

4.2 Possible Equations of State

The difficulties listed above suggest that we take another way for the probe. Some lessons on the thermodynamics of crystals are also beneficial. The Debye theory [12, 13] for crystals was well-known. In addition, drawn from the Landau school [6] to study Bose liquid, and a fluid phonon concept is developed. By learning the Debye [12, 13] theories, Fan and Fan [14] suggested an equation of state for soft-matter quasicrystals based on the four phonon model [15], the form of the equation is simple but the coefficients are complex which concern some macro- and micro-structure constants. For this reason, further discussion is omitted.

The difficulty lets us return to the classical thermodynamics [16], which is well-known for conventional liquids and solids, the first-order approximate equation of state is

$$V(T, p) = V_0(T_0, 0)[1 + \alpha(T - T_0) - \kappa_T p] \tag{4.2.1}$$

because soft matter is an intermediate phase between conventional liquid and solid, this equation can be used to describe soft matter. In (4.2.1) the initial state pressure is $p_0 = 0$, the initial state temperature is T_0. It is the temperature at which the vapor pressure is equal to zero. In addition, κ_T denotes the coefficient of an isothermal compression. Suppose the physical process in our studied soft matter is in an isothermal environment, that is $T = T_0$. The equation of state of soft matter may be approximated by the following relation

$$V(p) = V_0(0)[1 - \kappa_T p] \tag{4.2.2}$$

Due to the isothermal process, the total mass of soft matter is constant, and the volume ratio and density ratio are reciprocal to each other. The pressure expression of the equation of state for soft matter can be written as follows

$$p = \frac{1}{\kappa_T}\left[1 - \frac{\rho_0(0)}{\rho(p)}\right] = \frac{1}{\kappa_T}\left(1 - \frac{\rho_0}{\rho}\right) \tag{4.2.3}$$

The so-called coefficient of isothermal compression κ_T is a key for the dynamics of soft-matter quasicrystals in this book. According to our systematical and large scale computations, we take

$$\kappa_T = 10^{-5}/\text{Pa} \qquad\qquad (4.2.4)$$

all computational results are best.

The Eq. (4.2.3) with the coefficient of isothermal compression κ_T given by (4.2.4) constitutes one of the basic equations in the governing equations of dynamics of soft-matter quasicrystals, helps the establishing of the dynamics system of the matter and development of the solving system of the dynamics theory.

4.3 Applications to Dynamics of Soft-Matter Quasicrystals

The soft-matter quasicrystals belong to a complex liquid, the matter distribution, deformation, and motion should be described through corresponding complex fluid dynamics or generalized dynamics, which constitutes the theme of this book. If the fluid is compressible and there is no equation of state, then the equation system of the generalized dynamics is not closed, i.e., the number of the field equations is less than that of the field variables, the problem has no physical meaning, and cannot be solved mathematically. We use the equation of state (4.2.3), and so the problem is consistent physically and mathematically, the computations can be done, but the accuracy is considerably good. The choice of the value of parameter κ_T strongly affects the computational results, the experimental verification is undertaken.

4.4 The Incompressible Model of Soft Matter

If we consider the soft matter as an incompressible complex fluid, then the above equation of state is not needed, as we treat $\rho = $ const. Then the equations of motion of soft-matter quasicrystals will be much simplified so that a very difficult problem in the soft matter study is excluded in form. Solving the relevant initial- and boundary-value problems of these equations will be much easier for incompressible complex fluids than those for the compressible ones. In addition, a transition from the compressible model to the incompressible model of soft matter needs a change in the Hamiltonian of the considered system, this is not only a thermodynamic question but also a dynamic question in our study. The corresponding detail will be discussed in Chaps. 7, 8, 9, 10 and 11.

References

1. Chaikin, J., Lubensky, T.M.: Principles of Condensed Matter Physics. Oxford University Press, New York (1995)
2. Landau, L.D., Lifshitz, E.M.: Classical Theory of Field. Phys-Math Press, Moscow, in Russian (1962)
3. Qian, X.S.: Physical Mechanics, 1st edn. Science Press, Beijing (1962); 2nd edn. Shanghai Jiaotong University Press, Shanghai, in Chinese (2011)
4. Landau, L.D.: Theory of superfluidity of He II. J Phys. USSR **5**, 71–90 (1941)
5. Pitaevskii, L., Stringari, S.: Bose-Einstein Condensation. Oxford, Clarendon Press (2003)
6. Dalfovo, F., Lastri, A., Pricaupenko, L., Stringari, S., Treiner, J.: Phys. Rev. B **52**, 1193–1200 (1995)
7. Lifshitz, E.M., Pitaevskii, L.P.: Statistical Physics, Part 2. Pergamon Press, Oxford (1980)
8. Wensink, H.H.: Equation of state of a dense columnar liquid crystal. Phys. Rev. Lett. **93**, 157801 (2004)
9. Xu, W.S., Li, Y.W., Sun, Z.Y., An, L.J.: Hard ellipses: equation of state, structure and self-diffusion, arXiv.org. Condens Matter. arXiv.1212.6497 (2012)
10. Orlenko, L.P. (ed.): Explosion Physics, Moscow, ISBN, FIZMATLIT, 3rd edn. Chinese translation by Sun C W. Science Press, Beijing (2011)
11. Alcubierre, M.: Introduction to 3+1 Numerical Relativity. Oxford University Press, New York (2008)
12. Debye, P.: Die Eigentuemlichkeit der spezifischenWaermenbeitiefenTemperaturen. Arch de Genéve **33**, 256–258 (1912)
13. Sommerfeld, A.: VorlesungenuebertheoretischePhysik, Band II, Mechanik der deformierbaren-Medien, Verlag Harri-Deutsch, Thun.Frankfort/M (1992)
14. Fan, L., Fan, T.Y.: Equation of state of structured liquid. First Annual Symposium on Frontiers of Soft Matter Science and Engineering, Dec 12, Beijing, China (2015)
15. Fan, T.Y., Sun, J.J.: Four phonon model for for studying thermodynamics of soft-matter quasicrystals. Phil. Mag. Lett. **94**, 112–117 (2014)
16. Wang, Z.C.: Thermodynamics and Statistics. Higher Education Press, Beijing, in Chinese (2014)

Chapter 5
Poisson Brackets and Derivation of Equations of Motion in Soft-Matter Quasicrystals

Previous chapters provided basic concepts of soft-matter quasicrystals. For practice, we need to establish the equations of motion of the matter, then one can give a quantitative description of their structures and dynamic properties.

We will introduce briefly the Poisson bracket method in Sects. 5.1, 5.2, 5.3 and 5.4, and apply it for the derivation of equations of motion of soft-matter quasicrystals in Sects. 5.5 and 5.6.

5.1 Brownian Motion and Langevin Equation

Einstein [1] studied the motion of Brownian particles in 1905, and he proposed the displacement of a particle following the equation

$$m\frac{\mathrm{d}^2x}{\mathrm{d}t^2} + a\frac{\mathrm{d}x}{\mathrm{d}t} = f \tag{5.1.1}$$

in which the first term denotes inertia force, the second term is a resistance force, and the right-hand of the equation represents a stochastic force. This equation is named the Langevin equation. Einstein solved the equation first theoretically, which was proved by Perrin's experiments later [2].

5.2 Extended Version of Langevin Equation

In Eq. (5.1.1), define $\frac{\mathrm{d}x}{\mathrm{d}t} = V$, then we have

$$\frac{\mathrm{d}V}{\mathrm{d}t} = -\frac{a}{m}V + \frac{1}{m}f \tag{5.2.1}$$

© The Author(s), under exclusive license to Springer Nature Singapore Pte Ltd. 2022
T.-Y. Fan et al., *Generalized Dynamics of Soft-Matter Quasicrystals*, Springer Series in Materials Science 260, https://doi.org/10.1007/978-981-16-6628-5_5

which can be extended as

$$\frac{\partial \psi(r, t)}{\partial t} = -\Gamma \psi(r, t) + F_s \tag{5.2.2}$$

where $\psi(r, t)$ is a mechanics quantity, Γ represents a resistant force, and F_s a stochastic force. The equation describes a stochastic process.

5.3   Multivariable Langevin Equation, Coarse-Graining

Ginzburg and Landau extended (5.2.1) to the multivariable case as

$$\frac{\partial \psi_\alpha(r, t)}{\partial t} = -\Gamma_{\alpha\beta} \frac{\delta H}{\delta \psi_\beta(r, t)} + (F_s)_\alpha \tag{5.3.1}$$

here the summation convention is applied, i.e., the repetition of suffix means taking summation, and $H = H[\psi(r, t)]$ denotes an energy function, which can also be named as Hamiltonian, $\frac{\delta H}{\delta \psi_\beta(r, t)}$ represents a variation of $H = H[\psi(r, t)]$ with respect to $\psi_\beta(r, t)$, and $\Gamma_{\alpha\beta}$ is the element of resistant matrix (or dissipation kinetic coefficient matrix), the definitions of other quantities are the same as before. Equation (5.3.1) is a Langevin equation with multivariable, which can be extended in more scenarios. Taking the macroscopic quantity $\psi_\alpha(r, t)$ as a thermodynamic average of microscopic quantity $\psi_\alpha^\mu(r, \{q^\alpha\}, \{p^\alpha\})$, i.e.,

$$\psi_\alpha(r, t)) = \langle \psi_\alpha^\mu(r, \{q^\alpha\}, \{p^\alpha\}) \rangle \tag{5.3.2}$$

where p^α, q^α are the canonic momentum and canonic coordinate, the microquantities will follow the microscopic Liouville equation

$$\frac{\partial \psi_\alpha^\mu}{\partial t} = \{H^\mu, \psi_\alpha^\mu\} \tag{5.3.3}$$

where $\{H^\mu, \psi_\alpha^\mu\}$ represents classical Poisson bracket which will be discussed in detail in the following, $H^\mu(\{q^\alpha\}, \{p^\alpha\})$ denotes the Hamiltonian of the microsystem. Equation (5.3.2) represents a coarse-graining treatment, indicating that the discussion here is macroscopic.

Equation (5.3.1) implies that

$$\frac{\partial \Psi_\alpha(r, t)}{\partial t} + \int \left(\{\Psi_\beta(r'), \Psi_\alpha(r)\} \frac{\delta H}{\delta \Psi_\beta(r', t)} \right) d^d r'$$

$$- \int \left(\frac{\delta\{\Psi_\beta(r'), \Psi_\alpha(r)\}}{\delta \Psi_\beta(r', t)} \right) d^d r' = -\Gamma_{\alpha\beta} \frac{\delta H}{\delta \Psi_\beta(r, t)} + (F_s)_\alpha \tag{5.3.4}$$

where $\{\Psi_\beta(r'), \Psi_\alpha(r)\}$ is the Poisson bracket. This is a generalized Langevin equation.

More detailed discussion on generalized Langevin equation can be found in monographs [3, 4].

5.4 Poisson Bracket Method in Condensed Matter Physics

In the following chapters, the equations of motion of soft-matter quasicrystals are derived by the Poisson bracket method individually, we first introduce the methodology, followed by an outline of the derivation in common.

Due to symmetry breaking, some equations' derivation of the hydrodynamics of certain substantive systems cannot be obtained directly by conventional conservation laws. The Poisson brackets in condensed matter physics become an important tool for theoretical derivation, which can simplify the calculation dramatically. The method is originated from Landau and his school in the former Soviet Union, refer to [5–10]. Physicists Martin et al. [11], Fleming and Cohen [12] in the USA developed the method for hydrodynamics of crystals and liquid crystals, but their derivations were still lengthy. Lubensky et al. [13, 14] further developed the approach in deriving the hydrodynamic equations of solid quasicrystals and made it systematically. Fan [15] gave an introduce the applications of the method in other problems.

Poisson brackets come from classical analytic mechanics. For two mechanical quantities f, g the quantity

$$\{f, g\} = \sum_i \left(\frac{\partial f}{\partial q_i} \frac{\partial g}{\partial p_i} - \frac{\partial f}{\partial p_i} \frac{\partial g}{\partial q_i} \right) \qquad (5.4.1)$$

is called classical Poisson bracket, where p_i, q_i denote the canonic momentum and canonic coordinate, respectively.

Relative to the classical Poisson bracket (5.4.1), a quantum Poisson bracket related to the commutator in quantum mechanics as

$$\left[\hat{A}, \hat{B} \right] = \hat{A}\hat{B} - \hat{B}\hat{A} \qquad (5.4.2)$$

is often used, here $\hat{A}, \hat{B}$ represent two operators, e.g., $\hat{A}$ represents coordinate operator x_α, $\hat{B}$ the momentum operator p_β, then

$$\left[x_\alpha, p_\beta \right] = i\hbar\delta_{\alpha\beta}, \quad \left[x_\alpha, x_\beta \right] = 0, \quad \left[p_\alpha, p_\beta \right] = 0 \qquad (5.4.3)$$

where $i = \sqrt{-1}$, $\hbar = h/2\pi$, h the Planck constant, $\delta_{\alpha\beta}$ unit tensor. Equation (5.4.3) is named quantum Poisson bracket. In quantum mechanics, mechanical quantities represent operators. Equation (5.4.3) holds for any operators, in general.

There is an inherent connection between the quantum Poisson bracket and classical Poisson bracket, i.e.,

$$\lim_{\hbar \to 0} \frac{i\left[\hat{A}\hat{B} - \hat{B}\hat{A}\right]}{\hbar} = \{A, B\} \tag{5.4.4}$$

This is a well-known result in quantum mechanics.

Landau [6] introduced the limit passing over (5.4.4) from quantum Poisson bracket to classical Poisson bracket in deriving the hydrodynamic equations of a superfluid. He took the expansions of mass density and momentum such as:

$$\hat{\rho}(r) = \sum_{\alpha} m_{\alpha} \delta(r_{\alpha} - r) \tag{5.4.5}$$

$$\hat{g}_k(r) = \sum_{\alpha} \hat{p}_k^{\alpha} \delta(r_{\alpha} - r) \tag{5.4.6}$$

whose quantum Poisson brackets are

$$\left[\hat{\rho}(r_1), \hat{\rho}(r_2)\right] = 0$$
$$\left[\hat{p}_k(r_1), \hat{\rho}(r_2)\right] = i\hbar\hat{\rho}(r_1)\nabla_k(r_1)\delta(r_1 - r_2)$$
$$\left[\hat{p}_k(r_1), \hat{p}_l(r_2)\right] = i\hbar\big(\hat{p}_l(r_1)\nabla_k(r_1) - \hat{p}_k(r_2)\nabla_k(r_2)\big)\delta(r_1 - r_2) \tag{5.4.7}$$

where $\nabla_k(r_1)$ represents derivative carrying out on coordinate r_1, and $\nabla_l(r_2)$ on coordinate r_2.

By using the limit passing over (5.4.4) from quantum to classical Poisson bracket, from (5.4.7) one can obtain the corresponding classical Poisson brackets:

$$\{p_k(r_1), \rho(r_2)\} = \rho(r_1)\nabla_k(r_1)\delta(r_1 - r_2)$$
$$\{p_k(r_1), p_l(r_2)\} = (p_l(r_1)\nabla_k(r_1) - p_k(r_2)\nabla_l(r_2))\delta(r_1 - r_2) \tag{5.4.8}$$

5.5 Application of Poisson Bracket to Quasicrystals

Lubensky et al. [13] extended the above discussion to solid quasicrystals. In Chap. 3, we introduced the phason elementary excitation beyond the phonon elementary excitation. The phason concept was originated from aperiodic crystals [13, 14]. Aperiodic crystals include incommensurate crystals and quasicrystals, and we here consider only quasicrystals, in which the phonon type displacements are called phonon field u_i, and phason type displacements are called phason field w_i, respectively.

Through the similar expansion of the displacement vectors u_i and w_i, we have

$$u_k(r) = \sum_\alpha u_k^\alpha \delta(r_\alpha - r) \qquad (5.5.1)$$

$$w_k(r) = \sum_\alpha w_k^\alpha \delta(r_\alpha - r) \qquad (5.5.2)$$

By using the limit passing over (5.4.4) from quantum Poisson bracket to classical Poisson bracket, from (5.5.1) and (5.5.2) one can find those corresponding classical Poisson brackets as follows

$$\{u_k(r_1), g_l(r_2)\} = (-\delta_{kl} + \nabla_l(r_1)u_k)\delta(r_1 - r_2) \qquad (5.5.3)$$

$$\{w_k(r_1), g_l(r_2)\} = (\nabla_l(r_1)w_k)\delta(r_1 - r_2) \qquad (5.5.4)$$

It is evident that (5.5.4) is quite different from (5.5.3), which leads to the dissipation equations of phason given in the subsequent discussion, which are quite different from equations of motion of phonons due to symmetry breaking. The relevant derivations were carried out by Lubensky et al. [13].

5.6 Equations of Motion of Soft-Matter Quasicrystals

In Chap. 3, the equations on elasticity and hydrodynamics of solid quasicrystals have provided us a basis for equations of motion of soft-matter quasicrystals.

In this section, we will give a general outline of deriving the equations of motion of soft-matter quasicrystals through Poisson brackets. This method can be applied in many general physics fields. The details for different quasicrystal systems will be discussed individually in Chaps. 7–11. Below we will introduce the application of the Poisson bracket method with generalized Langevin equation.

5.6.1 Generalized Langevin Equation

Besides Poisson brackets, some other formulas will be taken as well in the derivation of hydrodynamic equations of quasicrystals, which are related to the Langevin equation or generalized Langevin equation, refer to Sects. 5.1, 5.2 and 5.3 of this chapter.

In the previous Sects. 5.1, 5.2 and 5.3, we introduced Langevin equation

$$\frac{\partial \Psi(r, t)}{\partial t} = -\Gamma \Psi(r, t) + F_s \qquad (5.6.1)$$

here, $\Psi(r, t)$ is a mechanical quantity, Γ represents a resistant force, F_s a stochastic force. We also know the equation with multi-variables

$$\frac{\partial \Psi_\alpha(r, t)}{\partial t} = -\Gamma_{\alpha\beta} \frac{\delta H}{\delta \Psi_\beta(r, t)} + (F_s)_\alpha \tag{5.6.2}$$

where $H = H[\Psi(r, t)]$ denotes an energy function Hamiltonian, $\frac{\delta H}{\delta \Psi_\beta(r,t)}$ represents a variation of $H = H[\Psi(r, t)]$ with respect to $\Psi_\beta(r, t)$, and $\Gamma_{\alpha\beta}$ is the element of resistant matrix (or dissipation kinetic coefficient matrix).

In d dimensional space, the partial derivative of macro-quantity $\Psi_\alpha(r, t)$ with time

$$\frac{\partial \Psi_\alpha(r, t)}{\partial t}$$

stands for

$$\frac{\partial \Psi_\alpha(r, t)}{\partial t} = - \int \left(\{\Psi_\beta(r'), \Psi_\alpha(r)\} \frac{\delta H}{\delta \Psi_\beta(r', t)} \right) d^d r'$$
$$+ \int \left(\frac{\delta \{\Psi_\beta(r'), \Psi_\alpha(r)\}}{\delta \Psi_\beta(r', t)} \right) d^d r' - \Gamma_{\alpha\beta} \frac{\delta H}{\delta \Psi_\beta(r, t)} + (F_s)_\alpha \tag{5.6.3}$$

where $d^d r' = dV$ represents the volume element of the integral. Combination with (5.4.8), (5.5.3) and (5.5.4), we will derive the hydrodynamic equations of quasicrystals from Eq. (5.6.3) in the next subsection. For the static case, the last term in (5.6.3) is omitted.

5.6.2 Derivation of Generalized Dynamic Equations of Soft-Matter Quasicrystals

The equation of mass conservation is the same as that of simple fluid, so we do not repeat it here.

First, we consider the equations of motion of phonons due to symmetry breaking. Put $\Psi_\alpha(r, t) = u_i(r, t)$, $\Psi_\beta(r', t) = g_j(r', t)$ in (5.6.3) and ignore the second and fourth terms in the right-hand side of the equation, then

$$\frac{\partial u_i(r, t)}{\partial t} = - \int \left(\{u_i(r'), g_j(r)\} \frac{\delta H}{\delta g_j(r', t)} \right) d^d r' - \Gamma_u \frac{\delta H}{\delta u_i(r, t)}$$

Substituting bracket (5.5.3) into the integral of right-hand side yields

$$\frac{\partial u_i(r, t)}{\partial t} = \int \left(-\delta_{ij} + \nabla_j(r) u_i \right) \delta(r - r') \frac{g_j(r')}{\rho(r')} d^d r' + \Gamma_u \frac{\delta H}{\delta u_i(r, t)}$$

$$= -V_j \nabla_j(r) u_i - \Gamma_u \frac{\delta H}{\delta u_i(r, t)} + V_i \tag{5.6.4}$$

where Γ_u denotes the phonon dissipation kinematic coefficient, and the Hamiltonian is defined as

$$H = H[\Psi(r, t)] = \int \frac{g^2}{2\rho} \mathrm{d}^d r + \int \left[\frac{1}{2} A\left(\frac{\delta\rho}{\rho_0}\right)^2 + B\left(\frac{\delta\rho}{\rho_0}\right) \nabla \cdot u \right] \mathrm{d}^d r + F_{\mathrm{el}}$$

$$= H_{\mathrm{kin}} + H_{\mathrm{density}} + F_{\mathrm{el}}$$

$$F_{\mathrm{el}} = F_u + F_w + F_{uw}, g = \rho V \tag{5.6.5}$$

and V represents the fluid velocity, A, B are constants describing density variation named as Lubensky-Ramaswamy-Toner (LRT) constants, the last term of (5.6.5) represents elastic energy, which consists of phonon, phason, and phonon-phason coupling parts

$$F_u = \int \frac{1}{2} C_{ijkl} \varepsilon_{ij} \varepsilon_{kl} \mathrm{d}^d r$$

$$F_w = \int \frac{1}{2} K_{ijkl} w_{ij} w_{kl} \mathrm{d}^d r$$

$$F_{uw} = \int \left(R_{ijkl} \varepsilon_{ij} w_{kl} + R_{klij} w_{ij} \varepsilon_{kl} \right) \mathrm{d}^d r \tag{5.6.6}$$

respectively, C_{ijkl} the phonon elastic constants, K_{ijkl} phason elastic constants, and R_{ijkl}, R_{klij} the phonon-phason coupling elastic constants, and the strain tensors ε_{ij}, w_{ij} are defined by

$$\varepsilon_{ij} = \frac{1}{2}\left(\frac{\partial u_i}{\partial x_j} + \frac{\partial u_j}{\partial x_i}\right), \quad w_{ij} = \frac{\partial w_i}{\partial x_j} \tag{5.6.7}$$

the associated stress tensors are related through the constitutive law for soft-matter quasicrystals including 5-, 8-, 10-, and 12-fold symmetry (the 7-, 9-, 14-, and 18-fold symmetry quasicrystals will be discussed in Chaps. 10 and 11, and are not included here)

$$\left.\begin{array}{l}
\sigma_{ij} = C_{ijkl}\varepsilon_{ik} + R_{ijkl} w_{kl}, \\
H_{ij} = K_{ijkl} w_{ij} + R_{klij}\varepsilon_{kl}, \\
p_{ij} = -p\delta_{ij} + \sigma'_{ij} = -p\delta_{ij} + \eta_{ijkl}\dot{\xi}_{kl}, \\
\varepsilon_{ij} = \frac{1}{2}\left(\frac{\partial u_i}{\partial x_j} + \frac{\partial u_j}{\partial x_i}\right), w_{ij} = \frac{\partial w_i}{\partial x_j}, \dot{\xi}_{ij} = \frac{1}{2}\left(\frac{\partial V_i}{\partial x_j} + \frac{\partial V_j}{\partial x_i}\right)
\end{array}\right\} \tag{5.6.8}$$

where η_{ijkl} denotes the viscosity coefficient tensor of fluid. Now consider the derivation of phason dissipation equations.

In (5.6.3) put $\Psi_\alpha(r, t) = w_i(r, t)$, $\Psi_\beta(r', t) = g_j(r', t)$, neglecting the second and fourth terms on the right-hand side, and substituting the Poisson bracket (5.6.4) into it leads to

$$\frac{\partial w_i(r, t)}{\partial t} = -\int \left(\{w_i(r'), g_j(r)\} \frac{\delta H}{\delta g_j(r', t)} \right) d^d r' - \Gamma_w \frac{\delta H}{\delta w_i(r, t)}$$

then

$$\frac{\partial w_i(r, t)}{\partial t} = \int \left(\nabla_j(r) w_i \right) \delta(r - r') \frac{g_j(r')}{\rho(r')} d^d r' - \Gamma_w \frac{\delta H}{\delta w_i(r, t)}$$

$$= -V_j \nabla_j(r) w_i - \Gamma_w \frac{\delta H}{\delta w_i(r, t)} \tag{5.6.9}$$

in which Γ_w denotes the phason dissipation coefficient, and Hamiltonian is defined by (5.6.5) and (5.6.6).

By comparing (5.6.4) and (5.6.9), it is found that the physical meanings of phonon and phason in a hydrodynamic sense are quite different. According to the explanation of Lubensky et al. [13], the phonon represents wave propagation, while phason represents diffusion.

Of course, the other difference between phonon and phason is that they belong to a different irreducible representation of the point group, which has been discussed in Chap. 3.

The momentum conservation equations are

$$\frac{\partial g_i(r, t)}{\partial t} = -\nabla_k(r)(V_k g_i) + \nabla_j(r)\left(-p\delta_{ij} + \eta_{ijkl}\nabla_k(r)V_l\right) - \left(\delta_{ij} - \nabla_i u_j\right)\frac{\delta H}{\delta u_j(r, t)}$$

$$+ \left(\nabla_i w_j\right)\frac{\delta H}{\delta w_j(r, t)} - \rho \nabla_i(r)\frac{\delta H}{\delta \rho(r, t)}, \quad g_j = \rho V_j \tag{5.6.10}$$

recall that η_{ijkl} denotes the viscosity coefficient tensor of fluid, and the fluid phonon stress tensor is

$$p_{ij} = -p\delta_{ij} + \sigma'_{ij} = -p\delta_{ij} + \eta_{ijkl}\dot{\xi}_{kl} \tag{5.6.11}$$

with the deformation velocity tensor

$$\dot{\xi}_{kl} = \frac{1}{2}\left(\frac{\partial V_k}{\partial x_l} + \frac{\partial V_l}{\partial x_k}\right) \tag{5.6.12}$$

Equations (5.6.10) can be understood as generalized Navier-Stokes equations. Equations (5.6.4), (5.6.9), (6.3.10), and mass density conservation equation

$$\frac{\partial \rho}{\partial t} + \nabla_k(\rho V_k) = 0 \tag{5.6.13}$$

together makes the equation set of motion of soft-matter quasicrystals.

The above hydrodynamic equations are similar to those obtained by Lubensky et al. for solid quasicrystals [13], which were introduced in Chap. 3 in brief. The difference between present results with those in Chap. 3 or given by Ref. [13] lies in the existence of fluid pressure p exception of field variables mass density ρ, velocities V_i (or momentums $g_i = \rho V_i$), phonon displacement u_i and phason displacement w_i. Due to these reasons, Eqs. (5.6.4), (5.6.9), (5.6.10), and (5.6.13) are not closed, additional equation of state needs to be supplemented, as mentioned in Chap. 4.

The publication of the work of Lubensky et al. opened the study of the hydrodynamics of quasicrystals. In addition, there are some concerns about their work [16, 17].

The results can be summarized as follows:

$$\frac{\partial \rho}{\partial t} + \nabla_k(\rho V_k) = 0 \tag{5.6.13}$$

$$\frac{\partial g_i(r, t)}{\partial t} = -\nabla_k(r)(V_k g_i) + \nabla_j(r)\big(-p\delta_{ij} + \eta_{ijkl}\nabla_k(r)V_l\big) - \big(\delta_{ij} - \nabla_i u_j\big)\frac{\delta H}{\delta u_i(r, t)}$$

$$+ \big(\nabla_i w_j\big)\frac{\delta H}{\delta w_i(r, t)} - \rho\nabla_i(r)\frac{\delta H}{\delta \rho(r, t)}, \ g_j = \rho V_j \tag{5.6.10}$$

$$\frac{\partial u_i(r, t)}{\partial t} = V_i - V_j\nabla_j(r)u_i - \Gamma_u\frac{\delta H}{\delta u_i(r, t)} \tag{5.6.4}$$

$$\frac{\partial w_i(r, t)}{\partial t} = -V_j\nabla_j(r)w_i - \Gamma_w\frac{\delta H}{\delta w_i(r, t)} \tag{5.6.9}$$

combining with the equation of state, which is discussed in Chap. 4

$$p = f(\rho) \tag{5.6.14}$$

we have built a closed consistent physical and mathematical equation system with the above five sets of equations.

More details on derivations and their simplifications of equations of motion for individual quasicrystal systems will be given in Chaps. 7–11, and refer to Refs. [18–20]. The applications of the results in hydrodynamics of solid and soft-matter quasicrystals can be found in Refs. [21–23].

5.7 Poisson Brackets Based on Lie Algebra

The derivation in Sect. 5.6 shows that the results of Poisson brackets (5.4.8), (5.5.3), and (5.5.4) are very important. These results can also be obtained and have more general significance mathematically through the Lie group, relevant introductions about these are listed below.

Lie group and point groups used frequently in the textbook are the same in satisfying four axioms of the group, but there are distinctions between them, where point group belongs to a discrete group, while Lie groups belong to a continuous group. In addition, the momentum operator is the generator of a group of movements, and the spin operator is the generator of a group of rotations in the spin space. There are inherent connections between quantum Poisson brackets and Lie group, so suggested as "group Poisson brackets" in Ref. [8].

Assume g be an element of group G, which is related to m real continuous parameters α_i, i.e.,

$$g(\alpha_i) \in G, \quad \alpha_i \in \mathbb{R}, \ i = 1, 2, \ldots, m \tag{5.7.1}$$

$\mathbb{R}$ represents real space.

Denting notation "$\cdot$" which connects two elements $a(\alpha_i)$ and $b(\beta_i)$, and gives another element $c(\gamma_i) \in G$:

$$c(\gamma_i) = a(\alpha_i) \cdot b(\beta_i), \quad i = 1, 2, \ldots, m \tag{5.7.2}$$

For continuous variational parameters, there is

$$\gamma_i = \varphi_i(\alpha_1, \alpha_2, \ldots, \alpha_m, \beta_1, \beta_2, \ldots, \beta_m) \tag{5.7.3}$$

If φ_i is a single-valued analytic function $\alpha_1, \alpha_2, \ldots, \alpha_m, \beta_1, \beta_2, \ldots, \beta_m$, then such continuous group is Lie group. The concept of single-valued analytic function can be found in any textbook on function theory.

One can take a parameter α_i and an identical element E (which is an element if $g_i \in G$, then $Eg_i = g_i$, $\alpha_i(E) = 0$. The infinitesimal generator L_i of the Lie group can be expressed by the following partial differential derivative

$$L_i = \frac{\partial a(\ldots, \alpha_i, \ldots)}{\partial \alpha_i}\bigg|_{\alpha_i=0} \tag{5.7.4}$$

and group element a can be expressed by the following expansion

$$a(\ldots, \alpha_i, \ldots) = E(\ldots, 0, \ldots) + \alpha_i L_i + O(\alpha_i^2) \tag{5.7.5}$$

The infinitesimal element of the Lie group presents important meaning in this kind of group. Lie group can be expressed by matrix. Assume matrix $D(A)$ is the expression of the group G. The parameter of the infinitesimal element $A(\alpha)$ is an infinitesimal quantity α_i. Matrix $D(A)$ can be expanded as following:

$$D(A) = 1 - i \sum_{j=1}^{N} \alpha_j I_j \tag{5.7.6}$$

in addition

$$I_j = i \frac{\partial D(A)}{\partial \alpha_j} \bigg|_{\alpha_j = 0} \tag{5.7.7}$$

N generators I_j are called the generators of expression $D(A)$. Lie algebra can be constructed through commutators between any two group generators

$$[L_i, L_j] = C_{ij}^k L_k, \quad i, j, k = 1, 2, \ldots, m \tag{5.7.8}$$

in which C_{ij}^k is called structure constant. There are three characters including asymmetry, linearity, and the LieJacobi identity as follows:

$$[L_i, L_j] = -[L_j, L_i] \tag{5.7.9}$$

$$[\alpha L_i + \beta L_j, L_k] = \alpha[L_i, L_k] + \beta[L_j, L_k], \quad \alpha, \beta \in \mathbb{R} \tag{5.7.10}$$

$$\big[L_i + [L_j, L_k]\big] + \big[L_k, [L_i, L_j]\big] + \big[L_j, [L_k, L_i]\big] = 0 \tag{5.7.11}$$

The coordinate transformation

$$x^k \rightarrow x^k + u^k(r) \tag{5.7.12}$$

in the theory of elasticity is called translational group or movement group, or infinitesimal movement group. Interestingly, $u^k(r)$ here presents evident physical meaning, represents displacements or phonons of the lattice. Note that, x^k here represents a contravariant vector and x_i is a covariant one. The close relationship between relevant physical quantities and group algebra is as mentioned previously, because the momentum operator is a generator of movement group, and the spin operator is a generator of a spin operator in spin space. Some connections between physical field variables $a, b, c, \ldots$ and elements of the transformation group $A, B, C, \ldots$ can be generated.

$$\{a, b, c, \ldots\} \rightarrow \{A, B, C, \ldots\} \tag{5.7.13}$$

An algebra element A can be expressed by the following linear form

$$A = \sum_{g \in G} A(g)g, \quad A(g) \in \mathbb{R} \tag{5.7.14}$$

and $A(g)$ denotes the coefficient of expansion (5.7.14).

Furthermore assuming A can be transformed like

$$A \rightarrow g A g^{-1} \tag{5.7.15}$$

in which, $g = 1 + \delta g$, and δg is an infinitesimal transformation, there is a linear approximation

$$A \rightarrow A + \delta A \tag{5.7.16}$$

and

$$\delta A = [\delta g, A] \tag{5.7.17}$$

the infinitesimal transformation δg takes the form

$$\delta g = \frac{i}{\hbar} \int \alpha^k(r) L^k(r) \mathrm{d}^d r \tag{5.7.18}$$

in which $\alpha^k(r)$ is the local infinitesimal "angular", $L^k(r)$ the generator of local transformation group, $i = \sqrt{-1}$, $\hbar = h/2\pi$, h the Planck constant.

For the movement group, taking $\alpha^k(r) = u^k(r)$, and the generator as the momentum, combining from (5.7.16) and (5.7.18), we have

$$\delta A(r) = \frac{i}{\hbar} \int \alpha^k(r') \big[L^k(r'), A(r) \big] \mathrm{d}^d r' \tag{5.7.19}$$

This equation shows that δA is the linear function of "angular" $\alpha^k(r)$ of infinitesimal local transformation, the corresponding variation is

$$\frac{\delta A(r)}{\delta \alpha^k(r')} = \frac{i}{\hbar} \big[L^k(r'), A(r) \big] \tag{5.7.20}$$

The limit passing over from quantum mechanics to classical mechanics is

$$\frac{\delta \hat{A}}{\delta \alpha} = \frac{i}{\hbar} \big[\hat{L}, \hat{A} \big] \rightarrow \frac{\delta A}{\delta \alpha} = \{L, A\} \tag{5.7.21}$$

Recall again that $\hat{L}, \hat{A}$ represent operators in quantum mechanics, L, A the field variables in classical mechanics. So that the right-hand side of (5.4.20) may be written as

$$\frac{\delta a}{\delta \alpha} = \{l, a\} \tag{5.7.22}$$

in which a can represent any field variables $a, b, c, \ldots$ in hydrodynamics, l the generator $l^k(r)$ corresponding to the group, so that from (5.7.22)

$$\frac{\delta a(r)}{\delta \alpha^k(r')} = \left\{l^k(r'), a(r)\right\} \tag{5.7.23}$$

Furthermore

$$\frac{\delta l^m(r)}{\delta \alpha^k(r')} = \left\{l^k(r'), l^m(r)\right\}, \{a, a\} = \{a, b\} = \{b, b\} = 0 \tag{5.7.24}$$

At the finite temperature, the Hamiltonian can be expressed by

$$H = \int \varepsilon(p, \rho, s) d^d r$$

$$d\varepsilon = V^k dp_k + \mu d\rho + T ds.$$

where ε denotes the energy density, the others are the same before, $p = (p_x, p_y, p_z)$ and ρ the momentum and mass density, s the entropy, $V = (V_x, V_y, V_z)$ the velocity, μ the chemical potential, T the absolute temperature, respectively, so

$$\delta p_k = -u^l \nabla_l p_k - p_k \nabla_l u^l - p_k \nabla_l u^l$$
$$\delta \rho = -u^l \nabla_l \rho - \rho \nabla_k u^k$$
$$\delta s = -u^l \nabla_l s - s \nabla_k u^k \tag{5.7.25}$$

From (5.7.24) and (5.7.25), one obtains

$$\{p_k(r_1), \rho(r_2)\} = \rho(r_1)\nabla_k(r_1)\delta(r_1 - r_2)$$
$$\{p_k(r_1), p_l(r_2)\} = (p_l(r_1)\nabla_k(r_1) - p_k(r_2)\nabla_k(r_2))\delta(r_1 - r_2) \tag{5.7.26}$$

This is identical to (5.4.8) given by Poisson bracket method of condensed matter physics, first resolved by Landau Ref. [6].

Applying the above results into quasicrystals, we have

$$\{u_k(r_1), g_l(r_2)\} = (-\delta_{kl} + \nabla_l(r_1)u_k)\delta(r_1 - r_2) \tag{5.7.27}$$

$$\{w_k(r_1), g_l(r_2)\} = (\nabla_l(r_1)w_k)\delta(r_1 - r_2) \tag{5.7.28}$$

They are identical to (5.5.3) and (5.5.4) given by Lubensky et al. [9], which were derived directly using the Poisson bracket method.

The above derivation shows the power of the Lie group method. As discussed in Ref. [8], introducing the Liouville equation, the equations of motion for some complex systems can be obtained, as derived in Sect. 5.6.

5.8 On Solving Governing Equations

Equations (5.6.13), (5.6.10), (5.6.4), (5.6.9), and (5.6.14) govern the generalized dynamics of soft-matter quasicrystals. For an individual quasicrystal system, substituting concrete constitutive law into the equations will reduce concrete final partial differential equations (be called final governing equations) with strong nonlinearity, they may be solved under appropriate initial and boundary conditions, which is a mathematical physics problem. These governing equations and their initial- and boundary-value problems will be discussed in Chaps. 7–11 which are novel ones in physics, chemistry, and materials science.

We must solve these equations to obtain information regarding the distribution, deformation, and motion in soft-matter quasicrystal applications. So far there is a lack of any commercial software for solving these equations in analytic or numerical methods. We have managed to computer some case studies. The success of the problem solving depends upon the correctness of the governing equations, the values of the initial and boundary conditions, and the suitable approach adopted. Otherwise, one cannot obtain a reasonable solution.

To judge the solution to be a success or a failure, besides the mathematical demonstration, one may compare the computational results with experimental observations although it is difficult to do so due to lack of these data up to now, or/and by comparing results from the classical fluid dynamics (in which the analytic solution with higher accuracy is more important), or/and by comparing with results from solid quasicrystals and crystals.

In the subsequent discussions in Chaps. 7 to 13, we will present some computational results. Among them, there are a few examples that have vigorously been checked through the above procedures. For such a new topic of generalized dynamics of soft-matter quasicrystals, the results should be validated for every case.

References

1. Einstein, A.: Ueber die von der molekularkinetischen Theorie der Waerme geforderte Bewegung von in ruhenden Fluessigkeiten suspendierten Teilchen. Ann d Phys **17**, 549–560 (1905)
2. Perrin, J.B.: The Atoms. Nabu Press, New York (2010) (English Trans. D.L. Hammick)
3. Forster, D.: Hydrodynamic fluctuation, broken symmetry and correlation functions. In: Benjamin, W.A. (ed.) Frontier in Physics, A Lecture Note and Reprint Series, vol. 47. Incorporated, Massachusetts (1975)
4. Chaikin, P., Lubensky, T.C.: Principles of Condensed Matter Physics. Cambridge University Press, Cambridge (1995)
5. Landau, L.D., Lifshitz, M.E.: Fluid Mechanics, Theory of Elasticity. Pergamon, Oxford (1998)
6. Landau, L.D.: The theory of superfluidity of heilium II. Zh. Eksp.Teor. Fiz, II, **592**, J. Phys. USSR **5**, 71–90 (1941)
7. Landau, L.D., Lifshitz, E.M.: Zur Theorie der Dispersion der magnetische Permeabilitaet der ferromagnetische Koerpern. Physik Zeitschrift fuer Sowjetunion **8**, 158–164 (1935)

8. Dzyaloshinskii, I.E., Volovick, G.E.: Poisson brackets in condensed matter physics. AnnPhys. (NY) **125**, 67–97 (1980)
9. Dzyaloshinskii, I.E., Volovick, G.E.: On the concept of local invariance in spin glass theory. J. de Phys. **39**, 693–700 (1978)
10. Volovick, G.E.: Additional localized degrees of freedom in spin glasses. Zh Eksp Teor Fiz **75**, 1102–1109 (1978)
11. Martin, P.C., Paron, O., Pershan, P.S.: Unified hydrodynamic theory for crystals, liquid crystals, and normal fluids. Phys. RevA **6**, 2401–2420 (1972)
12. Fleming, P.D., Cohen, C.: Hydrodynamics of solids. Phys. Rev. B **13**, 500–516 (1976)
13. Lubensky, T.C., Ramaswamy, S., Toner, J.: Hydrodynamics of icosahedral quasicrystals. Phys. Rev. B **32**, 7444–7411 (1985)
14. Lubensky, T.C.: Symmetry, elasticity and hydrodynamics of quasiperioic structures. In: Jaric, M.V. (ed.) Aperiodic Crystals, vol. I., pp. 199–280. Academic Press, Boston (1988)
15. Fan, T.Y.: Poisson bracket method and it applications to quasicrystals, liquid crystals and a class of soft matter. Acta Mechanica Sinica **45**, 548–559 in Chinese (2013)
16. Rochal, S.B., Lorman, V.L.: Minimal model of the phonon-phason dynamics in icosahedral quasicrystals and its application to the problem of internal friction in the i-AlPdMn alloy. Phys. Rev. B **66**, 144204 (2002)
17. Coddens, G.: On the problem of the relation between phason elasticity and phason dynamics in quasicrystals. Euro. Phys. J. B **54**, 37–65 (2006)
18. Fan, T.Y.: Equation system of generalized hydrodynamics of soft-matter quasicrystals. Appl. Math. Mech. **37**, 331–347 (2016) in Chinese; arXiv:1908.06425[cond-mat.soft] 15 Oct (2019)
19. Fan, T.Y.: Generalized hydrodynamics of soft-matter second kind two-dimensional quasicrystals. Appl. Math. Mech. **38**, 189–199 in Chinese (2017); arXiv:1908.06430[cond-mat.soft] 15 Oct (2019)
20. Fan, T.Y., Tang, Z.Y.: Three-dimensional generalized dynamics of soft-matter quasicrystals. Appl. Math. Mech. **38**, 1195–1207 (2017); Advances in Materials Science and Engineering, Vol. 2020, Article 1D4875854 (2020)
21. Cheng, H., Fan, T.Y., Wei, H.: Solution of hydrodynamics of 5- and 10-fold symmetry quasicrystals. Appl. Math. Mech. **37**, 1393–1404 (2016)
22. Cheng, H., Fan, T.Y., Wei, H.: Characters of deformation and motion of possible soft-matter quasicrystals with 5- and 10-fold symmetries. unpublished work (2016)
23. Wang, F., Cheng, H., Fan, T.Y., Hu, H.Y.: A stress analysis of some fundamental specimens of soft-matter quasicrystals with 8-fold symmetry based on generalized dynamics. Adv. Mater. Sci. Eng. Vol **2019**, Article 1D 8789151 (2019)

Chapter 6
Oseen Theory and Oseen Solution

In previous chapters, we introduced the physics and mathematics background for studying soft-matter quasicrystals. Like general soft matter, the soft-matter quasicrystals are complex liquids or structured liquids, so the knowledge on conventional liquid dynamics provides the base for further study on soft-matter quasicrystals. In this chapter, we will focus on basic knowledge about liquid dynamics especially the Oseen theory [1, 2].

6.1 Navier–Stokes Equations

The liquid dynamics is special for fluid dynamics, whose basic equations are the Navier–Stokes equations. In the two-dimensional case, the Navier–Stokes equations can be written as,

$$\left.\begin{array}{l} \frac{\partial \rho}{\partial t} + \nabla \cdot (\rho \mathbf{V}) = 0 \\ \frac{\partial (\rho V_x)}{\partial t} + \frac{\partial (V_x \rho V_x)}{\partial x} + \frac{\partial (V_y \rho V_x)}{\partial y} = -\frac{\partial p}{\partial x} + \eta \nabla^2 V_x + \frac{1}{3}\eta \frac{\partial}{\partial x} \nabla \cdot \mathbf{V} \\ \frac{\partial (\rho V_y)}{\partial t} + \frac{\partial (V_x \rho V_y)}{\partial x} + \frac{\partial (V_y \rho V_y)}{\partial y} = -\frac{\partial p}{\partial y} + \eta \nabla^2 V_y + \frac{1}{3}\eta \frac{\partial}{\partial y} \nabla \cdot \mathbf{V} \\ p = f(\rho) \end{array}\right\} \qquad (6.1.1)$$

in which, $\mathbf{V}$ denotes the velocity field, and $\nabla^2 = \frac{\partial^2}{\partial x^2} + \frac{\partial^2}{\partial y^2}, \nabla = \mathbf{i}\frac{\partial}{\partial x} + \mathbf{j}\frac{\partial}{\partial y}, \eta$ the fluid dynamic viscosity, ρ the mass density, p the pressure, respectively. In (6.1.1), the first equation is mass conservation, the second and third equations are momentum conservation in x, and the y component, the fourth one is the equation of state. In the left-hand side of the second and third equations, terms $\frac{\partial (\rho V_x)}{\partial t}, \frac{\partial (\rho V_y)}{\partial t}$ may be considered as the local derivatives, while terms $\frac{\partial (V_x \rho V_x)}{\partial x}, \frac{\partial (V_y \rho V_x)}{\partial y}, \frac{\partial (V_x \rho V_y)}{\partial x}, \frac{\partial (V_y \rho V_y)}{\partial y}$ the convective derivatives according to Sommerfeld's point of view [3]. The equations are the well-known Navier–Stokes equations.

© The Author(s), under exclusive license to Springer Nature Singapore Pte Ltd. 2022
T.-Y. Fan et al., *Generalized Dynamics of Soft-Matter Quasicrystals*, Springer Series in Materials Science 260, https://doi.org/10.1007/978-981-16-6628-5_6

6.2 Stokes Approximation

If the velocities are small, Stokes suggested the terms concerning convective derivatives in the second and third equations of (6.1.1) can be negligible, so (6.1.1) reduces to a as simplified version

$$\left.\begin{array}{l} \dfrac{\partial \rho}{\partial t} + \nabla \cdot (\rho \mathbf{V}) = 0 \\[4pt] \dfrac{\partial (\rho V_x)}{\partial t} = -\dfrac{\partial p}{\partial x} + \eta \nabla^2 V_x + \frac{1}{3}\eta \dfrac{\partial}{\partial x} \nabla \cdot \mathbf{V} \\[4pt] \dfrac{\partial (\rho V_y)}{\partial t} = -\dfrac{\partial p}{\partial y} + \eta \nabla^2 V_y + \frac{1}{3}\eta \dfrac{\partial}{\partial y} \nabla \cdot \mathbf{V} \\[4pt] p = f(\rho) \end{array}\right\} \tag{6.2.1}$$

This is the Stokes approximation, and Eq. (6.2.1) are called the Stokes equations.

In Chap. 1, we mentioned that in soft matter, the smaller fluid velocities and the greater viscosity is, which belong to the motion of the lower Reynolds number, and the Stokes approximation is useful.

6.3 Stokes Paradox

The three-dimensional Stokes equations are similar to those of (6.2.1) (refer to Chaps. 7 and 9). Its applications to flow past obstacles in incompressible fluid (i.e., $\rho = $ const) are very successful. However, the application of two-dimensional Stokes equations (6.2.1) to the problem of flow past a circular cylinder or other two-dimensional obstacles is failed at all. So far we do not have solutions for these problems. This is the well-known Stokes paradox.

6.4 Oseen Modification

To overcome the difficulty due to the Stokes paradox, Oseen analyzed the Navier–Stokes equations and found the source which leads to divergence of solutions at infinity and suggested making some modifications to the Stokes equations such as

$$\left.\begin{array}{l} \dfrac{\partial \rho}{\partial t} + \nabla \cdot (\rho \mathbf{V}) = 0 \\[4pt] \dfrac{\partial (\rho V_x)}{\partial t} + \dfrac{\partial (U_x \rho V_x)}{\partial x} + \dfrac{\partial (U_y \rho V_x)}{\partial y} = -\dfrac{\partial p}{\partial x} + \eta \nabla^2 V_x + \frac{1}{3}\eta \dfrac{\partial}{\partial x} \nabla \cdot \mathbf{V} \\[4pt] \dfrac{\partial (\rho V_y)}{\partial t} + \dfrac{\partial (U_x \rho V_y)}{\partial x} + \dfrac{\partial (U_y \rho V_y)}{\partial y} = -\dfrac{\partial p}{\partial y} + \eta \nabla^2 V_y + \frac{1}{3}\eta \dfrac{\partial}{\partial y} \nabla \cdot \mathbf{V} \\[4pt] p = f(\rho) \end{array}\right\} \tag{6.4.1}$$

in which U_x and U_y are given values of corresponding velocities in boundary conditions, this means that in momentum conservation equations, part of velocity components are replaced by known functions, Eq. (6.4.1) are called Oseen equations. According to this modification, people solved successfully the flow past cylinder and other two-dimensional obstacles. In the next section, we give an example of the applications.

6.5 Oseen Steady Solution of the Flow of Incompressible Fluid Past Cylinder

In the steady-state and incompressible fluid, the Oseen equations (6.4.1) can be simplified further as

$$\left.\begin{array}{l} \nabla \cdot \mathbf{V} = 0 \\ \rho\frac{\partial(U_x V_x)}{\partial x} + \rho\frac{\partial(U_y V_x)}{\partial y} = -\frac{\partial p}{\partial x} + \eta\nabla^2 V_x \\ \rho\frac{\partial(U_x V_y)}{\partial x} + \rho\frac{\partial(U_y V_y)}{\partial y} = -\frac{\partial p}{\partial y} + \eta\nabla^2 V_y \end{array}\right\} \qquad (6.5.1)$$

These equations are simplified because U_x and U_y are known functions already.

Suppose a slow flow along the direction x with velocity $U_x = U_\infty = \text{const}$, $U_y = 0$ shown by Fig. 6.1. In this case the equations (6.5.1) can be further simplified as

$$\left.\begin{array}{l} \nabla \cdot \mathbf{V} = 0 \\ \rho U_\infty\frac{\partial V_x}{\partial x} = -\frac{\partial p}{\partial x} + \eta\nabla^2 V_x \\ \rho U_\infty\frac{\partial V_y}{\partial x} = -\frac{\partial p}{\partial y} + \eta\nabla^2 V_y \end{array}\right\} \qquad (6.5.2)$$

From the second and third equations of (6.5.2), we can obtain

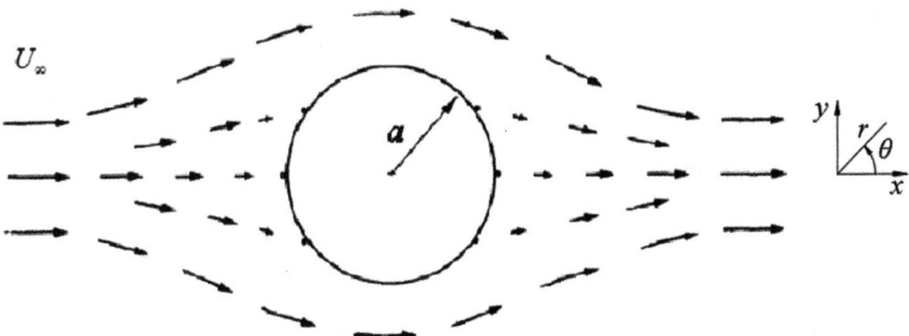

Fig. 6.1 Flow past cylinder with radius a

$$U_\infty \frac{\partial}{\partial x} \nabla \cdot \mathbf{V} = -\frac{\nabla^2 p}{\rho} + \frac{\eta}{\rho} \nabla^2 \nabla \cdot \mathbf{V} \tag{6.5.3}$$

Combining (6.5.3) with the first one of (6.5.2) yields

$$\nabla^2 p = 0 \tag{6.5.4}$$

This shows pressure is a harmonic function.
Assume

$$\mathbf{V} = \nabla \varphi + \mathbf{V}_2 \tag{6.5.5}$$

where $\mathbf{V}_2$ is unknown vector and φ satisfies

$$\nabla^2 \varphi = 0, \quad \forall \varphi \tag{6.5.6}$$

Substituting (6.5.5) into the second equation of (6.5.2) one can get

$$U_\infty \frac{\partial}{\partial x}\left(\frac{\partial \varphi}{\partial x} + V_{2x}\right) = -\frac{1}{\rho}\frac{\partial p}{\partial x} + \frac{\eta}{\rho}\frac{\partial}{\partial x}\left(\nabla^2 \varphi + \nabla^2 V_{2x}\right)$$

after simple treatment, one finds

$$\frac{\partial}{\partial x}\left(U_\infty \frac{\partial \varphi}{\partial x} + \frac{p}{\rho}\right) + U_\infty \frac{\partial V_{2x}}{\partial x} = \frac{\eta}{\rho}\nabla^2 V_{2x} \tag{6.5.7}$$

Because of the arbitrary property of function $\varphi(x, y)$, put

$$U_\infty \frac{\partial \varphi}{\partial x} + \frac{p}{\rho} = 0,$$

i.e.,

$$p = -\rho U_\infty \frac{\partial \varphi}{\partial x} \tag{6.5.8}$$

Substituting this formula into (6.5.7) yields

$$\frac{\partial V_{2x}}{\partial x} = \frac{\eta}{\rho U_\infty}\nabla^2 V_{2x} \tag{6.5.8a}$$

Similarly, the third equation of (6.5.2) is reduced to

$$\frac{\partial V_{2y}}{\partial x} = \frac{\eta}{\rho U_\infty}\nabla^2 V_{2y} \tag{6.5.8b}$$

In addition, insert (6.5.5) into the first one of (6.5.2) leads to

$$\nabla \cdot \mathbf{V}_2 = 0 \tag{6.5.8c}$$

Now we have

$$\left.\begin{aligned}
\nabla \cdot \mathbf{V}_2 &= 0 \\
\left(\tfrac{1}{2k}\nabla^2 - \tfrac{\partial}{\partial x}\right) V_{2x} &= 0 \\
\left(\tfrac{1}{2k}\nabla^2 - \tfrac{\partial}{\partial x}\right) V_{2y} &= 0
\end{aligned}\right\} \tag{6.5.9}$$

in which

$$\frac{1}{2k} \equiv \frac{\eta}{\rho U_\infty} \tag{6.5.10}$$

The parameter $2k$ carries an important constant, as

$$2ka = \frac{\rho U_\infty a}{\eta} = \text{Re}$$

is the well-known Reynolds number, a normalized number, here a is the characteristic size of the flow field. In the following, this number will be discussed frequently.

Equations (6.5.9) show our problem is reduced to determine function $\mathbf{V}_2$ only. Let's introduce a function $\chi(x, y)$ such as

$$V_{2x} = -\chi + \frac{1}{2k}\frac{\partial \chi}{\partial x}, \quad V_{2y} = \frac{1}{2k}\frac{\partial \chi}{\partial y} \tag{6.5.11}$$

and let $\chi(x, y)$ satisfy (6.5.9) and obtain

$$\left.\begin{aligned}
\frac{\partial}{\partial x}\left(V_{2x} - \frac{1}{2k}\frac{\partial}{\partial x} + \frac{1}{2k}\nabla^2\right)\chi &= 0 \\
\frac{\partial}{\partial y}\left(\frac{\partial}{\partial x} - \frac{1}{2k}\nabla^2\right)\chi &= 0
\end{aligned}\right\} \tag{6.5.12}$$

then we assume

$$\left(\frac{\partial}{\partial x} - \frac{1}{2k}\nabla^2\right)\chi = 0 \tag{6.5.13}$$

So that there are the solution expressions such as

$$\begin{cases}
V_x = -\chi + \frac{1}{2k}\frac{\partial \chi}{\partial x} + \frac{\partial \varphi}{\partial x} \\
V_y = \frac{1}{2k}\frac{\partial \chi}{\partial y} + \frac{\partial \varphi}{\partial y} \\
p = -\rho U_\infty \frac{\partial \varphi}{\partial x}
\end{cases} \tag{6.5.14}$$

in which function φ satisfies (6.5.6), and function χ satisfies (6.5.13) respectively.

For the case of flow past a circular cylinder, as shown in Fig. 6.1, there are the following boundary conditions in the polar coordinate system (r, θ) with $x = r \cos \theta$, $y = r \sin \theta$

$$\begin{cases} r = \sqrt{x^2 + y^2 + z^2} \to \infty : \\ V_r = U_\infty \cos \theta, \ V_\theta = -U_\infty \sin \theta, \\ r = a : \\ V_r = V_\theta = 0, \end{cases} \qquad (6.5.15)$$

According to (6.5.14) the velocity vector can be expressed by

$$\mathbf{V} = -\chi \mathbf{i} + \nabla \left(\varphi + \frac{1}{2k} \chi \right) \qquad (6.5.16)$$

In the polar coordinate system, we have the velocity components as below

$$\begin{aligned} V_r &= -\chi \cos \theta + \frac{1}{2k} \frac{\partial \chi}{\partial r} + \frac{\partial \varphi}{\partial r} \\ V_\theta &= \chi \sin \theta + \frac{1}{2kr} \frac{\partial \chi}{\partial \theta} + \frac{1}{r} \frac{\partial \varphi}{\partial \theta} \end{aligned} \qquad (6.5.17)$$

where function φ satisfies (6.5.6) and function χ can be expressed by

$$\chi = -U_\infty + e^{kx} Y \qquad (6.5.18)$$

and Y satisfies the Helmholtz equation

$$\left(\nabla^2 - k^2 \right) Y = 0 \qquad (6.5.19)$$

Based on the theory of partial differential equations, equation of (6.5.6) has a singular fundamental solution

$$\varphi = \ln r \qquad (6.5.20)$$

The general solution of Eq. (6.5.6) can be expressed by

$$\varphi = \sum_{n=0}^{\infty} A_n \frac{\partial^n}{\partial x^n} \ln r \qquad (6.5.21)$$

where

$$\frac{\partial^n}{\partial x^n} \ln r = (-1)^{n-1} (n-1)! \frac{\cos n\theta}{r^n} \qquad (n \geq 1)$$

So that

$$\varphi = A_0 \ln r + \sum_{n=1}^{\infty} A_n (-1)^{n-1} (n-1)! \frac{\cos n\theta}{r^n} \tag{6.5.22}$$

Considering the function Y connected only with radical variable r, (6.5.19) is reduced to

$$Y'' + \frac{1}{r} Y' - k^2 Y = 0 \tag{6.5.23}$$

which has solution

$$Y = C_1 I_0(kr) + C_2 K_0(kr)$$

where $I_0(kr)$ and $K_0(kr)$ are modified Bessel functions of the first and second kind with zero order, C_1, C_2 are arbitrary constants. Because $I_0(kr) \to \infty$ as $r \to \infty$ we must take $C_1 \equiv 0$. Similar to (6.5.21), the general solution of (6.5.18) is

$$\chi = -U_\infty + e^{kx} \sum_{n=0}^{\infty} B_n \frac{\partial^n}{\partial x^n} K_0(kr) \tag{6.5.24}$$

Here B_n = conts to be determined. In the following, we take the approximate expression of $K_0(kr)$ such as

$$K_0(kr) \approx -\ln\left(\frac{1}{2}\gamma_0 kr\right), \quad \frac{\partial}{\partial x} K_0(kr) \approx -\frac{x}{r^2} = -\frac{\cos\theta}{r} \tag{6.5.25}$$

where

$$\gamma_0 = 1.7811 = e^{\gamma}, \quad \gamma = 0.5772 \quad \text{(the Euler number)} \tag{6.5.26}$$

so other $B_n = 0$, as $n > 1$. In this case,

$$\chi = -U_\infty - B_0 \left[\ln\left(\frac{1}{2}\gamma_0 kr\right) - kr \cos\theta \ln\left(\frac{1}{2}\gamma_0 kr\right) \right] - \frac{B_1 \cos\theta}{r} \tag{6.5.27}$$

According to (6.5.17), (6.5.22), and (6.5.27), the velocity field is determined with some constants

$$\begin{cases} V_r = \frac{A_0}{r} - \frac{A_1 \cos\theta}{r^2} U_\infty \cos\theta - \frac{1}{2} B_0 \left[\frac{1}{kr} + \cos\theta - \cos\theta \ln\left(\frac{1}{2}\gamma_0 kr\right) \right] + \frac{B_1 \cos\theta}{2kr^2} \\ V_\theta = -\frac{A_1 \sin\theta}{r^2} - U_\infty \sin\theta - B_0 \frac{\sin\theta}{2} \ln\left(\frac{1}{2}\gamma_0 kr\right) + \frac{B_1 \sin\theta}{2kr^2} \end{cases}$$

$$\tag{6.5.28}$$

From the boundary conditions (6.5.14) and solution (6.5.28).

we find at last

$$
\begin{cases}
A_0 = \dfrac{2U_\infty}{k\left[1-2\ln\left(\frac{1}{2}ka\right)-2\gamma\right]} = \dfrac{4\eta}{\rho\left[1-2\ln\left(\frac{1}{2}ka\right)-2\gamma\right]} \\[2mm]
B_0 = -\dfrac{4U_\infty}{1-2\ln\left(\frac{1}{2}ka\right)-2\gamma} \\[2mm]
A_1 - \dfrac{B_1}{2k} = -\dfrac{U_\infty a^2}{1-2\ln\left(\frac{1}{2}ka\right)-2\gamma}
\end{cases}
\tag{6.5.29}
$$

At last, we obtain the velocity field around the cylinder as follows

$$
\begin{cases}
V_r(r,\theta) = \dfrac{U_\infty \cos\theta}{1-2\ln\left(\frac{1}{2}ka\right)-2\gamma}\left(-1+\dfrac{a^2}{r^2}+2\ln\dfrac{r}{a}\right) \\[2mm]
V_\theta(r,\theta) = -\dfrac{U_\infty \sin\theta}{1-2\ln\left(\frac{1}{2}ka\right)-2\gamma}\left(1-\dfrac{a^2}{r^2}+2\ln\dfrac{r}{a}\right)
\end{cases}
\tag{6.5.30}
$$

The solution holds only in the near field around the cylinder and in the case of $Re < 10$.

At the area far from the cylinder the solution is

$$
\begin{cases}
V_r(r,\theta) = A_0\left[\dfrac{1}{r}-\sqrt{\dfrac{\pi k}{2r}}\,e^{-kr(1-\cos\theta)}(1+\cos\theta)\right] \\[2mm]
V_\theta(r,\theta) = A_0\sqrt{\dfrac{\pi k}{2r}}\,e^{-kr(1-\cos\theta)}\sin\theta
\end{cases}
\tag{6.5.31}
$$

in which A_0 is defined by (6.5.29).

The above derivation was introduced in Ref. [4].

From the above results, we can obtain the stresses, i.e.,

$$
p_{ij} = -p\delta_{ij} + \sigma'_{ij}
$$

in which σ'_{ij} is the viscous stress tensor, according to the generalized Newton's law

$$
\sigma'_{ij} = 2\eta\left(\dot{\xi}_{ij} - \frac{1}{3}\dot{\xi}_{kk}\delta_{ij}\right), \quad \dot{\xi}_{kk} = \dot{\xi}_{xx} + \dot{\xi}_{yy} + \dot{\xi}_{zz}
$$

$$
\left(= \dot{\xi}_{rr} + \dot{\xi}_{\theta\theta} + \dot{\xi}_{zz} \quad \text{for cylinder coordinate}\right)
$$

$$
\dot{\xi}_{ij} = \frac{1}{2}\left(\frac{\partial V_i}{\partial x_j} + \frac{\partial V_j}{\partial x_i}\right)
$$

and concretely

$$
p_{rr} = -p + \sigma'_{rr} = -p + 2\eta\left(\dot{\xi}_{rr} - \frac{1}{3}\dot{\xi}_{kk}\right)
$$

$$
p_{\theta\theta} = -p + \sigma'_{\theta\theta} = -p + 2\eta\left(\dot{\xi}_{\theta\theta} - \frac{1}{3}\dot{\xi}_{kk}\right)
$$

$$
p_{r\theta} = p_{\theta r} = \sigma'_{r\theta} = 2\eta\dot{\xi}_{r\theta}
$$

and

$$\dot{\xi}_{rr} = \frac{\partial V_r}{\partial r},$$

$$\dot{\xi}_{\theta\theta} = \frac{1}{r}\frac{\partial V_\theta}{\partial \theta} + \frac{V_r}{r},$$

$$\dot{\xi}_{r\theta} = \dot{\xi}_{\theta r} = \frac{1}{2}\left(\frac{1}{r}\frac{\partial V_r}{\partial \theta} + \frac{\partial V_\theta}{\partial r} - \frac{V_\theta}{r}\right)$$

Substituting the previous results, we have

$$\begin{cases}
p = -\rho U_\infty \frac{\partial \varphi}{\partial x} \approx -\rho U_\infty A_0 \frac{\cos\theta}{r} \\
p_{rr} = -p + \sigma'_{rr} = -p + 2\eta\left(\dot{\xi}_{rr} - \frac{1}{3}\dot{\xi}_{kk}\right) \\
\quad = -\rho U_\infty A_0 \frac{\cos\theta}{r} + 2\eta\left[\frac{\partial V_r}{\partial r} - \frac{1}{3}\left(\frac{\partial V_r}{\partial r} + \frac{1}{r}\frac{\partial V_\theta}{\partial \theta} + \frac{V_r}{r}\right)\right] \\
p_{\theta\theta} = -p + \sigma'_{\theta\theta} = -p + 2\eta\left(\dot{\xi}_{\theta\theta} - \frac{1}{3}\dot{\xi}_{kk}\right) \\
\quad = -\rho U_\infty A_0 \frac{\cos\theta}{r} + 2\eta\left[\frac{1}{r}\frac{\partial V_\theta}{\partial \theta} + \frac{V_r}{r} - \frac{1}{3}\left(\frac{\partial V_r}{\partial r} + \frac{1}{r}\frac{\partial V_\theta}{\partial \theta} + \frac{V_r}{r}\right)\right] \\
p_{r\theta} = p_{\theta r} = \sigma'_{r\theta} = \sigma'_{\theta r} = 2\eta\dot{\xi}_{r\theta} = \eta\left(\frac{1}{r}\frac{\partial V_r}{\partial \theta} + \frac{\partial V_\theta}{\partial r} - \frac{V_\theta}{r}\right)
\end{cases} \tag{6.5.32}$$

The Oseen solution offers a complete description of the velocity field of flow around the obstacle and the field far away from the obstacle successfully overcomes the difficulty due to the Stokes paradox. In the meantime, the viscous stress field around the cylinder is also determined. Of course, it is only an approximate solution [3].

Landau et al. pointed out the flow past obstacle is not completely steady motion, the inertia effect should be considered [5].

6.6   The Reference Meaning of Oseen Theory and Oseen Solution to the Study in Soft Matter

The Oseen theory overcomes the difficulty of the Stokes paradox in two-dimensional viscous and incompressible flow. Although the Oseen solution provides only an approximate analytic solution for flow past cylinder in a viscous and incompressible fluid, it is significant to the study on dynamics of soft-matter quasicrystals. Soft-matter quasicrystals present high viscosity and low motion velocity, which can be treated as Oseen flow approximation. On the other hand, soft matter is compressible [6–9], which is different from the conventional incompressible fluid. Nevertheless, the Oseen solution can be referenced by the soft-matter solutions, and provide a testing stage to check the accuracy. Of course, the Oseen solution is not the unique standard for checking the soft-matter solutions.

References

1. Oseen, C.W.: Ueber die Stokes'sche Formel und uebereineverwandte Aufgabe in der Hydrodynamik. Ark Math Astronom Fys **6** (2010)
2. Oseen, C.W.: NeuereMethoden und Ergibnisse in der Hydrodynamik. AkademischeVerlagsgesellschaft, Leipzig (1927)
3. Sommerfeld, A.: VorlesungenuebertheoretischePhysik, Band II, Mechanik der deformierbarenMedien, Verlag Harri Deutsch, Thun. Frankfurt/M (1992)
4. Sleozkin, N.A.: Incompressible Viscous Fluid Dynamics. Gostehizdat Press, Moscow. In: Russian (1959); Kochin, N.E., Kibel'i, I.A., Roze, N.V.: Theoretical Hydrodynamics. Government Press of Phys-Math Literature, Moscow (in Russian) (1953)
5. Landau, L.D., Lifshitz, E.M.: Fluid Mechanics. Pergamon, Oxford (1980)
6. Fan, T.Y.: Equation system of generalized dynamics of soft-matter quasicrystals. Appl. Math. Mech. **37**, 331–344 (2016), in Chinese; arXiv:1908.06425[cond-mat.soft] 15 Oct 2019
7. Fan, T.Y.: Generalized hydrodynamics of second kind of two-dimensional soft-matter quasicrystals. Appl. Math. Mech. **38**, 189–199, in Chinese (2017); arXiv:1908.06430[cond-mat.soft] 15 Oct 2019
8. Cheng, H., Fan, T.Y., Wei, H.: Characters of deformation and motion of soft-matter quasicrystals of possible 5- and 10-fold symmetry (unpublished work) (2016)
9. Wang, F., Cheng, H., Fan, T.Y., Hu, H.Y.: A stress analysis on some fundamental specimens of 8-fold symmetry quasicrystals in soft matter based on generalized dynamics. Adv. Mater. Sci. Eng. Vol 2019, Article 1D 8789151 (2019)

Chapter 7
Dynamics of Soft-Matter Quasicrystals with 12-Fold Symmetry

The discussion in the first 6 chapters provides preparation for the subsequent study.

We aim to explore the structures and dynamic properties of soft-matter quasicrystals. The quantitative analysis lies in the dynamic equation system of the matter, which has been summarized by Fan in Refs [1–3] through analyzing the discoveries of Zeng et al. [4], Takano [5], Hayashida et al. [6], Talapin et al. [7], Fischer et al. [8] and Cheng et al. [9], etc. The setting up of the equation system has used the hydrodynamics of Lubensky et al. [10] on solid quasicrystals, but the fluid phonon concept including the relevant field variables and equation of state, in addition, are also introduced.

The quantitative analysis requires solving the initial-boundary value problems of these equations. This chapter will discuss some solutions of initial-boundary value problems for the plane field of two-dimensional soft-matter quasicrystals with 12-fold symmetry, which might be one of the most important soft-matter quasicrystals at present. In addition, the three-dimensional equations are presented as well. Due to the observation of 10-fold symmetry quasicrystals in the soft matter recently, the importance of the 12-fold symmetry ones may be reduced, although the detailed information on 10-fold symmetry quasicrystals of soft matter has not been reported yet.

In the previous chapters, we pointed out that the compressible model of soft matter is significant, in the meantime, the incompressible model of the matter is considered as well, which could be treated much easier because it does not need to use the equation of state.

T.-Y. Fan et al., *Generalized Dynamics of Soft-Matter Quasicrystals*, Springer Series in Materials Science 260, https://doi.org/10.1007/978-981-16-6628-5_7

7.1 Two-Dimensional Governing Equations of Soft-Matter Quasicrystals of 12-Fold Symmetry

According to the Landau principle of symmetry breaking and elementary excitation, the soft-matter quasicrystals comprise elementary excitations–phonons, phasons, and fluid phonon and the corresponding field variables are u_i, w_i and V_i, respectively. The soft-matter quasicrystals observed so far are two-dimensional quasicrystals. For simplicity, for 12-fold symmetry quasicrystals we here first only consider their plane fields in quasiperiodic symmetry plane, i.e., the xy—plane, if z—axis is 12-fold symmetry axis, there are spatial self variables $x_1 = x, x_2 = y$ only in the field variables and field equations.

Deformation and motion of soft-matter quasicrystals follow the laws of mass and momentum conservations and rules of symmetry breaking of phonons and phasons. This suggests hydrodynamics or generalized dynamics of soft-matter quasicrystals, which was discussed in Chap. 5, here we give a more detailed discussion.

Comparing the hydrodynamics given in Chap. 5 for soft-matter quasicrystals and that in Chap. 3 for solid quasicrystals, there are three differences between them [1] which should be emphasized again:

(1)   The solid viscosity constitutive equation in Ref [10] $\sigma'_{ij} = \eta_{ijkl}\dot{\xi}_{kl}$, $\dot{\xi}_{kl} = \frac{1}{2}\left(\frac{\partial V_k}{\partial x_l} + \frac{\partial V_l}{\partial x_k}\right)$ is replaced by the fluid constitutive equation $p_{ij} = -p\delta_{ij} + \sigma'_{ij} = -p\delta_{ij} + \eta_{ijkl}\dot{\xi}_{kl}$, $\dot{\xi}_{kl} = \frac{1}{2}\left(\frac{\partial V_k}{\partial x_l} + \frac{\partial V_l}{\partial x_k}\right)$; in which p is the fluid pressure, in addition, the constitutive laws for phonons and phasons can draw from those of solid quasicrystals given in Chaps. 3 and 5

$$\sigma_{ij} = C_{ijkl}\varepsilon_{kl} + R_{ijkl}w_{kl}$$
$$H_{ij} = K_{ijkl}w_{kl} + R_{klij}\varepsilon_{kl}$$

in addition, adding the fluid constitutive law

$$p_{ij} = -p\delta_{ij} + \sigma'_{ij} = -p\delta_{ij} + \eta_{ijkl}\dot{\xi}_{kl}$$

in which

$$\varepsilon_{ij} = \frac{1}{2}\left(\frac{\partial u_i}{\partial x_j} + \frac{\partial u_j}{\partial x_i}\right), w_{ij} = \frac{\partial w_i}{\partial x_j}, \dot{\xi}_{ij} = \frac{1}{2}\left(\frac{\partial V_i}{\partial x_j} + \frac{\partial V_j}{\partial x_i}\right)$$

u_i denotes phonon displacement vector, σ_{ij} the phonon stress tensor, ε_{ij} the phonon strain tensor; w_i the phason displacement vector, H_{ij} the phason stress tensor, w_{ij} the phason strain tensor; V_i the fluid phonon velocity vector, p_{ij} the fluid stress tensor, p the fluid pressure, η_{ijkl} the fluid viscosity coefficient tensor, $\dot{\xi}_{ij}$ the fluid deformation rate tensor; C_{ijkl}, K_{ijkl} and R_{ijkl} the phonon, phason, and phonon-phason coupling elastic constant tensors, respectively. In addition, for 12-fold symmetry quasicrystals $R_{ijkl} = 0$ (because of decoupling between phonons and phasons in elasticity). For simplicity, in the following, we only discuss the simplest fluid, i.e., $p_{ij} = -p\delta_{ij} + \sigma'_{ij} = -p\delta_{ij} + 2\eta(\dot{\xi}_{ij} - \frac{1}{3}\dot{\xi}_{kk}\delta_{ij}) + \eta'\dot{\xi}_{kk}\delta_{ij}$, $\dot{\xi}_{kk} = \dot{\xi}_{11} + \dot{\xi}_{22} + \dot{\xi}_{33}$, $\dot{\xi}_{ij} = \frac{1}{2}\left(\frac{\partial V_i}{\partial x_j} + \frac{\partial V_j}{\partial x_i}\right)$, in which η is so-called the first viscosity coefficient, η' the second one, which is omitted because it is too small (the details can be found in Chap. 5);

(2)   An equation of state $p = f(\rho)$ is supplemented and in solid quasicrystals, there is no need for the equation. The equation of state belongs to thermodynamics of soft matter;

(3)   Related to point two, unlike in the solid quasicrystals, the dynamics equations of soft-matter quasicrystals cannot be linearized in general, because the equation of state is nonlinear.

Regarding the above listed constitutive equations, for 12-fold symmetry quasicrystals we have

$$\left.\begin{array}{l} \sigma_{ij} = \dfrac{\partial f_{\text{def}}}{\partial \varepsilon_{ij}} = C_{ijkl}\varepsilon_{kl} \\[2mm] H_{ij} = \dfrac{\partial f_{\text{def}}}{\partial w_{ij}} = K_{ijkl}w_{kl} \end{array}\right\} \tag{7.1.1'}$$

due to decoupling (the phonon-phason coupling elastic tensor vanishes) and the concrete form of the deformation energy density f_{def} depends upon the symmetry groups of the quasicrystals. From data of solid quasicrystals, in two-dimensional quasicrystals of 12-fold symmetry, there are 2 Laue classes and 7 point groups (refer to [11]). We here consider point group $12mm$, for which in the quasiperiodic plane the elastic deformation energy density is

$$\begin{aligned} f_{\text{def}}(\mathbf{u}, \mathbf{w}) = f_{\text{def}}(\varepsilon_{ij}, w_{ij}) &= \frac{1}{2}L(\nabla \cdot \mathbf{u})^2 \\ &+ M\varepsilon_{ij}\varepsilon_{ij} + \frac{1}{2}K_1 w_{ij}w_{ij} + \frac{1}{2}K_2(w_{21}^2 + w_{12}^2 \\ &+ 2w_{11}w_{22}) + \frac{1}{2}K_3(w_{21} + w_{12})^2 = F_u + F_w, \\ &(x = x_1, y = x_2, i = 1, 2, j = 1, 2) \end{aligned} \tag{7.1.1''}$$

So that we can obtain the constitutive law based on $(7.1.1')$ as below

$$
\left.
\begin{aligned}
\sigma_{xx} &= L(\varepsilon_{xx} + \varepsilon_{yy}) + 2M\varepsilon_{xx} \\
\sigma_{yy} &= L(\varepsilon_{xx} + \varepsilon_{yy}) + 2M\varepsilon_{yy} \\
\sigma_{xy} &= \sigma_{yx} = 2M\varepsilon_{xy} \\
H_{xx} &= K_1 w_{xx} + K_2 w_{yy} \\
H_{yy} &= K_1 w_{yy} + K_2 w_{xx} \\
H_{xy} &= K_1 w_{xy} - K_2 w_{yx} \\
H_{yx} &= K_1 w_{yx} - K_2 w_{xy} \\
p_{xx} &= -p + \sigma'_{xx} = -p + 2\eta(\dot{\xi}_{xx} - \tfrac{1}{3}\dot{\xi}_{kk}) \\
p_{yy} &= -p + \sigma'_{yy} = -p + 2\eta(\dot{\xi}_{yy} - \tfrac{1}{3}\dot{\xi}_{kk}) \\
p_{xy} &= p_{yx} = \sigma'_{xy} = \sigma'_{yx} = 2\eta\dot{\xi}_{xy} \\
\dot{\xi}_{kk} &= \dot{\xi}_{xx} + \dot{\xi}_{yy}
\end{aligned}
\right\}
\qquad (7.1.2)
$$

in which there are no coupling elastic constants. Hence, the equations of motion of the plane field of soft-matter quasicrystals with 12-fold symmetry is obtained as follows by omitting the terms of $\nabla_i u_j \frac{\delta H}{\delta u_j}$ and $\nabla_i w_j \frac{\delta H}{\delta w_j}$ (refer to Chap. 5):

$$
\left.
\begin{aligned}
&\frac{\partial \rho}{\partial t} + \nabla \cdot (\rho \mathbf{V}) = 0 \\
&\frac{\partial(\rho V_x)}{\partial t} + \frac{\partial(V_x \rho V_x)}{\partial x} + \frac{\partial(V_y \rho V_x)}{\partial y} = -\frac{\partial p}{\partial x} + \eta\nabla^2 V_x + \tfrac{1}{3}\eta\frac{\partial}{\partial x}\nabla \cdot \mathbf{V} \\
&\quad + M\nabla^2 u_x + (L + M - B)\frac{\partial}{\partial x}\nabla \cdot \mathbf{u} - (A - B)\frac{1}{\rho_0}\frac{\partial \delta\rho}{\partial x} \\
&\frac{\partial(\rho V_y)}{\partial t} + \frac{\partial(V_x \rho V_y)}{\partial x} + \frac{\partial(V_y \rho V_y)}{\partial y} = -\frac{\partial p}{\partial y} + \eta\nabla^2 V_y + \tfrac{1}{3}\eta\frac{\partial}{\partial y}\nabla \cdot \mathbf{V} \\
&\quad + M\nabla^2 u_y + (L + M - B)\frac{\partial}{\partial y}\nabla \cdot \mathbf{u} - (A - B)\frac{1}{\rho_0}\frac{\partial \delta\rho}{\partial y} \\
&\frac{\partial u_x}{\partial t} + V_x\frac{\partial u_x}{\partial x} + V_y\frac{\partial u_x}{\partial y} = V_x + \Gamma_{\mathbf{u}}\left[M\nabla^2 u_x + (L + M)\frac{\partial}{\partial x}\nabla \cdot \mathbf{u}\right] \\
&\frac{\partial u_y}{\partial t} + V_x\frac{\partial u_y}{\partial x} + V_y\frac{\partial u_y}{\partial y} = V_y + \Gamma_{\mathbf{u}}\left[M\nabla^2 u_y + (L + M)\frac{\partial}{\partial y}\nabla \cdot \mathbf{u}\right] \\
&\frac{\partial w_x}{\partial t} + V_x\frac{\partial w_x}{\partial x} + V_y\frac{\partial w_x}{\partial y} = \Gamma_{\mathbf{w}}\left[K_1\nabla^2 w_x + (K_2 + K_3)\frac{\partial}{\partial y}\left(\frac{\partial w_x}{\partial y} + \frac{\partial w_y}{\partial x}\right)\right] \\
&\frac{\partial w_y}{\partial t} + V_x\frac{\partial w_y}{\partial x} + V_y\frac{\partial w_y}{\partial y} = \Gamma_{\mathbf{w}}\left[K_1\nabla^2 w_y + (K_2 + K_3)\frac{\partial}{\partial x}\left(\frac{\partial w_x}{\partial y} + \frac{\partial w_y}{\partial x}\right)\right] \\
&p = f(\rho)
\end{aligned}
\right\}
\qquad (7.1.3)
$$

where $\nabla^2 = \frac{\partial^2}{\partial x^2} + \frac{\partial^2}{\partial y^2}$, $\nabla = \mathbf{i}\frac{\partial}{\partial x} + \mathbf{j}\frac{\partial}{\partial y}$, $\mathbf{V} = \mathbf{i}V_x + \mathbf{j}V_y$, $\mathbf{u} = \mathbf{i}u_x + \mathbf{j}u_y$ and $L = C_{12}$, $M = (C_{11} - C_{12})/2$ the phonon elastic constants, K_1, K_2, K_3 the phason elastic constants, η the fluid dynamic viscosity, Γ_u and Γ_w the phonon and phason dissipation coefficients, A and B the material constants due to variation of mass density, namely LRT constants, respectively.

Equation (7.1.3) is the final governing equations of two-dimensional dynamics of soft-matter quasicrystals of 12-fold symmetry with field variables $u_x, u_y, w_x, w_y, V_x, V_y, \rho$ and p. The number of the field variables is 8 and the number of field equations is 8 too. Among them: (7.1.3a) is the mass conservation. Equations (7.1.3b) and (7.1.3c) are the momentum conservation equations or generalized Navier–Stokes equations, (7.1.3d) and (7.1.3e) are the equations of motion of phonons due to the symmetry breaking, (7.1.3f) and (7.1.3g) are the phason dissipation equations and (7.1.3h) is the equation of state. The equations are consistent to be mathematical solvability. If there is a lack of the equation of state, the equation system is not closed and has no meaning mathematically and physically. This shows the equation of state is necessary.

It is evident that the equation set is nonlinear and cannot be linearized in general, in which there are wave equations as well as diffusion equations. Because of the feature, the Fourier transform cannot be used and the discussion in the frequency domain after the Fourier transform is difficult to carry out. Related to this, the spectrum behavior of the equations has not been directly discussed here. For the subsequent discussion, the wave speeds of the system are important, which can be approximately obtained, i.e.,

$$c_1 = \sqrt{\frac{A + L + 2M - 2B}{\rho}}, \quad c_2 = c_3 = \sqrt{\frac{M}{\rho}}, \quad c_4 = \sqrt{\left(\frac{\partial p}{\partial \rho}\right)_s} \tag{7.1.4}$$

they are speeds of phonon longitudinal wave, phonon transverse waves (for the plane field there is only one transverse wave mode), and fluid phonon (longitudinal) wave, respectively.

Based on the final governing Eq. (7.1.3) we will discuss some solutions of soft-matter quasicrystals with 12-fold symmetry. Due to the complexity of the equations, solving the initial-boundary value problems of the equations is very difficult. In the following, we will introduce some results, in which most of them are approximate.

7.2 Simplification of Governing Equations

7.2.1 Steady Dynamic Problem of Soft-Matter Quasicrystals with 12-Fold Symmetry

We first consider the static case.

In Eq. (7.1.3), if term $\frac{\partial Q}{\partial t} = 0$, the problem is reduced to a steady problem, which is interesting in some cases. One example will be introduced in Sect. 7.3. Furthermore, we omit the terms of convective derivatives on the left-hand side in (7.1.3), so

$$\left.\begin{aligned}
&\nabla \cdot (\rho \mathbf{V}) = 0 \\
&-\frac{\partial p}{\partial x} + \eta \nabla^2 (\rho V_x) - (A - B)\frac{1}{\rho_0}\frac{\partial \delta \rho}{\partial x} = 0 \\
&-\frac{\partial p}{\partial y} + \eta \nabla^2 (\rho V_y) - (A - B)\frac{1}{\rho_0}\frac{\partial \delta \rho}{\partial y} = 0 \\
&p = f(\rho)
\end{aligned}\right\} \tag{7.2.1}$$

and the equations of phonons are related to fluid velocity field as below

$$\left.\begin{array}{l} V_x + \Gamma_u\left[M\nabla^2 u_x + (L+M)\frac{\partial}{\partial x}\nabla\cdot\mathbf{u}\right] = 0 \\ V_y + \Gamma_u\left[M\nabla^2 u_y + (L+M)\frac{\partial}{\partial y}\nabla\cdot\mathbf{u}\right] = 0 \end{array}\right\} \tag{7.2.2}$$

but the equations of phasons are independent of the fluid field as well as phonon field

$$\left.\begin{array}{l} K_1\nabla^2 w_x + (K_2+K_3)\frac{\partial}{\partial y}\left(\frac{\partial w_x}{\partial y}+\frac{\partial w_y}{\partial x}\right) = 0 \\ K_1\nabla^2 w_y + (K_2+K_3)\frac{\partial}{\partial x}\left(\frac{\partial w_x}{\partial y}+\frac{\partial w_y}{\partial x}\right) = 0 \end{array}\right\} \tag{7.2.3}$$

These equations are hydrostatic equations which are also meaningful in some cases.

If omitting velocity terms in Eq. (7.2.2), then it can be reduced to biharmonic equations

$$\nabla^2\nabla^2 F = 0 \quad (\mathbf{V}=0) \tag{7.2.2'}$$

in which

$$u_x = (L+M)\frac{\partial^2 F}{\partial x\partial y}, u_y = -(L+2M)\frac{\partial^2 F}{\partial x^2} - M\frac{\partial^2 F}{\partial y^2} \tag{7.2.2''}$$

At meantime the Eq. (7.2.3) will be reduced as

$$\nabla^2\nabla^2 G = 0 \quad (\mathbf{V}=0) \tag{7.2.3'}$$

if omitting velocity terms and

$$w_x = (K_1+K_2)\frac{\partial^2 G}{\partial x\partial y}, w_y = -K_1\frac{\partial^2 G}{\partial x^2} - (K_1+K_2+K_3)\frac{\partial^2 G}{\partial y^2} \tag{7.2.3''}$$

The complex representations of solutions of Eqs. (7.2.2$'$) and (7.2.3$'$) are well-known as follows

$$F = \mathrm{Re}\left[\bar{z}f_1(z) + \int f_2(z)\mathrm{d}z\right] \tag{7.2.4}$$

$$G = \mathrm{Re}\left[\bar{z}g_1(z) + \int g_2(z)\mathrm{d}z\right] \tag{7.2.5}$$

where Re denotes the real part of complex function, $f_1(z)$, $f_2(z)$ and $g_1(z)$, $g_2(z)$ are analytic functions of a complex variable $z = x + iy$, and $\bar{z} = x - iy$ with $i = \sqrt{-1}$, respectively.

7.2.2 *Pure Fluid Dynamics*

In the absence of phonon and phason fields, one may omit the inertia terms and the small quantities if introduce the flow function $U(x, y)$

$$V_x = \frac{\partial U}{\partial x}, \quad V_y = -\frac{\partial U}{\partial y} \tag{7.2.4}$$

in this case, equation (7.1.3) is reduced to

$$\nabla^2 \nabla^2 U = 0 \tag{7.2.5}$$

We have a complex representation of the solution of equation (7.2.5)

$$U = \mathrm{Re}\left[\bar{z}\phi_1(z) + \int \psi_1(z)\mathrm{d}z \right] \tag{7.2.6}$$

where Re denotes the real part of complex functions, $\phi_1(z)$ and $\psi_1(z)$ are two analytic functions of a complex variable $z = x + iy$.

Although the pure fluids do not represent soft matter, pure fluid dynamics can be used to determine the limits for soft matter dynamics. In this context, Eq. (7.2.5) can be viewed as similar to (7.2.2′); however, the solutions of (7.2.5) are meaningless, which can lead to the so-called Stokes paradox. This insight is beneficial, and we will build upon it when describing soft-matter quasicrystals throughout the subsequent sections of this chapter.

7.3 Dislocation and Solution

It is well-known that the dislocation problem is one of the important problems for soft-matter quasicrystals as well as solid ones.

As a zero-order approximation, in Eq. (7.2.2) the fluid velocity field is omitted, i.e., it becomes to Eq. (7.2.2′), the corresponding dislocation can be easily solved, then we obtain the dislocation solution [11] under the dislocation conditions $\int_\Gamma \mathrm{d}u_x = b_1^{\|}, \int_\Gamma \mathrm{d}u_y = b_2^{\|}$

$$\left.\begin{aligned}
u_x &= \frac{b_1^{\|}}{2\pi}\left(\arctan\frac{y}{x} + \frac{L+M}{L+2M}\frac{xy}{r^2}\right) \\
&\quad + \frac{b_2^{\|}}{2\pi}\left(\frac{M}{L+2M}\ln\frac{r}{r_0} + \frac{L+M}{L+2M}\frac{y^2}{r^2}\right) \\
u_y &= -\frac{b_1^{\|}}{2\pi}\left(\frac{M}{L+2M}\ln\frac{r}{r_0} + \frac{L+M}{L+2M}\frac{x^2}{r^2}\right) \\
&\quad + \frac{b_2^{\|}}{2\pi}\left(\arctan\frac{y}{x} - \frac{L+M}{L+2M}\frac{xy}{r^2}\right) \\
u_z &= 0
\end{aligned}\right\} \tag{7.3.1}$$

in which $L = C_{12}$, $M = (C_{11} - C_{12})/2$ are the phonon elastic constants, $b_1^\parallel$, $b_2^\parallel$ the phonon Burgers vector components.

The dislocation solution for the phason field can be found by solving Eq. (7.2.3$'$)

Under dislocation conditions, $\int_\Gamma dw_x = b_1^\perp$, $\int_\Gamma dw_y = b_2^\perp$ one can obtain the dislocation solution of the phason field

$$
\left.
\begin{aligned}
w_x &= \frac{b_1^\perp}{2\pi}\left(\arctan\frac{y}{x} - \frac{(K_1+K_2)(K_2+K_3)}{2K_1(K_1+K_2+K_3)}\frac{xy}{r^2}\right) \\
&+ \frac{b_2^\perp}{4\pi}\left(-\frac{K_2(K_1+K_2+K_3)-K_1K_3}{K_1(K_1+K_2+K_3)}\ln\frac{r}{r_0} + \frac{(K_1+K_2)(K_2+K_3)}{K_1(K_1+K_2+K_3)}\frac{y^2}{r^2}\right) \\
w_y &= \frac{b_1^\perp}{4\pi}\left(\frac{K_2(K_1+K_2+K_3)-K_1K_3}{K_1(K_1+K_2+K_3)}\ln\frac{r}{r_0} - \frac{(K_1+K_2)(K_2+K_3)}{K_1(K_1+K_2+K_3)}\frac{x^2}{r^2}\right) \\
&+ \frac{b_2^\perp}{2\pi}\left(\arctan\frac{y}{x} + \frac{(K_1+K_2)(K_2+K_3)}{2K_1(K_1+K_2+K_3)}\frac{xy}{r^2}\right)
\end{aligned}
\right\}
\tag{7.3.2}
$$

where K_1, K_2, K_3 are the phason elastic constants, $b_1^\perp$, $b_2^\perp$ the phason Burgers vector components, r_0 the dislocation core size. The detail can be referred to Fan [11].

7.3.1 Limitation of Zero-Order Solution of Dislocation, Possible Modification Considering the Fluid Effect

For soft-matter quasicrystals, due to the existence of fluid phonon, the fluid effect on the dislocations should be considered. The above solutions are only elastic solutions or say they are only zero-order approximate solutions of the dislocation problem. By considering the fluid effect, the strict analytic solution of dislocation in quasicrystals of soft matter is not available at present even if in the near future. An approximate solution is constructed, but the results are not so satisfactory. The discussion will be introduced in Sects. 7.9 and 8.4.

7.4 Generalized Oseen Approximation Under the Condition of Lower Reynolds Number

The motion of soft matter is with the feature of small Reynolds number as mentioned in Chap. 1, because the fluid velocity is small and viscosity is great, so that

$$
\text{Re} = \frac{\rho U a}{\eta} = 0.0001 \sim 1
\tag{7.4.1}
$$

where ρ the mass density, η the viscosity, U the characteristic velocity, a the characteristic size in the fluid field.

In Chap. 6 we have in detail discussed the Oseen [13, 14] theory for classical fluid. Taking the Oseen modification to the present complex fluid, the equations (7.3.1) are changed as below

$$
\left.
\begin{aligned}
&\frac{\partial \rho}{\partial t} + \nabla \cdot (\rho \mathbf{V}) = 0 \\
&\frac{\partial(\rho V_x)}{\partial t} + \frac{\partial(U_x \rho V_x)}{\partial x} + \frac{\partial(U_y \rho V_x)}{\partial y} = -\frac{\partial p}{\partial x} + \eta \nabla^2 V_x + \frac{1}{3}\eta \frac{\partial}{\partial x} \nabla \cdot \mathbf{V} \\
&\quad + M\nabla^2 u_x + (L + M - B)\frac{\partial}{\partial x}\nabla \cdot \mathbf{u} - (A - B)\frac{1}{\rho_0}\frac{\partial \delta\rho}{\partial x} \\
&\frac{\partial(\rho V_y)}{\partial t} + \frac{\partial(U_x \rho V_y)}{\partial x} + \frac{\partial(U_y \rho V_y)}{\partial y} = -\frac{\partial p}{\partial y} + \eta \nabla^2 V_y + \frac{1}{3}\eta \frac{\partial}{\partial y} \nabla \cdot \mathbf{V} \\
&\quad + M\nabla^2 u_y + (L + M - B)\frac{\partial}{\partial y}\nabla \cdot \mathbf{u} - (A - B)\frac{1}{\rho_0}\frac{\partial \delta\rho}{\partial y} \\
&\frac{\partial u_x}{\partial t} + U_x \frac{\partial u_x}{\partial x} + U_y \frac{\partial u_x}{\partial y} = V_x + \Gamma_{\mathbf{u}}\left[M\nabla^2 u_x + (L + M)\frac{\partial}{\partial x}\nabla \cdot \mathbf{u}\right] \\
&\frac{\partial u_y}{\partial t} + U_x \frac{\partial u_y}{\partial x} + U_y \frac{\partial u_y}{\partial y} = V_y + \Gamma_{\mathbf{u}}\left[M\nabla^2 u_y + (L + M)\frac{\partial}{\partial y}\nabla \cdot \mathbf{u}\right] \\
&\frac{\partial w_x}{\partial t} + U_x \frac{\partial w_x}{\partial x} + U_y \frac{\partial w_x}{\partial y} = \Gamma_{\mathbf{w}}\left[K_1\nabla^2 w_x + (K_2 + K_3)\frac{\partial}{\partial y}\left(\frac{\partial w_x}{\partial y} + \frac{\partial w_y}{\partial x}\right)\right] \\
&\frac{\partial w_y}{\partial t} + U_x \frac{\partial w_y}{\partial x} + U_y \frac{\partial w_y}{\partial y} = \Gamma_{\mathbf{w}}\left[K_1\nabla^2 w_y + (K_2 + K_3)\frac{\partial}{\partial x}\left(\frac{\partial w_x}{\partial y} + \frac{\partial w_y}{\partial x}\right)\right] \\
&p = f(\rho)
\end{aligned}
\right\} \quad (7.4.2)
$$

we call (7.4.2) as generalized Oseen equations of soft-matter quasicrystals of 12-fold symmetry, in which U_x, U_y represent known velocities (given in boundary conditions) and the others are defined as previously.

7.5 Steady Dynamic Equations Under Oseen Modification in Polar Coordinate System

In the next section, we will discuss a flow of soft-matter quasicrystals past a circular cylinder, in which the polar coordinate system will be used. We here list the version of generalized Oseen equations in polar coordinate system. In the polar coordinate system (r, θ), there are phonon strain components $\varepsilon_{rr}, \varepsilon_{\theta\theta}, \varepsilon_{r\theta} = \varepsilon_{\theta r}$ with the phonon stress components $\sigma_{rr}, \sigma_{\theta\theta}, \sigma_{r\theta} = \sigma_{\theta r}$ following the generalized Hooke's law of phonons

$$
\left.
\begin{aligned}
&\varepsilon_{rr} = \frac{\partial u_r}{\partial r},\ \varepsilon_{\theta\theta} = \frac{1}{r}\frac{\partial u_\theta}{\partial r} + \frac{u_r}{r},\ \varepsilon_{r\theta} = \varepsilon_{\theta r} = \frac{1}{2}\left[\frac{1}{r}\frac{\partial u_r}{\partial \theta} + \frac{\partial u_\theta}{\partial r} - \frac{u_\theta}{r}\right] \\
&\sigma_{rr} = (L + 2M)\varepsilon_{rr} + L\varepsilon_{\theta\theta},\ \sigma_{\theta\theta} = (L + 2M)\varepsilon_{\theta\theta} + L\varepsilon_{rr},\ \sigma_{r\theta} = \sigma_{\theta r} = 2M\varepsilon_{r\theta}
\end{aligned}
\right\}
$$
$$(7.5.1)$$

and the fluid phonon deformation velocity components $\dot{\xi}_{rr}, \dot{\xi}_{\theta\theta}, \dot{\xi}_{r\theta} = \dot{\xi}_{\theta r}$ and the fluid stress components $p_{rr}, p_{r\theta}, p_{r\theta} = p_{\theta r}$ including fluid pressure p following the generalized Newton's law

$$\left.\begin{array}{l}
\dot{\xi}_{rr} = \frac{\partial V_r}{\partial r}, \dot{\xi}_{\theta\theta} = \frac{1}{r}\frac{\partial V_\theta}{\partial r} + \frac{V_r}{r}, \dot{\xi}_{r\theta} = \dot{\xi}_{\theta r} = \frac{1}{2}\left(\frac{1}{r}\frac{\partial V_r}{\partial\theta} + \frac{\partial V_\theta}{\partial r} - \frac{V_\theta}{r}\right) \\[4pt]
p_{rr} = -p + \sigma'_{rr}, \; p_{\theta\theta} = -p + \sigma'_{\theta\theta}, \; p_{r\theta} = p_{\theta r} = \sigma'_{r\theta} = \sigma'_{\theta r} \\[4pt]
\sigma'_{rr} = 2\eta\dot{\xi}_{rr} + \frac{2}{3}\eta\dot{\xi}_{kk}, \; \sigma'_{\theta\theta} = 2\eta\dot{\xi}_{\theta\theta} \\[4pt]
+ \frac{2}{3}\eta\dot{\xi}_{kk}, \; \sigma'_{r\theta} = \sigma'_{\theta r} = 2\eta\dot{\xi}_{r\theta}, \; \dot{\xi}_{kk} = \dot{\xi}_{rr} + \dot{\xi}_{\theta\theta}
\end{array}\right\}
\qquad (7.5.2)$$

and the phason strain components $w_{rr}, w_{\theta\theta}, w_{r\theta}, w_{\theta r}$ and the phason stress components $H_{rr}, H_{\theta\theta}, H_{r\theta}, H_{\theta r}$ obeying the generalized Hooke's law of phasons stress and the phason strain components and the generalized Hooke's law of phasons

$$\left.\begin{array}{l}
w_{rr} = \frac{\partial w_r}{\partial r}, \; w_{\theta\theta} = \frac{1}{r}\frac{\partial w_\theta}{\partial r} + \frac{w_r}{r}, \; w_{r\theta} = \frac{1}{r}\frac{\partial w_r}{\partial\theta}, \; w_{\theta r} = \frac{\partial w_\theta}{\partial r} - \frac{w_\theta}{r} \\[4pt]
H_{rr} = K_1 w_{rr} + K_2 w_{\theta\theta}, \; H_{\theta\theta} = K_1 w_{\theta\theta} + K_2 w_{rr}, \\[4pt]
H_{r\theta} = (K_1 + K_2 + K_3)w_{r\theta} + K_3 w_{\theta r} \\[4pt]
H_{\theta r} = (K_1 + K_2 + K_3)w_{\theta r} + K_3 w_{r\theta}
\end{array}\right\}
\qquad (7.5.3)$$

So that we have the governing equations

$$\left.\begin{array}{l}
\nabla \cdot (\rho\mathbf{V}) = 0 \\[4pt]
\frac{\partial(U_r\rho V_r)}{\partial r} + \frac{\partial(U_\theta\rho V_\theta)}{r\partial\theta} = -\frac{\partial p}{\partial r} + \eta\nabla^2 V_r + \frac{1}{3}\eta\frac{\partial}{\partial r}\nabla\cdot\mathbf{V} \\[4pt]
+ M\nabla^2 u_r + (L + M - B)\frac{\partial}{\partial r}\nabla\cdot\mathbf{u} - (A - B)\frac{1}{\rho_0}\frac{\partial\delta\rho}{\partial r} \\[4pt]
\frac{\partial(U_r\rho V_\theta)}{r\partial\theta} + \frac{\partial(U_\theta\rho V_\theta)}{r\partial\theta} = -\frac{\partial p}{r\partial\theta} + \eta\nabla^2 V_\theta + \frac{1}{3}\eta\frac{\partial}{r\partial\theta}\nabla\cdot\mathbf{V} \\[4pt]
+ M\nabla^2 u_\theta + (L + M - B)\frac{\partial}{r\partial\theta}\nabla\cdot\mathbf{u} - (A - B)\frac{1}{\rho_0}\frac{\partial\delta\rho}{r\partial\theta} \\[4pt]
U_r\frac{\partial u_r}{\partial r} + U_\theta\frac{\partial u_\theta}{r\partial\theta} = V_x + \Gamma_\mathbf{u}\left[M\nabla^2 u_r + (L + M)\frac{\partial}{\partial r}\nabla\cdot\mathbf{u}\right] \\[4pt]
U_r\frac{\partial u_\theta}{\partial r} + U_\theta\frac{\partial u_\theta}{r\partial\theta} = V_y + \Gamma_\mathbf{u}\left[M\nabla^2 u_\theta + (L + M)\frac{\partial}{r\partial\theta}\nabla\cdot\mathbf{u}\right] \\[4pt]
U_r\frac{\partial w_r}{\partial r} + U_\theta\frac{\partial w_\theta}{r\partial\theta} = \Gamma_\mathbf{w}\left\{\left[K_1\left(\frac{\partial^2}{\partial r^2} + \frac{\partial}{r\partial r} - \frac{1}{r^2}\right) + (K_1 + K_2 + K_3)\frac{\partial^2}{r^2\partial\theta^2}\right]w_r\right. \\[4pt]
\left. + \left(K_2\frac{\partial^2}{r\partial r\partial\theta} + K_1\frac{\partial}{r^2\partial\theta}\right)w_\theta\right\} \\[4pt]
U_r\frac{\partial w_\theta}{\partial r} + U_\theta\frac{\partial w_\theta}{r\partial\theta} = \Gamma_\mathbf{w}\left\{\left[K_1\frac{\partial^2}{\partial r^2} + K_2\left(\frac{\partial^2}{r\partial r\partial\theta} + \frac{\partial}{r\partial r} + 2\frac{\partial}{r^2\partial\theta} - \frac{1}{r^2}\right)\right]w_r\right. \\[4pt]
\left. + \left[K_1\left(\frac{\partial^2}{r^2\partial\theta^2} + 3\frac{\partial}{r^2\partial\theta}\right) + K_2\frac{\partial}{r^2\partial\theta} + 4K_3\frac{\partial}{r^2\partial\theta}\right]w_\theta\right\} \\[4pt]
p = f(\rho)
\end{array}\right\}$$

$$(7.5.4)$$

in which $\nabla^2 = \frac{\partial^2}{\partial r^2} + \frac{\partial}{r\partial r} + \frac{\partial^2}{r^2\partial\theta^2}$, $\nabla = \mathbf{i}_r\frac{\partial}{\partial r} + \mathbf{i}_\theta\frac{\partial}{r\partial\theta}$, $\mathbf{V} = \mathbf{i}_r V_r + \mathbf{i}_\theta V_\theta$, $\mathbf{u} = \mathbf{i}_r u_r + \mathbf{i}_\theta u_\theta$, $\nabla \cdot \mathbf{a} = \frac{\partial}{r\partial r}(r a_r) + \frac{\partial a_\theta}{r\partial\theta}$.

7.6 Flow Past a Circular Cylinder

This section reports the formulation and results on the flow of soft-matter quasicrystals with 12-fold symmetry past a circular cylinder. Here we only consider the steady motion of low Reynolds number of structured viscous liquid with compressibility.

The equation system of the generalized dynamics of soft-matter quasicrystals is given by (7.5.4), in which the elementary excitations—phonons, phasons, and fluid phonon and their interaction play a central role. The equation of state is a key for the analysis. The computation verifies the equations and the formulation and reveals some significant behavior of the flow.

7.6.1 Two-Dimensional Flow Past Obstacle

This section discusses an application of the mathematical theory of dynamics of soft-matter quasicrystals for solving flow past a cylinder. The classical analytic Oseen solution offered in Chap. 6 for conventional liquid is significant which provides a comparison example for the present discussion.

Suppose a slow flow along direction x with velocity U_∞ shown by Fig. 7.1 the pressure p_∞ at infinity is omitted here and a circular cylinder with radius a in an infinite soft-matter quasicrystal. We have the boundary conditions in circular cylindrical coordinate system (r, θ)

$$\begin{cases} r = \sqrt{x^2 + y^2} \to \infty : \\ V_r = U_\infty \cos\theta, \ V_\theta = -U_\infty \sin\theta, \ \sigma_{rr} = \sigma_{\theta\theta} = 0, \ H_{rr} = H_{\theta\theta} = 0, \\ r = a : \\ V_r = V_\theta = 0, \ \sigma_{rr} = \sigma_{r\theta} = 0, \ H_{rr} = H_{r\theta} = 0 \end{cases} \tag{7.6.1}$$

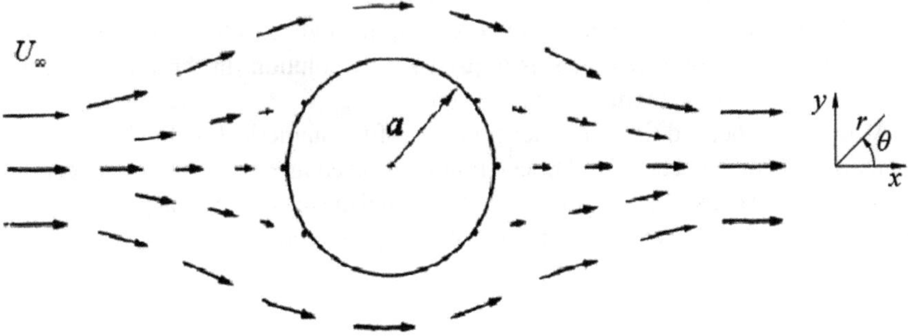

Fig. 7.1 Flow of soft-matter quasicrystal past cylinder with radius a

7.6.2　Quasi-Steady Analysis—Numerical Solution by Finite Difference Method

Due to the complexity of Eq. (7.5.4) and boundary conditions (7.6.1), any analytic solution is not available at present. We have to use numerical methods to solve the boundary value problem and the finite difference method in polar coordinate system is used, whose grid is shown in Fig. 7.2. Since one can only take finite

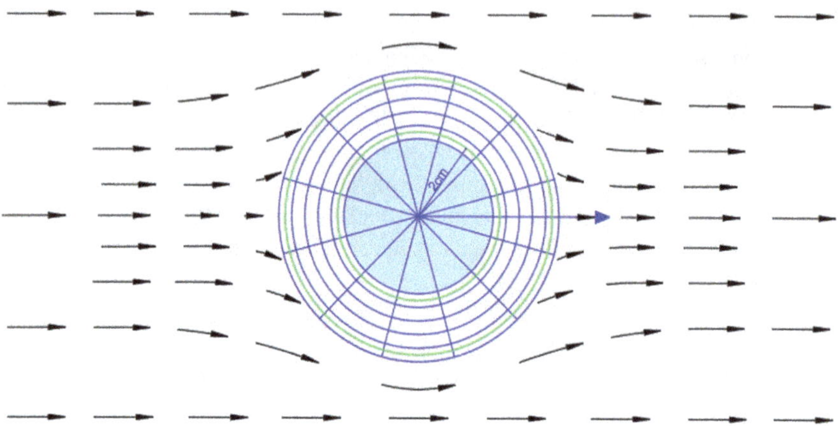

Fig. 7.2　Finite difference grid in polar coordinate system

"computational infinity" in the numerical method and different "computational infinity" will lead to different computational results, this indicates the problem is time-dependent. The present treatment is a quasi-steady solution. In the following, we take the "computational infinity" at $r_\infty = 6a$.

At first, we checked the computer program of the numerical method of ours by comparison with the generalized Oseen problem solved in Ref [15]. As the checking is satisfactory, we just continue the computation and the results and analysis are given by Cheng and Fan [16] and listed in the following subsections.

7.6.3　Numerical Results and Analysis

In the numerical analysis, we must give a concrete equation of state and here we adopt the following equation

$$p = f(\rho) = \frac{1}{\kappa_T}\left(1 - \frac{\rho_0}{\rho}\right) \tag{7.6.2}$$

this means that the calculation holds for quasicrystals in certain soft matter only, refer to Chap. 4.

In the computation the following material constants (part of which come from a Chap. 1 and most among the others are estimated values) and relevant physical and geometry parameters.

$U_\infty = 0.01\,\text{m/s}$, $\rho_0 = 1.5\,\text{g/cm}^3$, $\eta = 1\,\text{Poise}$, $r/a = 1.55$, $a = 1\,\text{cm}$, $L = 10\,\text{MPa}$, $M = 4\,\text{MPa}$, $K_1 = 0.5\text{L}$, $K_2 = -0.1\text{L}$, $K_3 = 0.05\text{L}$, $\Gamma_u = 4.8 \times 10^{-17}\text{m}^3 \cdot \text{s/kg}$, $\Gamma_w = 4.8 \times 10^{-19}\text{m}^3 \cdot \text{s/kg}$, $A \sim 0.2\,\text{MPa}$, $B \sim 0.2\,\text{MPa}$ and the phonon-phason coupling constant $R = 0$ is used and the computation is stable.

A part of numerical results obtained is listed in the following through a series of illustrations. We find that among factors influencing the computational results, the Reynolds number $\text{Re} = \frac{\rho U_\infty a}{\eta}$ is the most important.

The relative variation of the mass density $\frac{\delta\rho}{\rho_0}$ is obtained, where $\delta\rho = \rho - \rho_0$. The computation result is

$$\left|\frac{\delta\rho}{\rho_0}\right| \sim 10^{-6} \tag{7.6.3}$$

which is quite large than that of solid quasicrystals, refer to Chap. 3 and it is confirmed that the compressibility of soft-matter quasicrystals should be considered.

The angular distribution of phonon stresses is shown in Figs. 7.3, 7.4 and 7.5. These results belong to soft-matter quasicrystals only. The conventional liquids do not contain this kind of these field variables, i.e., there are no phonon elastic stresses (Figs. 7.6, 7.7, 7.8, 7.9, 7.10 and 7.11).

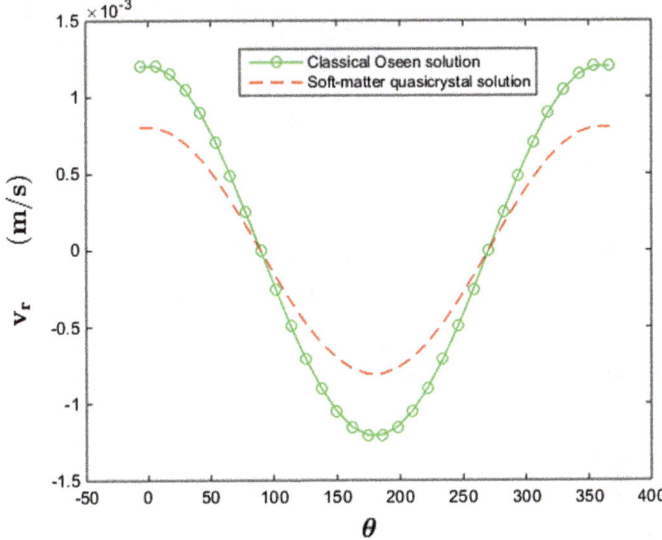

Fig. 7.3 Angular distribution of the radial velocity of fluid phonon at $r = 1.55a$ and comparison with that of the classical Oseen solution

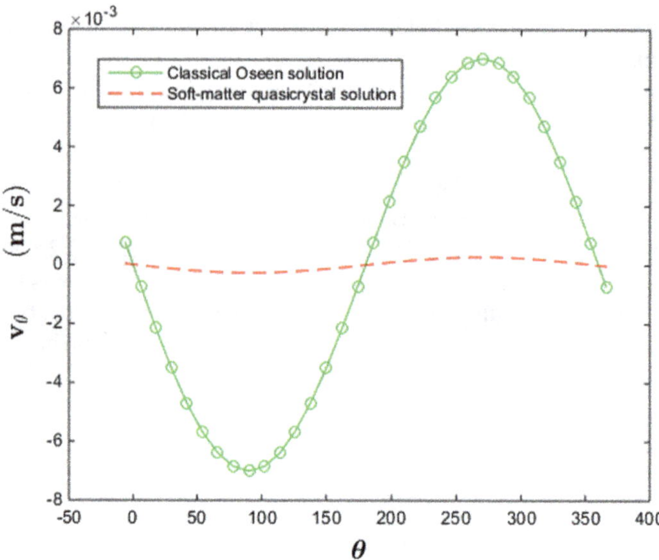

Fig. 7.4 Angular distribution of the circumferential velocity of fluid phonon at $r = 1.55a$ and comparison with that of the classical Oseen solution

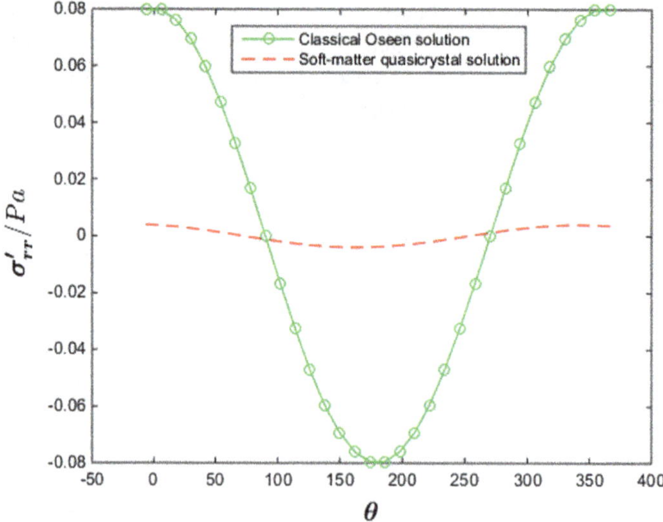

Fig. 7.5 Angular distribution of radial normal viscous stress of fluid phonon at $r = 1.55a$ and comparison with that of the classical Oseen solution

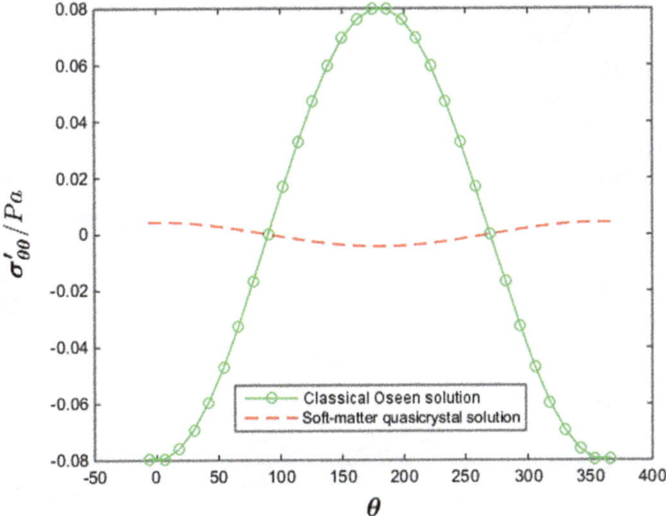

Fig. 7.6   Angular distribution of circumference normal viscous stress of fluid phonon at $r = 1.55a$ and comparison with that of the classical Oseen solution

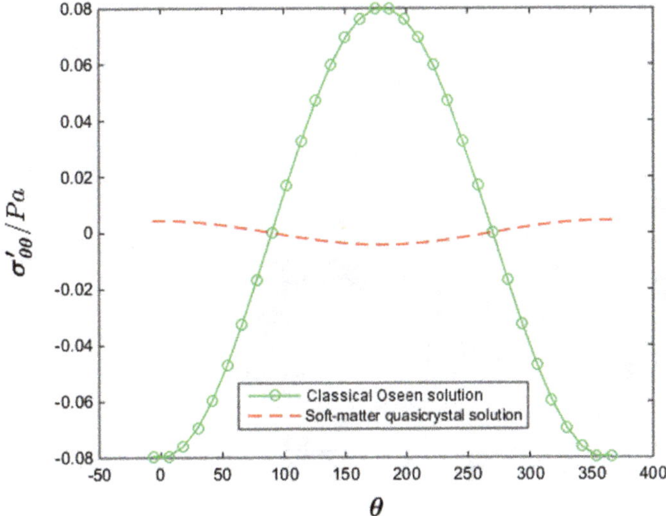

Fig. 7.7   Angular distribution of radial normal viscous stress of fluid phonon at $r = 1.55a$ and comparison with that of the classical Oseen solution

Although there is no phonon field in simple fluid, the results given by Figs. 7.12, 7.13, 7.14, 7.15 and 7.16 cannot be compared with the classical Oseen solution, it is evident these results are significant, which are induced by the fluid motion and shows the strong coupling between phonons and fluid phonon. Especially that the phonon stresses are greater much more than those of the fluid phonon stress field, this shows the phonon elementary excitation plays a more important role in the

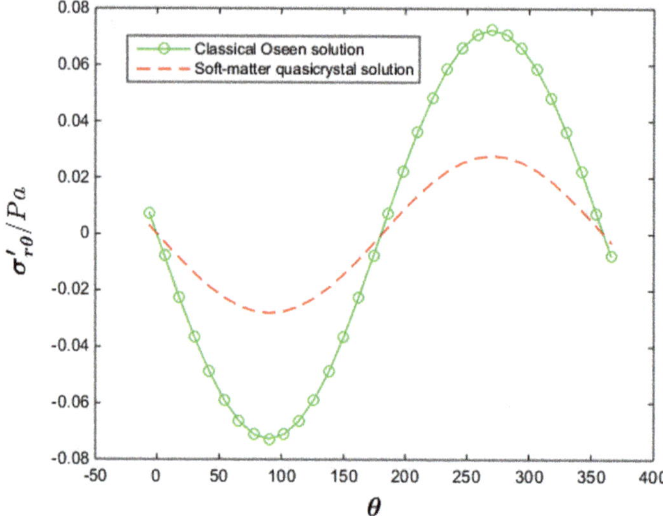

Fig. 7.8 Angular distribution of radial shear stress of fluid phonon at $r = 1.55a$ and comparison with that of the classical Oseen solution

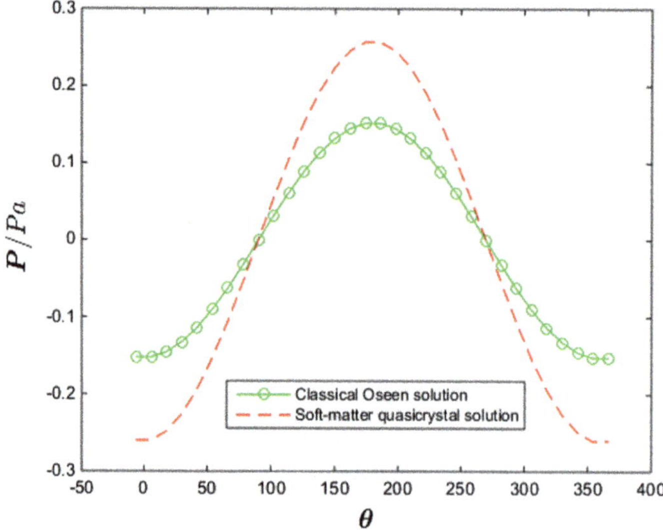

Fig. 7.9 Angular distribution of fluid pressure at $r = 1.55a$ and comparison with that of the classical Oseen solution

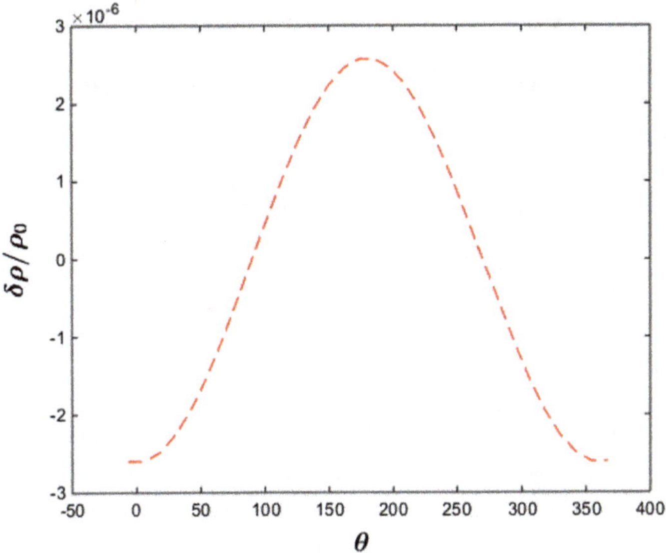

Fig. 7.10 Angular distribution of the relative variation of mass density at $r = 1.55a$

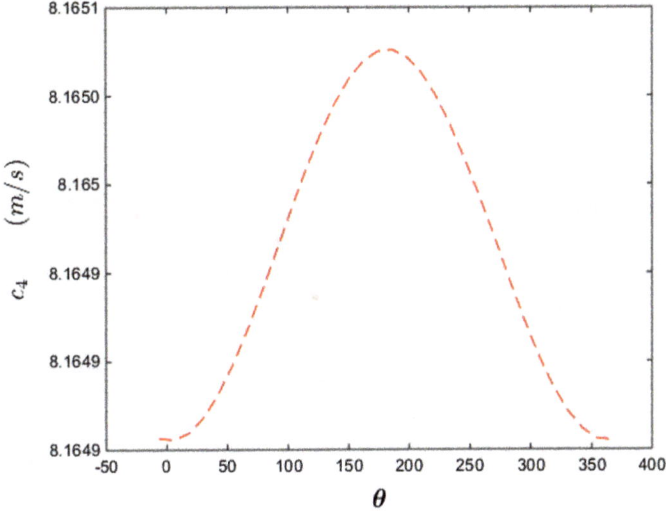

Fig. 7.11 Angular distribution of the fluid acoustic speed at $r = 1.55a$

flow motion. However, due to decoupling between phonons and phasons, we find that the coupling between the fluid phonon and phasons does not exist, at least the coupling is very weak if it exists so that during the flow past a cylinder, we could not obtain any nonzero solution on phason field (but this is in 12-fold and 18-fold

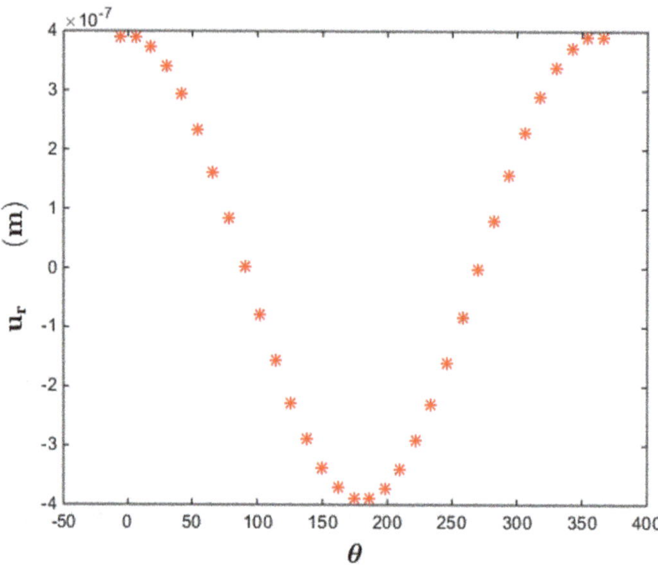

Fig. 7.12 The phonon radial displacement at $r = 1.55a$ versus angular θ

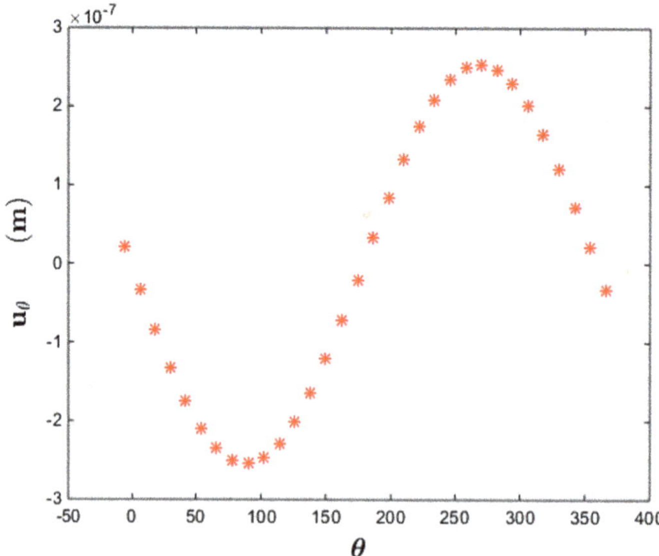

Fig. 7.13 The phonon circumference displacement at $r = 1.55a$ versus angular θ

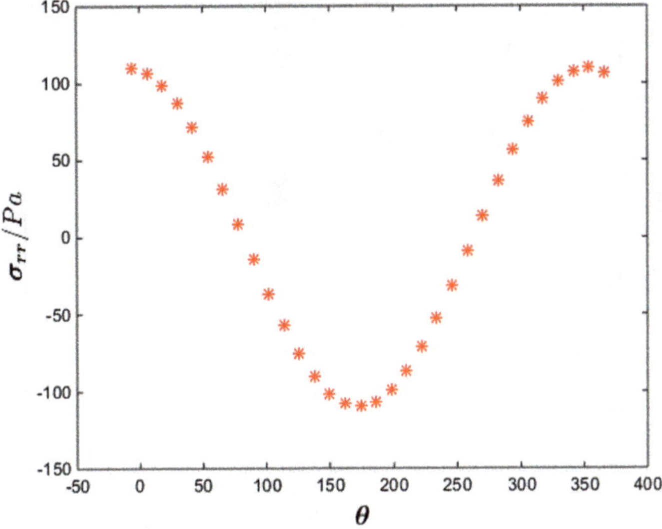

Fig. 7.14 The phonon radial normal stress at $r = 1.55a$ versus angular θ

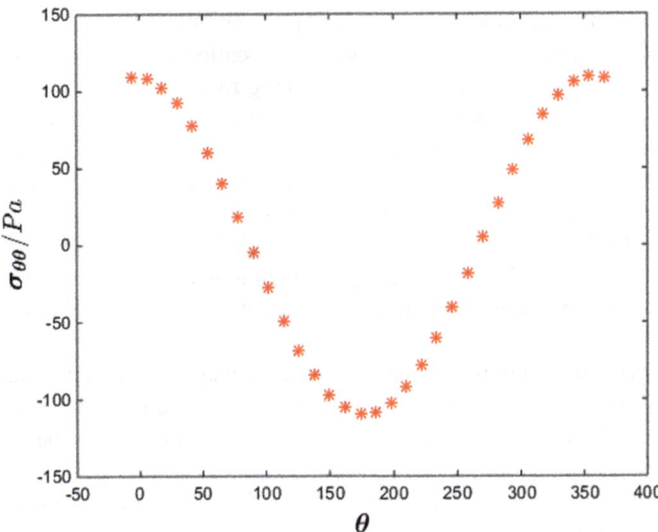

Fig. 7.15 The phonon circumference normal stress at $r = 1.55a$ versus angular θ

quasicrystals due to the decoupling between phonon and phason elasticity, while for other quasicrystals, e.g., decagonal or octagonal quasicrystals, the phason field is nonvarnished, this will be reported in Chap. 9).

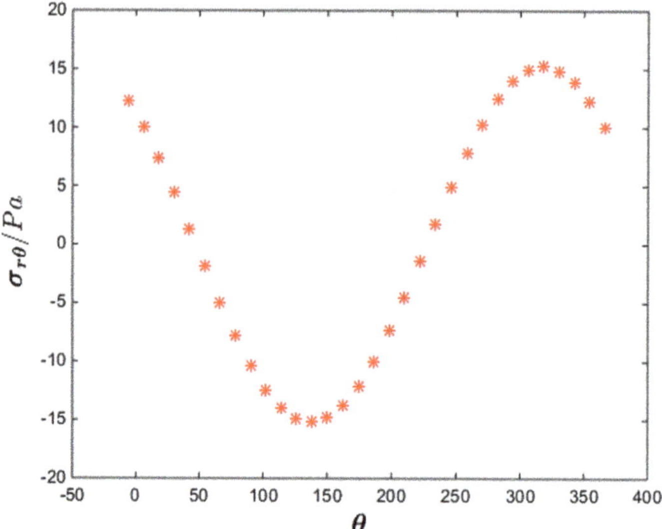

Fig. 7.16 The phonon shear stress at $r = 1.55a$ versus angular θ

Due to the limitation of space, we cannot include more illustrations. The results show the differences in solutions between conventional liquids and soft-matter quasicrystals are quite large because they belong to two different kinds of materials. This explores the importance of degrees of freedom of phonons and phasons in quasicrystals, although the results of phasons cannot be included here (their values are too small) in the pictures due to the decoupling between phonons and phasons in the 12-fold symmetry quasicrystals. In contrast to 12-fold symmetry quasicrystals, the 10- and 8-fold symmetry quasicrystals (although the 8-fold symmetry quasicrystals are not observed in the soft matter yet, which may be found in the near future) due to the coupling between phonons and phasons, the solution of phasons is strongly exhibited.

Another evident feature is that the Reynolds number plays an important role.

The equation of state is important too, if there is no equation then the basic equation set is not closed and there are no solutions at all. The including of the equation of state is necessary.

The viscous fluid stress components and the fluid pressure are compared from the Oseen solutions of the classical incompressible viscous fluid, whose derivation refers to Chap. 6 which are introduced from references [12, 17]:

$$V_r^{\text{classical}} = \frac{U_\infty \cos\theta}{1 - 2\ln\left(\frac{1}{2}ka\right) - 2\gamma}\left(-1 + \frac{a^2}{r^2} + 2\ln\frac{r}{a}\right)$$

$$V_\theta^{\text{classical}} = -\frac{U_\infty \sin\theta}{1 - 2\ln\left(\frac{1}{2}ka\right) - 2\gamma}\left(1 - \frac{a^2}{r^2} + 2\ln\frac{r}{a}\right) \tag{7.6.4}$$

with

$$2k = \rho U_\infty/\eta, \quad \gamma = 0.5772 \tag{7.6.5}$$

The comparison is meaningful, although only a motion of steady-state and incompressible simple liquid is discussed in the classical Oseen solution, belong to the linear regime but it is an analytic solution, presents very important reference significance. Our solution is a numerical solution for such a complex system described by Eq. (7.5.4), which are complex liquids with degrees of freedom of phonons, phanons, and compressible fluid, presents highly nonlinearity, our solution could be checked by the Oseen solution given in Sect. 6.5. To some extent, this gives a vigorous verification of the generalized dynamics of soft-matter quasicrystals by the classical fluid dynamics. It is well-known, for classical viscous fluid dynamics of two-dimension the classical Oseen solution is precious.

Exactly speaking, the numerical solution in the section depends upon the equation of state (7.6.2), which needs to be verified by experiments. Due to the difficulty, the experiments are undertaken still.

The further analysis, refer to Chap. 9, where we shall provide some results on phason field in the generalized Oseen flow of soft-matter quasicrystals of 8-fold symmetry where the phonon-phason is coupled, which is more interesting.

7.7 Three-Dimensional Equations of Generalized Dynamics of Soft-Matter Quasicrystals with 12-Fold Symmetry

The discussion of Sects. 7.1 to 7.6 is only concerned with the planar field, we did not concern the three-dimensional problem of the dynamics, whose version is listed in this section.

At first, we list the three-dimensional constitutive laws on phonons, phasons, and fluid phonon, respectively, as follows [3]

$$\left. \begin{aligned}
\sigma_{xx} &= C_{11}\varepsilon_{xx} + C_{12}\varepsilon_{yy} + C_{13}\varepsilon_{zz} \\
\sigma_{yy} &= C_{12}\varepsilon_{xx} + C_{11}\varepsilon_{yy} + C_{13}\varepsilon_{zz} \\
\sigma_{zz} &= C_{13}\varepsilon_{xx} + C_{13}\varepsilon_{yy} + C_{33}\varepsilon_{zz} \\
\sigma_{yz} &= \sigma_{zy} = 2C_{44}\varepsilon_{yz} \\
\sigma_{zx} &= \sigma_{xz} = 2C_{44}\varepsilon_{zx} \\
\sigma_{xy} &= \sigma_{yx} = 2C_{66}\varepsilon_{xy}
\end{aligned} \right\} \tag{7.7.1a}$$

$$\left.\begin{aligned}
H_{xx} &= K_1 w_{xx} + K_2 w_{yy}\\
H_{yy} &= K_2 w_{xx} + K_1 w_{yy}\\
H_{yz} &= K_4 w_{yz}\\
H_{xy} &= (K_1 + K_2 + K_3)w_{xy} + K_3 w_{yx}\\
H_{xz} &= K_4 w_{xz}\\
H_{yx} &= K_3 w_{xy} + (K_1 + K_2 + K_3)w_{yx}
\end{aligned}\right\} \tag{7.7.1b}$$

$$\left.\begin{aligned}
p_{xx} &= -p + 2\eta\dot{\xi}_{xx} - \tfrac{2}{3}\eta\dot{\xi}_{kk}\\
p_{yy} &= -p + 2\eta\dot{\xi}_{yy} - \tfrac{2}{3}\eta\dot{\xi}_{kk}\\
p_{zz} &= -p + 2\eta\dot{\xi}_{zz} - \tfrac{2}{3}\eta\dot{\xi}_{kk}\\
p_{yz} &= 2\eta\dot{\xi}_{yz},\ p_{zx} = 2\eta\dot{\xi}_{zx},\ p_{xy} = 2\eta\dot{\xi}_{xy}\\
\dot{\xi}_{kk} &= \dot{\xi}_{xx} + \dot{\xi}_{yy} + \dot{\xi}_{zz}
\end{aligned}\right\} \tag{7.7.1c}$$

then the equations of dynamics of soft-matter quasicrystals of 12-fold symmetry are as follows

$$\left.\begin{aligned}
&\frac{\partial\rho}{\partial t} + \nabla\cdot(\rho\mathbf{V}) = 0\\
&\frac{\partial(\rho V_x)}{\partial t} + \frac{\partial(V_x\rho V_x)}{\partial x} + \frac{\partial(V_y\rho V_x)}{\partial y} + \frac{\partial(V_z\rho V_x)}{\partial z} = -\frac{\partial p}{\partial x} + \eta\nabla^2 V_x\\
&+ \tfrac{1}{3}\eta\frac{\partial}{\partial x}\nabla\cdot\mathbf{V} + \left(C_{66}\frac{\partial^2}{\partial y^2} + C_{44}\frac{\partial^2}{\partial z^2}\right)u_x + (C_{12} + C_{66})\frac{\partial^2 u_y}{\partial x\partial y}\\
&+ (C_{13} + C_{44} - C_{11})\frac{\partial^2 u_z}{\partial x\partial z} + (C_{11} - B)\frac{\partial}{\partial x}\nabla\cdot\mathbf{u} - (A - B)\frac{1}{\rho_0}\frac{\partial\delta\rho}{\partial x}\\
&\frac{\partial(\rho V_y)}{\partial t} + \frac{\partial(V_x\rho V_y)}{\partial x} + \frac{\partial(V_y\rho V_y)}{\partial y} + \frac{\partial(V_z\rho V_y)}{\partial z} = -\frac{\partial p}{\partial y} + \eta\nabla^2 V_y\\
&+ \tfrac{1}{3}\eta\frac{\partial}{\partial y}\nabla\cdot\mathbf{V} + (C_{12} + C_{66})\frac{\partial^2 u_x}{\partial x\partial y} + \left(C_{66}\frac{\partial^2}{\partial x^2} + C_{11}\frac{\partial^2}{\partial y^2} + C_{44}\frac{\partial^2}{\partial z^2}\right)u_y\\
&+ (C_{13} + C_{44})\frac{\partial^2 u_z}{\partial y\partial z} + (C_{11} - B)\frac{\partial}{\partial y}\nabla\cdot\mathbf{u} - (A - B)\frac{1}{\rho_0}\frac{\partial\delta\rho}{\partial y}\\
&\frac{\partial(\rho V_z)}{\partial t} + \frac{\partial(V_x\rho V_z)}{\partial x} + \frac{\partial(V_y\rho V_z)}{\partial y} + \frac{\partial(V_z\rho V_z)}{\partial z} = -\frac{\partial p}{\partial z} + \eta\nabla^2 V_z\\
&+ \tfrac{1}{3}\eta\frac{\partial}{\partial z}\nabla\cdot\mathbf{V} + \left(C_{44}\frac{\partial^2}{\partial x^2} + C_{44}\frac{\partial^2}{\partial y^2} + (C_{33} - C_{13} - C_{44})\frac{\partial^2}{\partial z^2}\right)u_z\\
&+ (C_{13} + C_{44} - B)\frac{\partial}{\partial z}\nabla\cdot\mathbf{u} - (A - B)\frac{1}{\rho_0}\frac{\partial\delta\rho}{\partial z}\\
&\frac{\partial u_x}{\partial t} + V_x\frac{\partial u_x}{\partial x} + V_y\frac{\partial u_x}{\partial y} + V_z\frac{\partial u_x}{\partial z} = V_x + \Gamma_{\mathbf{u}}\left[\left(C_{11}\frac{\partial^2}{\partial x^2} + C_{66}\frac{\partial^2}{\partial y^2} + C_{44}\frac{\partial^2}{\partial z^2}\right)u_x\right.\\
&+ (C_{11} - C_{66})\frac{\partial^2 u_y}{\partial x\partial y} + (C_{13} + C_{44})\frac{\partial^2 u_z}{\partial x\partial z}\bigg]\ \frac{\partial u_y}{\partial t} + V_x\frac{\partial u_y}{\partial x} + V_y\frac{\partial u_y}{\partial y} + V_z\frac{\partial u_y}{\partial z} = V_y\\
&+ \Gamma_{\mathbf{u}}\left[(C_{11} - C_{66})\frac{\partial^2 u_x}{\partial x\partial y} + \left(C_{66}\frac{\partial^2}{\partial x^2} + C_{11}\frac{\partial^2}{\partial y^2} + C_{44}\frac{\partial^2}{\partial z^2}\right)u_y + (C_{13} + C_{44})\frac{\partial^2 u_z}{\partial y\partial z}\right]\\
&\frac{\partial u_z}{\partial t} + V_x\frac{\partial u_z}{\partial x} + V_y\frac{\partial u_z}{\partial y} + V_z\frac{\partial u_z}{\partial z} = V_z + \Gamma_{\mathbf{u}}\left[(C_{13} + C_{44})\left(\frac{\partial^2 u_x}{\partial x\partial z} + \frac{\partial^2 u_y}{\partial y\partial z}\right)\right.\\
&+ \left(C_{44}\frac{\partial^2}{\partial x^2} + C_{44}\frac{\partial^2}{\partial y^2} + C_{33}\frac{\partial^2}{\partial z^2}\right)u_z\bigg]\ \frac{\partial w_x}{\partial t} + V_x\frac{\partial w_x}{\partial x} + V_y\frac{\partial w_x}{\partial y} + V_z\frac{\partial w_x}{\partial z}\\
&= \Gamma_{\mathbf{w}}\left[K_1\nabla_1^2 w_x + (K_2 + K_3)\frac{\partial^2 w_x}{\partial y^2} + K_4\frac{\partial^2 w_x}{\partial z^2} + 2K_3\frac{\partial^2 w_y}{\partial x\partial y}\right]\ \frac{\partial w_y}{\partial t} + V_x\frac{\partial w_y}{\partial x}\\
&+ V_y\frac{\partial w_y}{\partial y} V_z\frac{\partial w_y}{\partial z} = \Gamma_{\mathbf{w}}\left[(K_2 + K_3)\frac{\partial^2 w_x}{\partial x\partial y} + K_3\frac{\partial^2 w_x}{\partial y\partial z} + K_1\nabla_1^2 w_y + (K_2 + K_3)\frac{\partial^2 w_y}{\partial x^2}\right.\\
&+ (K_1 + K_2 + K_3))\frac{\partial^2 w_y}{\partial x\partial z}\bigg]\\
&p = f(\rho)
\end{aligned}\right\} \tag{7.7.2}$$

in which $\nabla^2 = \frac{\partial^2}{\partial x^2} + \frac{\partial^2}{\partial y^2} + \frac{\partial^2}{\partial z^2}$, $\nabla_1^2 = \frac{\partial^2}{\partial x^2} + \frac{\partial^2}{\partial y^2}$, $\nabla = \mathbf{i}\frac{\partial}{\partial x} + \mathbf{j}\frac{\partial}{\partial y} + \mathbf{k}\frac{\partial}{\partial z}$, $\mathbf{V} = \mathbf{i}V_x + \mathbf{j}V_y + \mathbf{k}V_z$, $\mathbf{u} = \mathbf{i}u_x + \mathbf{j}u_y + \mathbf{k}u_z$ and $C_{11}, C_{12}, C_{13}, C_{33}, C_{44}, C_{66} = (C_{11} - C_{12})/2$ the phonon elastic constants, K_1, K_2, K_3, K_4 the phason elastic constants, η the fluid dynamic viscosity and Γ_u and Γ_w the phonon and phason dissipation coefficients, A and B the material constants due to variation of mass density, i.e., the LRT constants, respectively.

The Eq. (7.7.2) is the final governing equations of dynamics of soft-matter quasicrystals of 12-fold symmetry in the three-dimensional case with field variables $u_x, u_y, u_z, w_x, w_y, V_x, V_y, V_z, \rho$ and p, the amount of the field variables is 10 and the amount of field equations is 10 too. Among them: (7.7.2a) is the mass conservation equation, (7.7.2b)–(7.7.2d) the momentum conservation equations or the generalized Navier–Stokes equations, (7.7.2e)–(7.7.2g) the equations of motion of phonons due to the symmetry breaking, (7.7.2h) and(7.7.2i) the phason dissipation equations and (7.7.2j) the equation of state, respectively. The equations are consistent to be mathematical solvability, if there is a lack of the equation of state, the equation system is not closed and has no meaning mathematically and physically. This shows the equation of state is necessary.

These equations reveal the nature of wave propagation of fields $\mathbf{u}$ and $\mathbf{V}$ with phonon wave speeds $c_1 = \sqrt{\frac{A + C_{11} - 2B}{\rho}}$, $c_2 = c_3\sqrt{\frac{C_{11} - C_{12}}{2\rho}}$ and fluid phonon wave speed $c_4 = \sqrt{\left(\frac{\partial p}{\partial \rho}\right)_s}$ and the nature of the diffusion of field $\mathbf{w}$ with major diffusive coefficient $D_1 = \Gamma_w K_1$ and other less important diffusive coefficients $D_2 = \Gamma_w K_2$, etc. from the viewpoint of hydrodynamics.

The detail of derivation was given by Ref [3]. Tang and Fan further derived the version of the equations in a spherical coordinate system and Cheng et al. gave a solution on the flow of the structured liquid past a sphere by using the finite difference method, whose detail is quite lengthy which can refer to Chap. 13 of this book.

7.8   Governing Equations of Generalized Dynamics of Incompressible Soft-Matter Quasicrystals of 12-Fold Symmetry

If we consider that the soft-matter quasicrystals are incompressible, i.e., the mass density is a constant, i.e.,

$$\rho = \text{const}$$

then $A = B = 0$, $\frac{\delta\rho}{\rho_0} = 0$, the Eq. (7.7.2) will be reduced to

$$
\left.
\begin{aligned}
&\nabla \cdot \mathbf{V} = 0 \\
&\rho\left(\frac{\partial V_x}{\partial t} + \frac{\partial(V_x V_x)}{\partial x} + \frac{\partial(V_y V_x)}{\partial y} + \frac{\partial(V_z V_x)}{\partial z}\right) = -\frac{\partial p}{\partial x} + \eta \nabla^2 V_x \\
&+ \left(C_{66}\frac{\partial^2}{\partial y^2} + C_{44}\frac{\partial^2}{\partial z^2}\right)u_x + (C_{12} + C_{66})\frac{\partial^2 u_y}{\partial x \partial y} + (C_{13} + C_{44} \\
&- C_{11})\frac{\partial^2 u_z}{\partial x \partial z} + C_{11}\frac{\partial}{\partial x}\nabla \cdot \mathbf{u}\, \rho\left(\frac{\partial V_y}{\partial t} + \frac{\partial(V_x V_y)}{\partial x} + \frac{\partial(V_y V_y)}{\partial y} + \frac{\partial(V_z V_y)}{\partial z}\right) = -\frac{\partial p}{\partial y} \\
&+ \eta \nabla^2 V_y + (C_{12} + C_{66})\frac{\partial^2 u_x}{\partial x \partial y} + \left(C_{66}\frac{\partial^2}{\partial x^2} + C_{11}\frac{\partial^2}{\partial y^2} + C_{44}\frac{\partial^2}{\partial z^2}\right)u_y + (C_{13} \\
&+ C_{44})\frac{\partial^2 u_z}{\partial y \partial z} + C_{11}\frac{\partial}{\partial y}\nabla \cdot \mathbf{u}\, \rho\left(\frac{\partial V_z}{\partial t} + \frac{\partial(V_x V_z)}{\partial x} + \frac{\partial(V_y V_z)}{\partial y} + \frac{\partial(V_z V_z)}{\partial z}\right) = -\frac{\partial p}{\partial z} \\
&+ \eta \nabla^2 V_z + \left(C_{44}\frac{\partial^2}{\partial x^2} + C_{44}\frac{\partial^2}{\partial y^2} + (C_{33} - C_{13} - C_{44})\frac{\partial^2}{\partial z^2}\right)u_z + (C_{13} + C_{44})\frac{\partial}{\partial z}\nabla \cdot \mathbf{u} \\
&\frac{\partial u_x}{\partial t} + V_x\frac{\partial u_x}{\partial x} + V_y\frac{\partial u_x}{\partial y} + V_z\frac{\partial u_x}{\partial z} = V_x + \Gamma_{\mathbf{u}}\left[\left(C_{11}\frac{\partial^2}{\partial x^2} + C_{66}\frac{\partial^2}{\partial y^2} + C_{44}\frac{\partial^2}{\partial z^2}\right)u_x\right. \\
&+ (C_{11} - C_{66})\frac{\partial^2 u_y}{\partial x \partial y} + (C_{13} + C_{44})\frac{\partial^2 u_z}{\partial x \partial z}\bigg]\frac{\partial u_y}{\partial t} + V_x\frac{\partial u_y}{\partial x} + V_y\frac{\partial u_y}{\partial y} + V_z\frac{\partial u_y}{\partial z} \\
&= V_y + \Gamma_{\mathbf{u}}\left[C_{11} - C_{66})\frac{\partial^2 u_x}{\partial x \partial y} + \left(C_{66}\frac{\partial^2}{\partial x^2} + C_{11}\frac{\partial^2}{\partial y^2} + C_{44}\frac{\partial^2}{\partial z^2}\right)u_y\right. \\
&+ (C_{13} + C_{44})\frac{\partial^2 u_z}{\partial y \partial z}\bigg]\frac{\partial u_z}{\partial t} + V_x\frac{\partial u_z}{\partial x} + V_y\frac{\partial u_z}{\partial y} + V_z\frac{\partial u_z}{\partial z} = V_z + \Gamma_{\mathbf{u}}[(C_{13} \\
&+ C_{44})\left(\frac{\partial^2 u_x}{\partial x \partial z} + \frac{\partial^2 u_y}{\partial y \partial z}\right) + \left(C_{44}\frac{\partial^2}{\partial x^2} + C_{44}\frac{\partial^2}{\partial y^2} + C_{33}\frac{\partial^2}{\partial z^2}\right)u_z\bigg]\frac{\partial w_x}{\partial t} + V_x\frac{\partial w_x}{\partial x} \\
&+ V_y\frac{\partial w_x}{\partial y} + V_z\frac{\partial w_x}{\partial z} = \Gamma_{\mathbf{w}}\left[K_1\nabla_1^2 w_x + (K_2 + K_3)\frac{\partial^2 w_x}{\partial y^2} + K_4\frac{\partial^2 w_x}{\partial z^2} + 2K_3\frac{\partial^2 w_y}{\partial x \partial y}\right] \\
&\frac{\partial w_y}{\partial t} + V_x\frac{\partial w_y}{\partial x} + V_y\frac{\partial w_y}{\partial y} V_z\frac{\partial w_y}{\partial z} = \Gamma_{\mathbf{w}}\left[(K_2 + K_3)\frac{\partial^2 w_x}{\partial x \partial y} + K_3\frac{\partial^2 w_x}{\partial y \partial z} + K_1\nabla_1^2 w_y\right. \\
&+ (K_2 + K_3)\frac{\partial^2 w_y}{\partial x^2} + (K_1 + K_2 + K_3))\frac{\partial^2 w_y}{\partial x \partial z}\bigg]
\end{aligned}
\right\}
$$

$$\tag{7.8.1}$$

So that the equation of state is not needed, the number of field variables is 9 and one of governing equations is 9 too. The initial and boundary conditions are similar to those discussed associated with Eq. (7.7.2). Solving the corresponding initial- and boundary value problems of (7.8.1) is similar to those associated with (7.7.2) but simpler.

The incompressible model is significant for simplifying computation however the simplification and approximation may lead to the loss of some intention and information to soft matter.

7.9 Conclusion and Discussion

This chapter discussed the two-dimensional dynamics of soft-matter quasicrystals of 12-fold symmetry, the generalized dynamics of a complex system is the basis of the study. The basic Eq. (7.1.3) are partly verified by the applications through the dislocation solution and solution of flow past obstacle.

The present work is a continuation, extension, and development of hydrodynamics of solid quasicrystals of Lubensky et al. [10] which was created by using the Poisson bracket method, referenced from which Fan [1] derived the dynamic equation system of soft-matter quasicrystals, in which the key lies in the introduction of fluid phonon which originated from the Landau school, and supplementation of the equation of state for the thermodynamic of compressible liquid model. The equations for the three-dimensional problem are discussed too. And for the incompressible complex fluid model, the equation of state is not needed. The computation shows that the equation system of soft-matter quasicrystals suggested in Ref [1] is valid, solvable, and effective.

The solution of dislocation is approximate.

In a two-dimensional case, to overcome the Stokes paradox, taking the generalized Oseen modification, shown by Eq. (7.4.1), is necessary. The computational results are satisfactorily exhibited basically in the previous sections, in which fruitful illustrations are displayed. Due to the lack of experimental data, the numerical solutions are verified through the classical Oseen solution, because the classical Oseen solution is a very significant work of traditional fluid dynamics. For three-dimensional problems, the Oseen modification is not necessary, so that the equations are taken the version of (7.7.2). In addition, the thermodynamic stability of 12-fold symmetry quasicrystals in soft matter will be discussed in Chap. 13, which shows the importance of three-dimensional theory as well.

The incompressible complex fluid model is introduced in Sect. 7.8.

Recently Ref [18] reported an opinion that the simulation found that the phason degrees of freedom of dodecagonal quasicrystals in smectic B do not exist from the molecular dynamic modeling in their work. This might be a very interesting problem for soft-matter quasicrystals.

The discussion above is focused on ideal soft-matter quasicrystals although the dislocations are considered. Apart from the micro-defects, there are macro-defects, for example, cracks in soft matter including soft-matter quasicrystals, refer to Refs [19–25]. These problems have not been discussed in the chapter due to the limitation of space. We confirm that the generalized dynamics of soft matter may present important application meaning in practice, an example on plastic crack will be discussed in Chap. 16 for smectic A liquid crystals, which was reported by Ref [21].

References

1. Fan, T.Y.: Equation system of generalized hydrodynamics of soft-matter quasicrystals. Appl. Math. Mech. **37**, 331–347 (2016), in Chinese (2016); arXiv: 1908.06425[cond-mat.soft]. 15 Oct 2019
2. Fan, T.Y.: Generalized hydrodynamics of second two-dimensional soft-matter quasicrystals. Appl. Math. Mech. **38**, 189–199 (2017), in Chinese (2017); arXiv: 1908.06430[cond-mat.soft], 15 Oct 2019
3. Fan, T.Y., Z.Y.: Three-dimensional generalized dynamics of soft-matter quasicrystals. Appl. Math. Mech. **38**, 2017, 1195–1207, in Chinese (2017); Adv. Mat. Sci. Eng. **2020**, Article 1D 4875854 (2020)
4. Zeng, X., Ungar, G., Liu, Y., Percec, V., Dulcey, A.S.E., Hobbs, J.K.: Supermolecular dentritic liquid quasicrystals. Nature **428**, 157–160 (2004)
5. Takano, K.: A mesoscopic Archimedian tiling having a complexity in polymeric stars. J. Polym. Sci. Pol. Phys. **43**, 2427–2432 (2005)

6. Hayashida, K., Dotera, T., Takano, A., Matsushita, Y.: Polymeric quasicrystal: mesoscopic quasicrystalline tiling in ABC star polymers. Phys. Rev. Lett. **98**, 195502 (2007)
7. Talapin, V.D., Shevechenko, E.V., Bodnarchuk, M.I., Ye, X.C., Chen, J., Murray, C.B.: Quasicrystalline order in self-assembled binary nanoparticle superlattices. Nature **461**, 964–967 (2009)
8. Fischer, S., Exner, A., Zielske, K., Perlich, J., Deloudi, S., Steuer, W., Linder, P., Foestor, S.: Colloidal quasicrystals with 12-fold and 18-fold symmetry. Proc. Nat. Acad. Sci **108**, 1810–1814 (2011)
9. Yue, K., Huang, M.J., Marson, R., He, J.L., Huang, J.H., Zhou, Z., Liu, C., Yan, X.S., Wu, K., Wang, J., Guo, Z.H., Liu, H., Zhang, W., Ni, P.H., Wesdemiotis, C., Wen-Bin Zhang, W.B., Sharon, Glotzer, S.C., Cheng, S.Z.D.: Geometry induced sequence of nanoscale Frank-Kasper and quasicrystal mesophases in giant surfactants. Proc. Nat. Acad. Sci. **113**, 1392–1400 (2016)
10. Lubensky, T.C., Ramaswamy, S., Toner, J.: Hydrodynamics of icosahedral quasicrystals. Phys. Rev. B **32**, 7444–7452 (1985)
11. Fan, T.Y.: Mathematical theory of elasticity of quasicrystals and its applications, Science Press, Beijing/Springer-Verlag, Heidelberg, 1st edition, 2010; 2nd edition, 2016, in which the more detailed discussion on symmetry groups of quasicrystals, refer to Hu, C.Z., Wang, R.H., Ding, D.H.: Symmetry groups, physical constant tensors, elasticity and dislocations. Rep. Prog. Phys. **63**, 1–39 (2000); the dislocation solution of phason field was obtained firstly by Ding, D.H., refer to Yang, S.H., Ding, D.H.: Foundation of Theory of Crystal Dislocations, vol. 2, Science Press. Beijing (1998), in Chinese
12. Sleozkin, N.A.: Incompressible Viscous Fluid Dynamics. Gostehizdat Press, Moscow, in Russian (1959)
13. Oseen, C.W.: Ueber die Stokes'sche Formel und ueber eine verwandte Aufgabe in der Hydrodynamik, Ark Math Astronom Fys. **6** (1910)
14. Oseen, C.W.: Neuere Methoden und Ergibnisse in der Hydrodynamik. Akademische Verlagsgesellschaft, Leipzig (1927)
15. Cheng, H., Fan, T.Y., Tang, Z.Y.: Flow of compressible viscous fluid past a circular cylinder, unpublished work (2016)
16. Cheng, H., Fan, T.Y.: Flow of soft-matter quasicrystals with 12-fold symmetry past a circular cylinder, unpublished work (2017)
17. Kochin, N.E., Kibel'I, I.A., Roze, N.V.: Theoretical Hydrodynamics. Government Press of Phys-Math Literature, Moscow, in Russian (1953)
18. Oleynikov, P., Dzugutov, M., Lidin, S.: A smectic quasicrystal. Soft Matter **12**, 8869–8876 (2016)
19. Brostow, W., Cunha, A.M., Quintanila, J., Simoes, R.: Crack formation and propagation of polymer-liquid crystals. Macromol. Theor. Simul. **11**, 308–312 (2002)
20. Fan, T.Y.: A model of crack in smectic a liquid crystals. Philol. Mag. Lett. **92**, 153–158 (2012)
21. Fan, T.Y., Tang, Z.Y.: A model of crack based on dislocations in smectic A liquid crystals. Chin. Phys. B. **20**, 106103 (2014)
22. Bohn, S., Pauchard, L., Couder, Y.: Hierarchical crack pattern as formed by successive domain divisions. I. Temporal and geometrical hierarchy. Phys. Rev. E **71**, 046214 (2005)
23. Tirumkudulu, M.S.: Cracking in drying latex films. Langmuir **21**, 4938–4948 (2005)
24. Yow, H.N., Goikoetra, M., Goehring, L., Routh, A.F.: Effect of film thickness and particle size on cracking stresses in drying latex films. J. Colloid. Interface Sci. **352**, 542–548 (2010)
25. van der Kooij, H.M., Sprakel, J.: Watching paint dry; more exciting than it seems. Soft Matter **11**, 6353–6359 (2015)

Chapter 8
Dynamics of 10-Fold Symmetrical Soft-Matter Quasicrystals

8.1 Statement on Soft-Matter Quasicrystals of 10-Fold Symmetries

In Chap. 7 we discussed the dynamics of soft-matter quasicrystals with 12-fold symmetry observed in liquid crystals, polymers, colloids, and so on. There are some other quasicrystals, e.g., 10-fold symmetry quasicrystals that have been observed but not yet reported, the symmetry of which is similar to that of the 12-fold symmetry quasicrystals, and they also belong to the first type of two-dimensional quasicrystals. This chapter discusses the soft-matter quasicrystals with 10-fold symmetry. The quasicrystal system exhibits some characteristics, for example, a strong coupling between phonons and phasons for these quasicrystals, which is very interesting. Apart from this, in the previous chapter, only steady dynamics were discussed. In this chapter, the transient dynamic analysis for some samples of 10-fold symmetry quasicrystals will be studied in detail to further show some dynamic behavior of the motion. For example, in the transient dynamics, the wave propagation in matter and response of matter to waves are very interesting, where there are very simple and intuitive response laws of motion providing a physical basis testing the validity of dynamic equations and correctness of formulation for solving initial- and boundary-value problems of the equations. With this validity and correctness already proved, we can just talk about the accuracy of the studied problems.

8.2 Two-Dimensional Basic Equations of Soft-Matter Quasicrystals of Point Groups 10, $\overline{10}$

The most evident difference of 10-fold symmetrical quasicrystals with those of 12- and 18-fold symmetrical quasicrystals of soft matter is the strong coupling between phonons and phasons, i.e., in the constitutive law

$$\left.\begin{aligned}
\sigma_{ij} &= C_{ijkl}\varepsilon_{kl} + R_{ijkl}w_{kl} \\
H_{ij} &= K_{ijkl}w_{kl} + R_{klij}\varepsilon_{kl} \\
\varepsilon_{ij} &= \frac{1}{2}\left(\frac{\partial u_i}{\partial x_j} + \frac{\partial u_j}{\partial x_i}\right), \; w_{ij} = \frac{\partial w_i}{\partial x_j} \\
p_{ij} &= -p\delta_{ij} + 2\eta(\dot{\xi}_{ij} - \frac{1}{3}\dot{\xi}_{kk}\delta_{ij}) \\
\dot{\xi}_{ij} &= \frac{1}{2}\left(\frac{\partial V_i}{\partial x_j} + \frac{\partial V_j}{\partial x_i}\right)
\end{aligned}\right\} \qquad (8.2.1)$$

the phonon-phason coupling constants

$$R_{ijkl} \neq 0 \qquad (8.2.2)$$

In addition the generalized Newton's law for the fluid field

$$p_{ij} = -p\delta_{ij} + 2\eta(\dot{\xi}_{ij} - \frac{1}{3}\dot{\xi}_{kk}\delta_{ij}), \quad \dot{\xi}_{ij} = \frac{1}{2}\left(\frac{\partial V_i}{\partial x_j} + \frac{\partial V_j}{\partial x_i}\right)$$

is important for the study of course.

The important meaning of soft-matter quasicrystals and the generalized dynamics have been analyzed systematically in Refs. [1–3].

A difficulty in the generalized dynamics due to lack of equation of state $p = f(\rho)$ has been preliminarily discussed by Refs. [1–3], but some progress is just promoted recently, please refer to Chap. 4.

The concrete form of constitutive law depends upon the symmetry groups of quasicrystals. We can draw the data on point groups and relevant elastic constitutive laws for phonons and phasons from the theory of solid quasicrystals [4], for example, the phonon and phason constitutive laws for point groups 5, $\bar{5}$, and 10, $\overline{10}$ present the following form in xy-plane and if z-axis is the 5- or 10-fold symmetry axis

$$\left.\begin{aligned}
\sigma_{xx} &= L(\varepsilon_{xx} + \varepsilon_{yy}) + 2M\varepsilon_{xx} + R_1(w_{xx} + w_{yy}) + R_2(w_{xy} - w_{yx}) \\
\sigma_{yy} &= L(\varepsilon_{xx} + \varepsilon_{yy}) + 2M\varepsilon_{yy} - R_1(w_{xx} + w_{yy}) - R_2(w_{xy} - w_{yx}) \\
\sigma_{xy} &= \sigma_{yx} = 2M\varepsilon_{xy} + R_1(w_{yx} - w_{xy}) + R_2(w_{xx} + w_{yy}) \\
H_{xx} &= K_1 w_{xx} + K_2 w_{yy} + R_1(\varepsilon_{xx} - \varepsilon_{yy}) + 2R_2\varepsilon_{xy} \\
H_{yy} &= K_1 w_{yy} + K_2 w_{xx} + R_1(\varepsilon_{xx} - \varepsilon_{yy}) + 2R_2\varepsilon_{xy} \\
H_{xy} &= K_1 w_{xy} - K_2 w_{yx} - 2R_1\varepsilon_{xy} + R_2(\varepsilon_{xx} - \varepsilon_{yy}) \\
H_{yx} &= K_1 w_{yx} - K_2 w_{xy} + 2R_1\varepsilon_{xy} - R_2(\varepsilon_{xx} - \varepsilon_{yy})
\end{aligned}\right\} \qquad (8.2.2')$$

in addition

$$\left.\begin{aligned}
p_{xx} &= -p + 2\eta(\dot{\xi}_{xx} - \tfrac{1}{3}\dot{\xi}_{kk}) \\
p_{yy} &= -p + 2\eta(\dot{\xi}_{yy} - \tfrac{1}{3}\dot{\xi}_{kk}) \\
p_{xy} &= p_{yx} = 2\eta\dot{\xi}_{xy} \\
\dot{\xi}_{kk} &= \dot{\xi}_{xx} + \dot{\xi}_{yy}
\end{aligned}\right\} \qquad (8.2.2'')$$

then the final governing equations in xy-plane are derived as follows (The systematic derivation details are given by Fan [1])

$$\left.\begin{aligned}
&\frac{\partial\rho}{\partial t} + \nabla\cdot(\rho V) = 0 \\[4pt]
\frac{\partial(\rho V_x)}{\partial t} + \frac{\partial(V_x\rho V_x)}{\partial x} + \frac{\partial(V_y\rho V_x)}{\partial y} &= -\frac{\partial p}{\partial x} + \eta\nabla^2 V_x + \frac{1}{3}\eta\frac{\partial}{\partial x}\nabla\cdot V \\
&\quad + M\nabla^2 u_x + (L + M - B)\frac{\partial}{\partial x}\nabla\cdot u \\
&\quad + R_1\left(\frac{\partial^2 w_x}{\partial x^2} + 2\frac{\partial^2 w_y}{\partial x\partial y} - \frac{\partial^2 w_x}{\partial y^2}\right) \\
&\quad - R_2\left(\frac{\partial^2 w_y}{\partial x^2} - 2\frac{\partial^2 w_x}{\partial x\partial y} - \frac{\partial^2 w_y}{\partial y^2}\right) \\
&\quad - (A - B)\frac{1}{\rho_0}\frac{\partial\delta\rho}{\partial x} \\[6pt]
\frac{\partial(\rho V_y)}{\partial t} + \frac{\partial(V_x\rho V_y)}{\partial x} + \frac{\partial(V_y\rho V_y)}{\partial y} &= -\frac{\partial p}{\partial y} + \eta\nabla^2 V_y + \frac{1}{3}\eta\frac{\partial}{\partial y}\nabla\cdot V \\
&\quad + M\nabla^2 u_y + (L + M - B)\frac{\partial}{\partial y}\nabla\cdot u \\
&\quad + R_1\left(\frac{\partial^2 w_y}{\partial x^2} - 2\frac{\partial^2 w_x}{\partial x\partial y} - \frac{\partial^2 w_y}{\partial y^2}\right) \\
&\quad + R_2\left(\frac{\partial^2 w_x}{\partial x^2} + 2\frac{\partial^2 w_y}{\partial x\partial y} - \frac{\partial^2 w_x}{\partial y^2}\right) \\
&\quad - (A - B)\frac{1}{\rho_0}\frac{\partial\delta\rho}{\partial y} \\[6pt]
\frac{\partial u_x}{\partial t} + V_x\frac{\partial u_x}{\partial x} + V_y\frac{\partial u_x}{\partial y} &= V_x \\
&\quad + \Gamma_u\Big[M\nabla^2 u_x + (L + M)\frac{\partial}{\partial x}\nabla\cdot u \\
&\quad + R_1\left(\frac{\partial^2 w_x}{\partial x^2} + 2\frac{\partial^2 w_y}{\partial x\partial y} - \frac{\partial w_x}{\partial y^2}\right) \\
&\quad - R_2\left(\frac{\partial^2 w_y}{\partial x^2} - 2\frac{\partial^2 w_x}{\partial x\partial y} - \frac{\partial^2 w_y}{\partial y^2}\right)\Big]
\end{aligned}\right\}$$

$$
\left.
\begin{aligned}
&\frac{\partial u_y}{\partial t} + V_x \frac{\partial u_y}{\partial x} + V_y \frac{\partial u_y}{\partial y} = V_y + \Gamma_u[M\nabla^2 u_y + (L+M)\frac{\partial}{\partial y}\nabla \cdot u \\
&\qquad\qquad + R_1(\frac{\partial^2 w_y}{\partial x^2} - 2\frac{\partial^2 w_x}{\partial x \partial y} - \frac{\partial^2 w_y}{\partial y^2}) \\
&\qquad\qquad + R_2(\frac{\partial^2 w_x}{\partial x^2} + 2\frac{\partial^2 w_y}{\partial x \partial y} - \frac{\partial^2 w_x}{\partial y^2})] \\
&\frac{\partial w_x}{\partial t} + V_x \frac{\partial w_x}{\partial x} + V_y \frac{\partial w_x}{\partial y} = \Gamma_w[K_1\nabla^2 w_x \\
&\qquad\qquad + R_1(\frac{\partial^2 u_x}{\partial x^2} - 2\frac{\partial^2 u_y}{\partial x \partial y} - \frac{\partial^2 u_x}{\partial y^2}) \\
&\qquad\qquad + R_2(\frac{\partial^2 u_y}{\partial x^2} + 2\frac{\partial^2 u_x}{\partial x \partial y} - \frac{\partial^2 u_y}{\partial y^2})] \\
&\frac{\partial w_y}{\partial t} + V_x \frac{\partial w_y}{\partial x} + V_y \frac{\partial w_y}{\partial y} = \Gamma_w[K_1\nabla^2 w_y \\
&\qquad\qquad + R_1(\frac{\partial^2 u_y}{\partial x^2} + 2\frac{\partial^2 u_x}{\partial x \partial y} - \frac{\partial^2 u_y}{\partial y^2}) \\
&\qquad\qquad - R_2(\frac{\partial^2 u_x}{\partial x^2} - 2\frac{\partial^2 u_y}{\partial x \partial y} - \frac{\partial^2 u_x}{\partial y^2})] \\
&p = f(\rho)
\end{aligned}
\right\}
\qquad (8.2.3)
$$

in which the z-axis is 5- or 10-fold symmetry axis, u_x, u_y, w_x, w_y, V_x, V_y the phonon displacement, phason displacement, velocity components, and $\nabla^2 = \frac{\partial^2}{\partial x^2} + \frac{\partial^2}{\partial y^2}$, $\nabla = \mathbf{i}\frac{\partial}{\partial x} + \mathbf{j}\frac{\partial}{\partial y}$, $\mathbf{V} = \mathbf{i}V_x + \mathbf{j}V_y$, $\mathbf{u} = \mathbf{i}u_x + \mathbf{j}u_y$, respectively.

Among a set of Eq. (8.2.3), the first equation represents mass conservation law, the second and third ones represent the momentum conservation law or the generalized Navier–Stokes equations, and the fourth to seventh ones describe symmetry breaking due to phonons and phasons, and $\Gamma_\mathbf{u}$, Γ_w represent phonon and phason dissipation coefficients, respectively. Considering a series of the equation of state, we present the most likely form. The eighth equation among (8.2, 3) is the equation of state given by (4.2.2), refer to Chap. 4.

The above model belongs to a compressible complex fluid model of soft-matter quasicrystals with 10-fold symmetry, in contrast to which there is another model, i.e., an incompressible complex fluid model, in which $\rho = $ const, so the equation of state is not needed, in the case the governing equations of generalized dynamics will be simplified and discussed concretely in Sect. 8.8.

8.3 Dislocations and Solutions

For this kind of soft-matter quasicrystals, like other kinds of crystalline or quasicrystalline matter, dislocation is one of the basic problems. Due to the coupling between phonons and phasons the 5- and 10-fold symmetry quasicrystals are more complex than those of 12-fold ones. At first, we consider the zero-order approximate solution, i.e., we study the static case and omit the effect of fluid, then Eqs. (8.2.3) are reduced to

$$M\nabla^2 u_x + (L+M)\frac{\partial}{\partial x}\nabla\cdot\mathbf{u} + R_1\left(\frac{\partial^2 w_x}{\partial x^2} + 2\frac{\partial^2 w_y}{\partial x\partial y} - \frac{\partial^2 w_x}{\partial y^2}\right) - R_2\left(\frac{\partial^2 w_y}{\partial x^2} - 2\frac{\partial^2 w_x}{\partial x\partial y} - \frac{\partial^2 w_y}{\partial y^2}\right) = 0$$

$$M\nabla^2 u_y + (L+M)\frac{\partial}{\partial y}\nabla\cdot\mathbf{u} + R_1\left(\frac{\partial^2 w_y}{\partial x^2} - 2\frac{\partial^2 w_x}{\partial x\partial y} - \frac{\partial^2 w_y}{\partial y^2}\right) + R_2\left(\frac{\partial^2 w_x}{\partial x^2} + 2\frac{\partial^2 w_y}{\partial x\partial y} - \frac{\partial^2 w_x}{\partial y^2}\right) = 0$$

$$K_1\nabla^2 w_x + R_1\left(\frac{\partial^2 u_x}{\partial x^2} - 2\frac{\partial^2 u_y}{\partial x\partial y} - \frac{\partial^2 u_x}{\partial y^2}\right) + R_2\left(\frac{\partial^2 u_y}{\partial x^2} + 2\frac{\partial^2 u_x}{\partial x\partial y} - \frac{\partial^2 u_y}{\partial y^2}\right) = 0$$

$$K_1\nabla^2 w_y + R_1\left(\frac{\partial^2 u_y}{\partial x^2} + 2\frac{\partial^2 u_x}{\partial x\partial y} - \frac{\partial^2 u_y}{\partial y^2}\right) - R_2\left(\frac{\partial^2 u_x}{\partial x^2} - 2\frac{\partial^2 u_y}{\partial x\partial y} - \frac{\partial^2 u_x}{\partial y^2}\right) = 0$$

$$(8.3.1)$$

and further is simplified by introducing displacement potential $F(x, y)$ [5] in which $F(x, y)$ is satisfied the quadruple harmonic equation such as

$$\nabla^2\nabla^2\nabla^2\nabla^2 F = 0 \tag{8.3.2}$$

which is more complex than those of equations of 12-fold symmetry quasicrystals discussed in Chap. 7.

Solving Eq. (8.3.2) under the dislocation conditions

$$\int_\Gamma du_x = b_1^\parallel, \quad \int_\Gamma du_y = b_2^\parallel, \quad \int_\Gamma dw_x = b_1^\perp, \quad \int_\Gamma dw_y = b_2^\perp, \tag{8.3.3}$$

and other relevant boundary conditions and through some complicated and lengthy analysis one obtained the solution of dislocation by Li and Fan [6], Li et al. [5], and beforehand Ding et al. [7] found the solution in a different angle:

$$u_x = \frac{b_1^\parallel}{2\pi}\left[\arctan\left(\frac{y}{x}\right) + \frac{c_1 - c_2}{c_1}\left(\frac{xy}{r^2}\right)\right]$$

$$+ \frac{c_1 b_1^\perp}{\pi c_0 e_1}\left\{\frac{R_1}{R}\left[\frac{xy}{r^2} - \left(\frac{c_1 - c_2}{c_1}\right)\frac{2xy^3}{r^4}\right]\right.$$

$$\left. + \frac{R_2}{R}\left[\frac{y^2}{r^2} + \left(\frac{c_1 - c_2}{c_1}\right)\frac{y^2(x^2 - y^2)}{r^4}\right]\right\} \tag{8.3.4a}$$

$$u_y = \frac{b_1^\parallel}{2\pi}\left[-\ln\frac{r}{a} + \frac{c_1 - c_2}{c_1}\left(\ln\frac{r}{a} + \frac{y^2}{r^2}\right)\right]$$

$$+ \frac{c_1 b_1^\perp}{\pi c_0 e_1}\left\{-\frac{R_1}{R}\left[\frac{y^2}{r^2} - \left(\frac{c_1 - c_2}{c_1}\right)\frac{y^2(x^2 - y^2)}{r^4}\right] + \frac{R_2}{R}\left[\frac{xy}{r^2} + \left(\frac{c_1 - c_2}{c_1}\right)\frac{2xy^3}{r^4}\right]\right\}$$

$$\tag{8.3.4b}$$

$$w_x = \frac{c_0 b_1^\parallel}{2\pi c_1}\left[\left(\frac{R_1}{R}\right)\frac{2x^3 y}{r^4} + \left(\frac{R_2}{R}\right)\frac{y^2(3x^2 + y^2)}{r^4}\right]$$

$$+ \frac{b_1^{\perp}}{2\pi} \left[\begin{array}{l} \arctan\left(\dfrac{y}{x}\right) + \left(\dfrac{R_1^2 - R_2^2}{e_1 R^2}\right) \dfrac{xy(3x^2 - y^2)(3y^2 - x^2)}{3r^6} \\[3mm] + \left(\dfrac{2R_1 R_2}{e_1 R^2}\right) \dfrac{y^2(3x^2 - y^2)^2}{3r^6} \end{array} \right] \tag{8.3.4c}$$

$$w_y = \frac{c_0 b_1^{\parallel}}{2\pi c_1} \left[\left(\frac{R_1}{R}\right) \frac{y^2(3x^2 + y^2)}{r^4} + \left(\frac{R_2}{R}\right) \frac{2x^3 y}{r^4} \right]$$

$$+ \frac{b_1^{\perp}}{2\pi e_1} \left[\begin{array}{l} e_2 \ln \dfrac{r}{a} + \left(\dfrac{R_1^2 - R_2^2}{R^2}\right) \dfrac{y^2(3x^2 - y^2)^2}{3r^6} \\[3mm] - \left(\dfrac{2R_1 R_2}{e_1 R^2}\right) \dfrac{xy(3x^2 - y^2)(3y^2 - x^2)}{3r^6} \end{array} \right] \tag{8.3.4d}$$

in which

$$e_1 = \frac{2c_1 c_2}{c_0 k_0}, \quad e_2 = \frac{c_1 c_2}{c_0 k_0}\left(\frac{c_1'}{c_1} + \frac{c_2'}{c_2}\right)$$

$$c_1' = (L + 2M)K_2 - R^2, \quad c_2' = MK_2 - R^2$$

$$c_0 = (L + 2M)R, \quad c_1 = (L + 2M)K_1 - R^2, \quad c_2 = MK_1 - R^2$$

$$k_0 = R(K_1 - K_2)$$

the phonon and phason solutions are coupled to each other, and are more complicated than those for 12- fold symmetry quasicrystals listed in Chap. 7.

8.4 Probe on Modification of Dislocation Solution by Considering the Fluid Effect

The solutions listed in Sects. 7.3 and 8.3 respectively discussed the phonon and phason fields induced by the dislocations, and in Sect. 8.3 the interaction between phonons and phasons is also explored. However, the effect of fluid phonon has not been described yet. In this section, a probe to explore the effect is suggested.

To study the fluid effect, one must consider the complete governing equations after Oseen's modification at least for the case of steady-state and low Reynolds number

$$\nabla \cdot (\rho \mathbf{V}) = 0$$

$$\frac{\partial(U_x \rho V_x)}{\partial x} + \frac{\partial(U_y \rho V_x)}{\partial y} = -\frac{\partial p}{\partial x} + \eta \nabla^2 V_x + \frac{1}{3}\eta \frac{\partial}{\partial x}\nabla \cdot \mathbf{V}$$

$$+ M\nabla^2 u_x + (L + M - B)\frac{\partial}{\partial x}\nabla \cdot \mathbf{u}$$

$$+ R_1\left(\frac{\partial^2 w_x}{\partial x^2} + 2\frac{\partial^2 w_y}{\partial x \partial y} - \frac{\partial^2 w_x}{\partial y^2}\right)$$

$$- R_2\left(\frac{\partial^2 w_y}{\partial x^2} - 2\frac{\partial^2 w_x}{\partial x \partial y} - \frac{\partial^2 w_y}{\partial y^2}\right)$$

$$- (A - B)\frac{1}{\rho_0}\frac{\partial \delta\rho}{\partial x}$$

$$\frac{\partial(U_x \rho V_y)}{\partial x} + \frac{\partial(U_y \rho V_y)}{\partial y} = -\frac{\partial p}{\partial y} + \eta \nabla^2 V_y + \frac{1}{3}\eta \frac{\partial}{\partial y}\nabla \cdot \mathbf{V}$$

$$+ M\nabla^2 u_y + (L + M - B)\frac{\partial}{\partial y}\nabla \cdot \mathbf{u}$$

$$+ R_1(\frac{\partial^2 w_y}{\partial x^2} - 2\frac{\partial^2 w_x}{\partial x \partial y} - \frac{\partial^2 w_y}{\partial y^2}) + R_2(\frac{\partial^2 w_x}{\partial x^2} + 2\frac{\partial^2 w_y}{\partial x \partial y} - \frac{\partial^2 w_x}{\partial y^2})$$

$$- (A - B)\frac{1}{\rho_0}\frac{\partial \delta\rho}{\partial y}$$

$$U_x\frac{\partial u_x}{\partial x} + U_y\frac{\partial u_x}{\partial y} = V_x + \Gamma_{\mathbf{u}}\left[M\nabla^2 u_x + (L + M)\frac{\partial}{\partial x}\nabla \cdot \mathbf{u}\right.$$

$$\left. + R_1\left(\frac{\partial^2 w_x}{\partial x^2} + 2\frac{\partial^2 w_y}{\partial x \partial y} - \frac{\partial w_x}{\partial y^2}\right) - R_2\left(\frac{\partial^2 w_y}{\partial x^2} - 2\frac{\partial^2 w_x}{\partial x \partial y} - \frac{\partial^2 w_y}{\partial y^2}\right)\right]$$

$$U_x\frac{\partial u_y}{\partial x} + U_y\frac{\partial u_y}{\partial y} = V_y + \Gamma_{\mathbf{u}}\left[M\nabla^2 u_y + (L + M)\frac{\partial}{\partial y}\nabla \cdot \mathbf{u}\right.$$

$$\left. + R_1\left(\frac{\partial^2 w_y}{\partial x^2} - 2\frac{\partial^2 w_x}{\partial x \partial y} - \frac{\partial^2 w_y}{\partial y^2}\right) + R_2\left(\frac{\partial^2 w_x}{\partial x^2} + 2\frac{\partial^2 w_y}{\partial x \partial y} - \frac{\partial^2 w_x}{\partial y^2}\right)\right]$$

$$U_x\frac{\partial w_x}{\partial x} + U_y\frac{\partial w_x}{\partial y} = \Gamma_{\mathbf{w}}\left[K_1\nabla^2 w_x\right.$$

$$\left. + R_1\left(\frac{\partial^2 u_x}{\partial x^2} - 2\frac{\partial^2 u_y}{\partial x \partial y} - \frac{\partial^2 u_x}{\partial y^2}\right) + R_2\left(\frac{\partial^2 u_y}{\partial x^2} + 2\frac{\partial^2 u_x}{\partial x \partial y} - \frac{\partial^2 u_y}{\partial y^2}\right)\right]$$

$$U_x\frac{\partial w_y}{\partial x} + U_y\frac{\partial w_y}{\partial y} = \Gamma_{\mathbf{w}}\left[K_1\nabla^2 w_y\right.$$

$$\left. + R_1\left(\frac{\partial^2 u_y}{\partial x^2} + 2\frac{\partial^2 u_x}{\partial x \partial y} - \frac{\partial^2 u_y}{\partial y^2}\right) - R_2\left(\frac{\partial^2 u_x}{\partial x^2} - 2\frac{\partial^2 u_y}{\partial x \partial y} - \frac{\partial^2 u_x}{\partial y^2}\right)\right]$$

$$p = f(\rho)$$

$$\tag{8.4.1}$$

in which U_x and U_y are known quantities in the fluid field even if there is no external velocity field. According to Witten [8], the flow is existed always in soft matter. We call U_x or/and U_y intra-velocities.

For a dislocation in the matter, there are in following boundary conditions

$$\left.\begin{array}{l} \sqrt{x^2 + y^2} \to \infty: \quad \sigma_{ij} = H_{ij} = p_{ij} = 0; \\[2mm] r = r_0: \quad V_r = V_\theta = 0. \quad \sigma_{rr} = \sigma_{r\theta} = 0, \quad H_{rr} = H_{r\theta} = 0; \\[2mm] \displaystyle\int_\Gamma du_x b_1^{\parallel}, \quad \int_\Gamma du_y = b_2^{\parallel}, \quad \int_\Gamma dw_x = b_1^{\perp}, \quad \int_\Gamma dw_y = b_2^{\perp} \end{array}\right\} \qquad (8.4.2)$$

where r_0 denotes the radius of the dislocation core.

The work in solving Eq. (8.4.1) under boundary conditions (8.4.2) is very hard. Many efforts have been devoted to constructing solutions but there is no satisfactory result and no strict analytic solution is available so far. The difficulty comes from the complexity of both the equations and boundary conditions. So, the fluid effect on dislocation has not been explored analytically yet.

8.5 Transient Dynamic Analysis

8.5.1 Specimen and Initial-Boundary Conditions

Equation (8.2.3) are the basis for the generalized dynamics of the plane field of soft-matter quasicrystals with 10-fold symmetries. If we want to obtain further information on the distribution, deformation, and motion of the material, we must solve the equations under appropriate initial- and boundary-value conditions. For this purpose, a specimen made by the matter should be optioned which is subjected to some initial- and boundary-value conditions. Here the specimen is shown by Fig. 8.1, concerning Cheng et al. [9], and the corresponding initial- and boundary-value conditions are as follows:

$$t = 0: V_x = V_y = 0, u_x = u_y = 0, w_x = w_y = 0, \ \rho = \rho_0; \qquad (8.5.1)$$

$$y = \pm H, |x| < W: V_x = V_y = 0, \sigma_{yy} = \sigma_0 g(t), \sigma_{yx} = 0, H_{yy} = H_{yx} = 0;$$
$$x = \pm W, |y| < H: V_x = V_y = 0, \sigma_{xx} = \sigma_{xy} = 0, H_{xx} = H_{xy} = 0.$$
$$(8.5.2)$$

In the present computation we take $2H = 0.01\,\mathrm{m}$, $2W = 0.01\,\mathrm{m}$, $\sigma_0 = 100\,\mathrm{Pa}$, $\rho_0 = 1.5 \times 10^3\,\mathrm{kg/m^3}, \eta = 0.1\,\mathrm{Pa\,s}, L = 10\,\mathrm{MPa}$, $M = 4\,\mathrm{MPa}$, $K_1 = 0.5\,\mathrm{L}, K_2 = -0.1\,\mathrm{L}, R = R_1 = 0.04\,\mathrm{M}, R_2 = 0, \Gamma_u = 4.8 \times 10^{-17}\,\mathrm{m^3\,s/kg}, \Gamma_w = 4.8 \times 10^{-19}\,\mathrm{m^3\,s/kg}, \mathrm{A} \sim 0.2\,\mathrm{MPa}, \mathrm{B} \sim 0.2\,\mathrm{MPa}$, in which $g(t)$ is a Heaviside function of time, and a part of these material constants are introduced from Chap. 1 of this book and drawn from Refs. [4, 8, 10], and the others are estimated.

The initial-boundary-value problems of (8.5.1), (8.5.2) of nonlinear partial differential Eq. (8.2.3) are consistent mathematically, but the existence and uniqueness of the solution have not been proved yet due to the complexity of the problem. We can

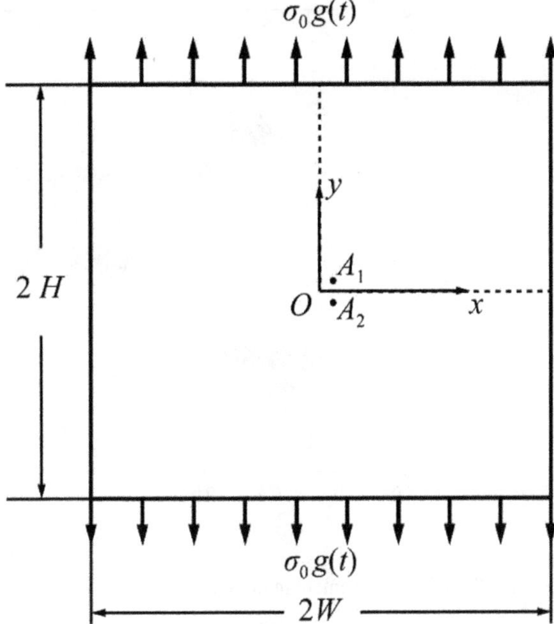

Fig. 8.1 Specimen of soft-matter quasicrystals of 5- or 10-fold symmetries under dynamic loading

solve it by numerical method and the stability and correctness of the solution can be verified by the numerical results only.

8.5.2 Numerical Analysis and Results

We here take the finite difference method to solve the problem, and a part of numerical results are given through the following illustrations, in which the computational point is $A_1(10\,\text{m}^{-4}, 10\,\text{m}^{-4})$ (or $A_2(10\,\text{m}^{-4}, 10\,\text{m}^{-4})$) (Figs. 8.2, 8.3, 8.4, 8.5, 8.6, 8.7, 8.8, 8.9, 8.10, 8.11, 8.12 and 8.13).

As pointed out at the beginning of the chapter the proof of existence and uniqueness of the solution of initial-boundary-value problem (8.5.1), (8.5.2) of Eq. (8.2.2) cannot be given at present, and the numerical solution presents high stability, which shows that the correctness of the equations and the formulation of corresponding initial-boundary-value problem in one direction at least. The correctness of the solution can also be checked by some verification physically. For example, from Figs. 8.2, 8.3, 8.4 and 8.5, the wave emanated from the upper or lower surface propagates to point $A_1(10^{-4}\,\text{m}, 10^{-4}\,\text{m})$ (or $A_2(10^{-4}\,\text{m}, 10^{-4}\,\text{m})$) experiences $t_0 = 4.07 \times 10^{-5}\,\text{s}$ which can be called response time of the matter at the location to the dynamic loading. The wave propagating distance is $H_0 = H - 10^{-4} = 0.0049\,\text{m}$, thus the speed of

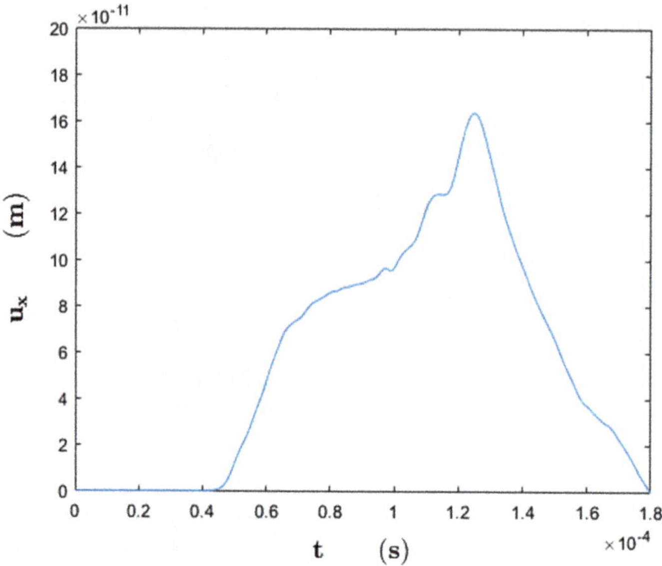

Fig. 8.2 Phonon displacement in direction x at point A_1(or A_2) versus time

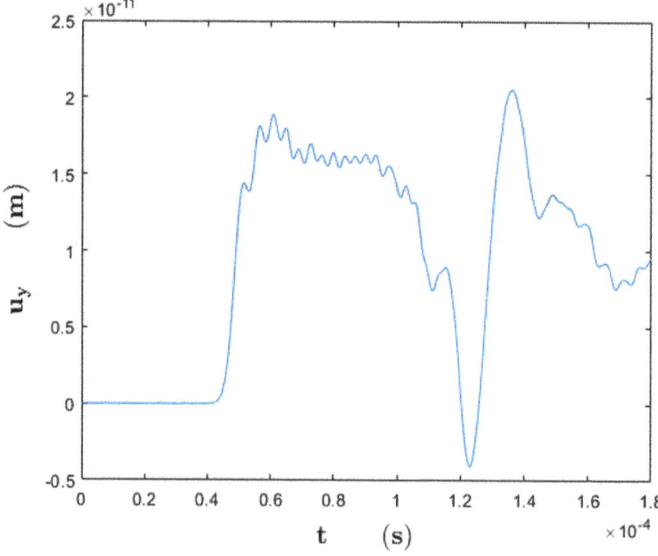

Fig. 8.3 Phonon displacement in direction y at point A_1(or A_2) versus time

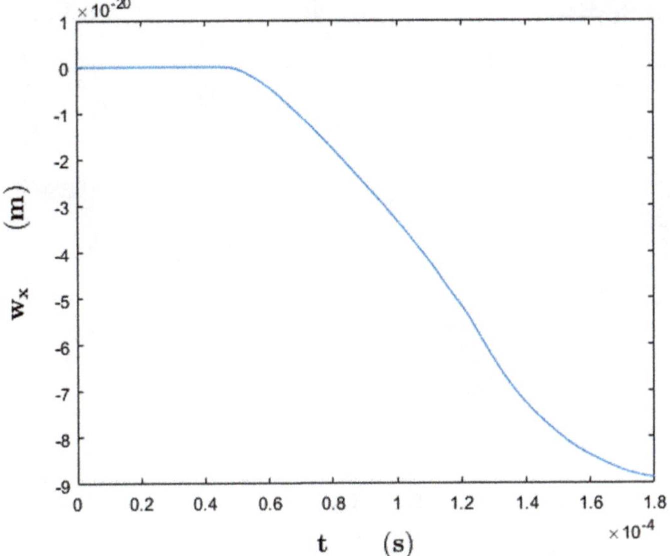

Fig. 8.4 Phason displacement in direction x at point A_1(or A_2) versus time

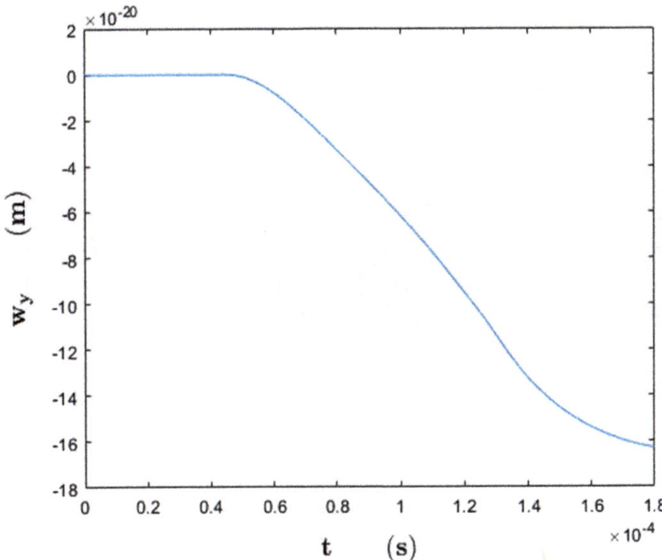

Fig. 8.5 Phason displacement in direction y at point A_1(or A_2) versus time

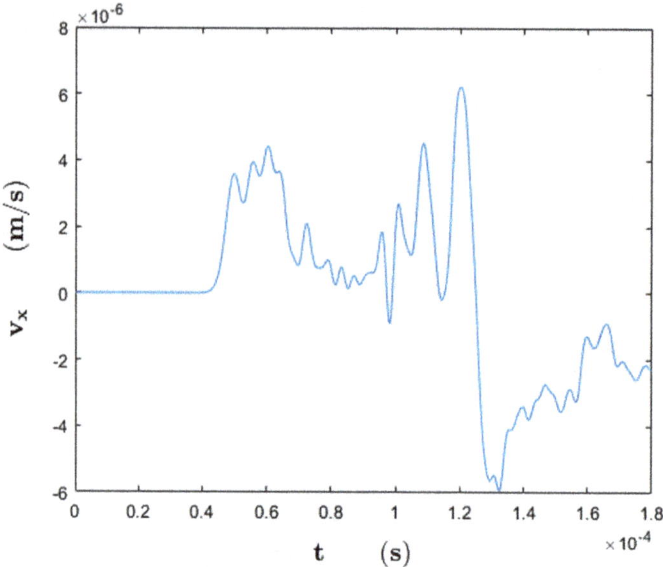

Fig. 8.6 Fluid phonon velocity in direction x at point A_1(or A_2) versus time

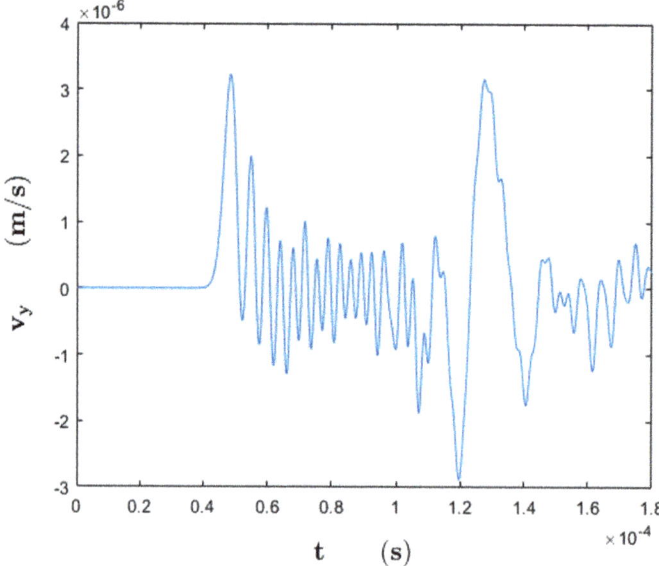

Fig. 8.7 Fluid phonon velocity in direction y at point A_1(or A_2) versus time

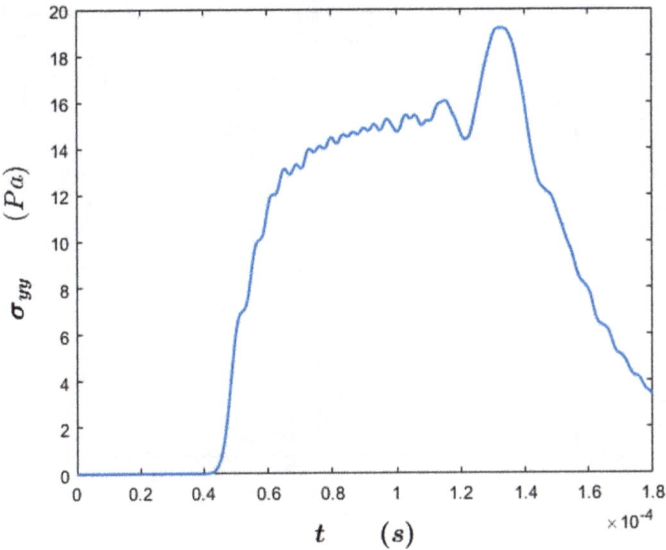

Fig. 8.8 Phonon normal stress in direction y at point A_1(or A_2) versus time

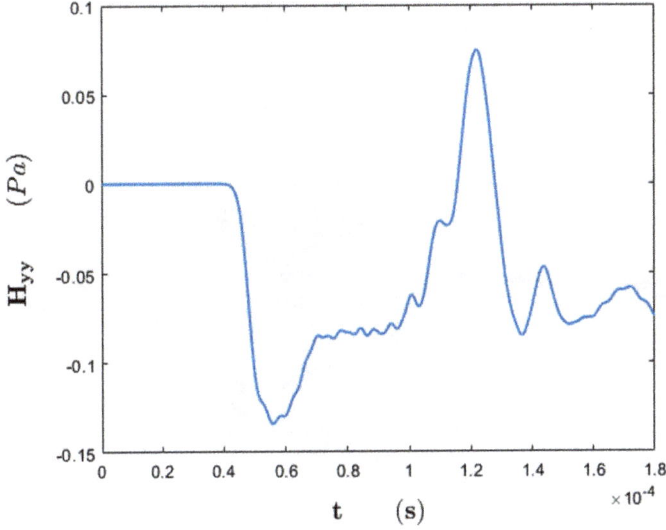

Fig. 8.9 Phason normal stress in direction y at point A_1(or A_2) versus time

the wave is $c = \frac{H_0}{t_0} = \frac{0.0049}{4.07 \times 10^{-5}} = 120.39\,\text{m/s}$. Also, we can see that, the density of soft-matter quasicrystals decreases to $\rho_{\min} = 1498\,\text{kg/m}^3$, the speed of elastic longitudinal wave $c_{1\,\max} = \sqrt{\frac{A+L+2M-2B}{\rho_{\min}}} = 109.6176\,\text{m/s}$, which is very close to

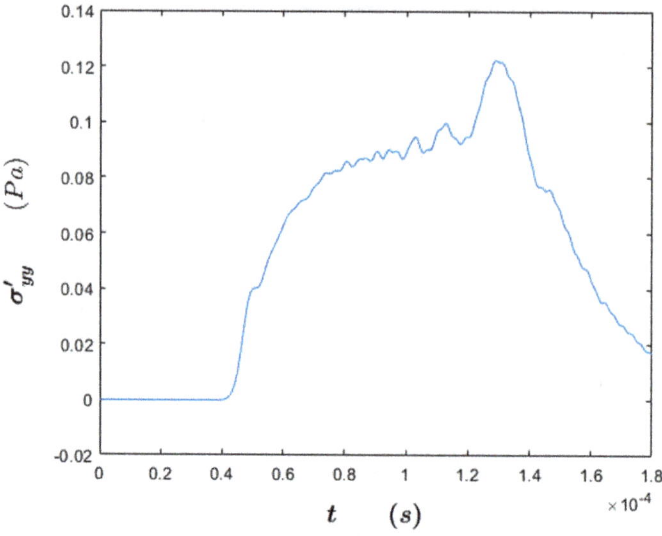

Fig. 8.10 Fluid phonon viscous normal stress in direction y at point A_1(or A_2) versus time

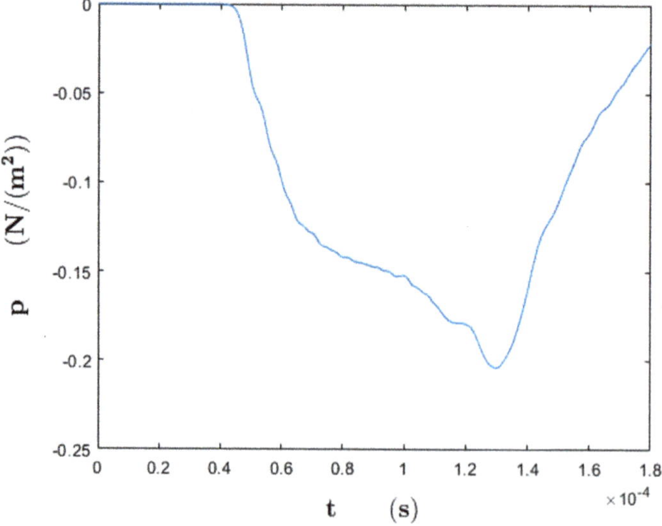

Fig. 8.11 Fluid pressure at point A_1(or A_2) versus time

the observed value c, and the error is $(c - c_{1\,\text{max}})/c = 6 \times 10^{-3}$, this indicates theoretical prediction is quite exact. Before the wave reaches the point, i.e., $t < t_0$ all field variables are equal to zero or their initial values (for displacements, velocities, stresses, mass density, and fluid pressure). The responses of the field variables appear

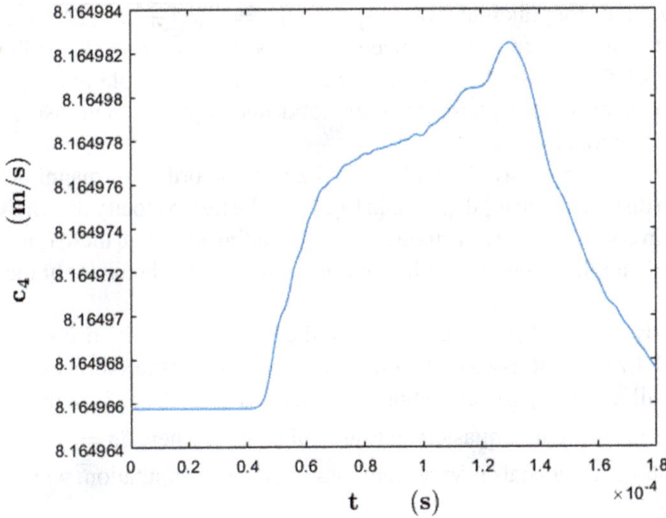

Fig. 8.12 Fluid phonon acoustic wave speed at point A_1 (or A_2) versus time

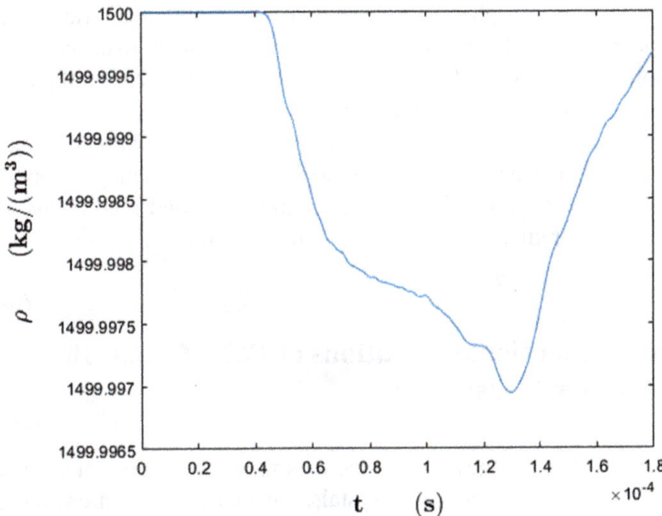

Fig. 8.13 Mass density at point A_1 (or A_2) versus time

only as $t > t_0$. This simple fact implies a "response law", and shows the importance of phonon (elasticity) for soft matter.

We have the longitudinal wave speed $c_1 = \sqrt{A + L + 2M - 2B/\rho} = 109.5445$ m/s, and two transverse speeds $c_2 = c_3 = \sqrt{M/\rho} = 51.6398$ m/s. The comparison of the results shows that c_1 plays the dominant role although the soft matter is an intermediate phase between solid and liquid, which also shows the importance of phonons.

Phonon displacement u_y shown by Fig. 8.2 is many orders of magnitude greater than phason displacement w_y depicted in Fig. 8.3. The fluid velocity drawn by Fig. 8.4 has arrived in considerable value induced by the loading although there is no action of any externally applied flow field, which explores soft-matter behavior of the material indeed.

In hydrodynamics of solid quasicrystals, the situation is quite different, because there are solid viscous stresses rather than fluid stresses, and the solid viscous stresses are very small according to our computation discussed in Chap. 3. In the meantime, $\left| \frac{\delta\rho}{\rho_0} \right| \sim 10^{-7}$ in soft-matter quasicrystals are quite large, where $\delta\rho = \rho - \rho_0$, and the value for solid quasicrystals is very small based on our computation, which was also discussed in Chap. 3.

Because the specimen has a limited size, the effect of the boundaries is evident. The wave shapes of solutions of phonon fields and fluid phonon fields describe the boundary effects, these also depicted the effect of interaction between phonons and fluid phonon. Although the phasons represent diffusion rather than wave propagation, the strong coupling effect between phonons and phasons leads to the configuration of numerical solutions of phason field that does not exhibit the shapes of pure diffusive solution of the classical diffusion equation in mathematical physics.

The 10-fold symmetry quasicrystals in the soft matter are observed but have not been openly reported, and we believe that the details of which may be reported soon. In this situation, we simplify data including some detailed predictions and avoid leading to some contradictions with the realistic experimental results.

8.6 Three-Dimensional Equations of Point Group 10 mm Soft-Matter Quasicrystals

The presentation offered in the previous sections shows the interest of the fivefold and 10-fold symmetrical soft-matter quasicrystals, especially as there are strong coupling effects between the phonons and phasons. However, the discussion is limited in the two-dimensional case of the quasiperiodicity, i.e., in the xy-plane. We now consider the three-dimensional dynamics of point group 10 mm quasicrystals in soft matter. At first, we list their constitutive laws [4, 11] as follows

$$\left.\begin{aligned}
\sigma_{xx} &= C_{11}\varepsilon_{xx} + C_{12}\varepsilon_{yy} + C_{13}\varepsilon_{zz} + R(w_{xx} + w_{yy}) \\
\sigma_{yy} &= C_{12}\varepsilon_{xx} + C_{11}\varepsilon_{yy} + C_{13}\varepsilon_{zz} - R(w_{xx} + w_{yy}) \\
\sigma_{zz} &= C_{13}\varepsilon_{xx} + C_{13}\varepsilon_{yy} + C_{33}\varepsilon_{zz} \\
\sigma_{yz} &= \sigma_{zy} = 2C_{44}\varepsilon_{yz} \\
\sigma_{zx} &= \sigma_{xz} = 2C_{44}\varepsilon_{zx} \\
\sigma_{xy} &= \sigma_{yx} = 2C_{66}\varepsilon_{xy} - R(w_{xy} - w_{yx}) \\
H_{xx} &= K_1 w_{xx} + K_2 w_{yy} + R(\varepsilon_{xx} - \varepsilon_{yy}) \\
H_{yy} &= K_2 w_{xx} + K_1 w_{yy} + R(\varepsilon_{xx} - \varepsilon_{yy}) \\
H_{yz} &= K_4 w_{yz} \\
H_{xy} &= K_1 w_{xy} - K_2 w_{yx} \\
H_{xz} &= K_4 w_{xz} \\
H_{yx} &= -K_2 w_{xy} + K_1 w_{yx} + 2R\varepsilon_{xy} \\
p_{xx} &= -p + 2\eta\dot{\xi}_{xx} - \frac{2}{3}\eta\dot{\xi}_{kk} \\
p_{yy} &= -p + 2\eta\dot{\xi}_{yy} - \frac{2}{3}\eta\dot{\xi}_{kk} \\
p_{zz} &= -p + 2\eta\dot{\xi}_{zz} - \frac{2}{3}\eta\dot{\xi}_{kk} \\
p_{yz} &= 2\eta\dot{\xi}_{yz} \\
p_{zx} &= 2\eta\dot{\xi}_{zx} \\
p_{xy} &= 2\eta\dot{\xi}_{xy} \\
\dot{\xi}_{kk} &= \dot{\xi}_{xx} + \dot{\xi}_{yy} + \dot{\xi}_{zz}
\end{aligned}\right\} \tag{8.6.1}$$

Therefore, we obtain the three-dimensional governing equations of 10-fold symmetry quasicrystals in soft matter

$$\left.\begin{aligned}
&\frac{\partial \rho}{\partial t} + \nabla \cdot (\rho \mathbf{V}) = 0 \\
&\frac{\partial(\rho V_x)}{\partial t} + \frac{\partial(V_x \rho V_x)}{\partial x} + \frac{\partial(V_y \rho V_x)}{\partial y} + \frac{\partial(V_z \rho V_x)}{\partial z} \\
&\quad = -\frac{\partial p}{\partial x} + \eta\nabla^2 V_x + \frac{1}{3}\eta\frac{\partial}{\partial x}\nabla \cdot \mathbf{V} \\
&\quad\quad + \left(C_{66}\frac{\partial^2}{\partial y^2} + C_{44}\frac{\partial^2}{\partial z^2}\right)u_x + (C_{12} + C_{66})\frac{\partial^2 u_y}{\partial x \partial y}
\end{aligned}\right\}$$

$$+ (C_{13} + C_{44} - C_{11})\frac{\partial^2 u_z}{\partial x \partial z} + (C_{11} - B)\frac{\partial}{\partial x}\nabla \cdot \mathbf{u}$$

$$+ R\frac{\partial}{\partial x}\nabla_1 \cdot \mathbf{w} - R\frac{\partial}{\partial y}\left(\frac{\partial w_x}{\partial y} - \frac{\partial w_y}{\partial x}\right)$$

$$- (A - B)\frac{1}{\rho_0}\frac{\partial \delta\rho}{\partial x}$$

$$\frac{\partial(\rho V_y)}{\partial t} + \frac{\partial(V_x \rho V_y)}{\partial x} + \frac{\partial(V_y \rho V_y)}{\partial y} + \frac{\partial(V_z \rho V_y)}{\partial z}$$

$$= -\frac{\partial p}{\partial y} + \eta \nabla^2 V_y + \frac{1}{3}\eta\frac{\partial}{\partial y}\nabla \cdot \mathbf{V}$$

$$+ (C_{12} + C_{66})\frac{\partial^2 u_x}{\partial x \partial y}$$

$$+ \left(C_{66}\frac{\partial^2}{\partial x^2} + C_{11}\frac{\partial^2}{\partial y^2} + C_{44}\frac{\partial^2}{\partial z^2}\right)u_y$$

$$+ (C_{13} + C_{44})\frac{\partial^2 u_z}{\partial y \partial z} + (C_{11} - B)\frac{\partial}{\partial y}\nabla \cdot \mathbf{u}$$

$$- R\frac{\partial}{\partial x}\left(\frac{\partial w_x}{\partial y} - \frac{\partial w_y}{\partial x}\right) - R\frac{\partial}{\partial y}\nabla_1 \cdot \mathbf{w}$$

$$- (A - B)\frac{1}{\rho_0}\frac{\partial \delta\rho}{\partial y}$$

$$\frac{\partial(\rho V_z)}{\partial t} + \frac{\partial(V_x \rho V_z)}{\partial x} + \frac{\partial(V_y \rho V_z)}{\partial y} + \frac{\partial(V_z \rho V_z)}{\partial z}$$

$$= -\frac{\partial p}{\partial z} + \eta \nabla^2 V_z + \frac{1}{3}\eta\frac{\partial}{\partial z}\nabla \cdot \mathbf{V}$$

$$+ \left(C_{44}\frac{\partial^2}{\partial x^2} + C_{44}\frac{\partial^2}{\partial y^2} + (C_{33} - C_{13} - C_{44})\frac{\partial^2}{\partial z^2}\right)u_z$$

$$+ (C_{13} + C_{44} - B)\frac{\partial}{\partial z}\nabla \cdot \mathbf{u} - (A - B)\frac{1}{\rho_0}\frac{\partial \delta\rho}{\partial z}$$

$$\frac{\partial u_x}{\partial t} + V_x\frac{\partial u_x}{\partial x} + V_y\frac{\partial u_x}{\partial y} + V_z\frac{\partial u_x}{\partial z}$$

$$= V_x + \Gamma_{\mathbf{u}}\left[\begin{array}{l}\left(C_{11}\frac{\partial^2}{\partial x^2} + C_{66}\frac{\partial^2}{\partial y^2} + C_{44}\frac{\partial^2}{\partial z^2}\right)u_x \\[2mm] +(C_{12} + C_{66})\frac{\partial^2 u_y}{\partial x \partial y} + (C_{13} + C_{44})\frac{\partial^2 u_z}{\partial x \partial z} \\[2mm] +R\frac{\partial}{\partial x}\nabla_1 \cdot \mathbf{w} - R\frac{\partial}{\partial y}\left(\frac{\partial w_x}{\partial y} - \frac{\partial w_y}{\partial x}\right)\end{array}\right]$$

$$
\begin{aligned}
&\frac{\partial u_y}{\partial t} + V_x \frac{\partial u_y}{\partial x} + V_y \frac{\partial u_y}{\partial y} + V_z \frac{\partial u_y}{\partial z} \\
&= V_y + \Gamma_{\mathbf{u}}
\begin{bmatrix}
(C_{12} + C_{66})\dfrac{\partial^2 u_x}{\partial x \partial y} \\[2mm]
+\left(C_{66}\dfrac{\partial^2}{\partial x^2} + C_{11}\dfrac{\partial^2}{\partial y^2} + C_{44}\dfrac{\partial^2}{\partial z^2}\right)u_y \\[2mm]
+(C_{13} + C_{44})\dfrac{\partial^2 u_z}{\partial y \partial z} - R\dfrac{\partial}{\partial x}\left(\dfrac{\partial w_x}{\partial y} - \dfrac{\partial w_y}{\partial x}\right) \\[2mm]
- R\dfrac{\partial}{\partial y}\nabla_1 \cdot \mathbf{w}
\end{bmatrix} \\[4mm]
&\frac{\partial u_z}{\partial t} + V_x \frac{\partial u_z}{\partial x} + V_y \frac{\partial u_z}{\partial y} + V_z \frac{\partial u_z}{\partial z} \\
&= V_z + \Gamma_{\mathbf{u}}
\begin{bmatrix}
(C_{13} + C_{44})\left(\dfrac{\partial^2 u_x}{\partial x \partial z} + \dfrac{\partial^2 u_y}{\partial y \partial z}\right) \\[2mm]
+\left(C_{44}\dfrac{\partial^2}{\partial x^2} + C_{44}\dfrac{\partial^2}{\partial y^2} + C_{33}\dfrac{\partial^2}{\partial z^2}\right)u_z
\end{bmatrix} \\[4mm]
&\frac{\partial w_x}{\partial t} + V_x \frac{\partial w_x}{\partial x} + V_y \frac{\partial w_x}{\partial y} + V_z \frac{\partial w_x}{\partial z} \\
&= \Gamma_{\mathbf{w}}
\begin{bmatrix}
K_1 \nabla_1^2 w_x + K_4 \dfrac{\partial^2 w_x}{\partial z^2} + K_2 \dfrac{\partial}{\partial y}\left(\dfrac{\partial}{\partial x} - \dfrac{\partial}{\partial y}\right)w_y \\[2mm]
+ R\dfrac{\partial}{\partial x}\left(\dfrac{\partial u_x}{\partial x} - \dfrac{\partial u_y}{\partial y}\right) - R\dfrac{\partial}{\partial y}\left(\dfrac{\partial u_x}{\partial y} + \dfrac{\partial u_y}{\partial x}\right)
\end{bmatrix} \\[4mm]
&\frac{\partial w_y}{\partial t} + V_x \frac{\partial w_y}{\partial x} + V_y \frac{\partial w_y}{\partial y} V_z \frac{\partial w_y}{\partial z} \\
&= \Gamma_{\mathbf{w}}
\begin{bmatrix}
K_1 \nabla_1^2 w_y + K_4 \dfrac{\partial^2 w_y}{\partial z^2} + R\dfrac{\partial}{\partial x}\left(\dfrac{\partial u_x}{\partial y} + \dfrac{\partial u_y}{\partial x}\right) \\[2mm]
+ R\dfrac{\partial}{\partial y}\left(\dfrac{\partial u_x}{\partial x} - \dfrac{\partial u_y}{\partial y}\right)
\end{bmatrix} \\[4mm]
&p = f(\rho)
\end{aligned}
\tag{8.6.2}
$$

in which $\nabla^2 = \frac{\partial^2}{\partial x^2} + \frac{\partial^2}{\partial y^2} + \frac{\partial^2}{\partial z^2}$, $\nabla_1^2 = \frac{\partial^2}{\partial x^2} + \frac{\partial^2}{\partial y^2}$, $\nabla = \mathbf{i}\frac{\partial}{\partial x} + \mathbf{j}\frac{\partial}{\partial y} + \mathbf{k}\frac{\partial}{\partial z}$, $\mathbf{V} = \mathbf{i}V_x + \mathbf{j}V_y + \mathbf{k}V_z$, $\mathbf{u} = \mathbf{i}u_x + \mathbf{j}u_y + \mathbf{k}u_z$, and $C_{11}, C_{12}, C_{13}, C_{33}, C_{44}, C_{66} = (C_{11} - C_{12})/2$ the phonon elastic constants, K_1, K_2, K_3, K_4 the phason elastic constants, R the phonon-phason coupling constant, η the fluid dynamic viscosity, and Γ_u Γ_w the phonon and phason dissipation coefficients, A and B the material constants due to variation of mass density, named LRT constants respectively.

Equation (8.6.2) are the final governing equations of dynamics of soft-matter quasicrystals of 10-fold symmetry in the three-dimensional case with fields variables $u_x, u_y, u_z, w_x, w_y, V_x, V_y, V_z, \rho$, and p, and the amount of the field variables and the field equations are 10, among which: (8.6.2a) is the mass conservation equation, (8.6.2b)–(8.6.2d) the momentum conservation equations or generalized Navier–Stokes equations, (8.6.2e)–(8.6.2g) the equations of motion of phonons due to the symmetry breaking, (8.6.2h) and (8.6.2i) the phason dissipation equations, and (8.6.2j) the equation of state, respectively. It is only when the equations are consistent that they are mathematically solvable; if the equation of state is missing, the system of equations is not closed and is mathematically and physically meaningless. This shows the equation of state is necessary.

These equations reveal the nature of wave propagation of fields $\mathbf{u}$ and $\mathbf{V}$ with phonon wave speeds $c_1 = \sqrt{A + C_{11} - 2B/\rho}$, $c_2 = c_3 = \sqrt{C_{11} - C_{12}/2\rho}$ and fluid phonon wave speed $c_4 = \sqrt{(\partial p/\partial \rho)_s}$ and the nature of the diffusion of the field $\mathbf{w}$ with major diffusive coefficient $D_1 = \Gamma_w K_1$ and other less important diffusive coefficients $D_2 = \Gamma_w K_2$, etc. from the viewpoint of hydrodynamics.

In contrast to the above model, which belongs to a compressible complex fluid model of soft-matter quasicrystals of 10-fold symmetry, there is another model, i.e., an incompressible complex fluid model, in which $\rho = const$, in the case, the equation of state is not required, so that the governing equations of generalized dynamics will be simplified, as will be discussed in Sect. 8.7.

8.7 Incompressible Complex Fluid Model of Soft-Matter Quasicrystals with 10-Fold Symmetry

Similar to Sect. 7.8, we here can discuss an incompressible complex fluid model of soft-matter quasicrystals with 10-fold symmetry, in this case $\rho = const$ then $A = B = 0$, $\frac{\delta \rho}{\rho_0} = 0$, and equation of state is not needed, and the Eq. (8.6.2) will be reduced to

$$\nabla \cdot \mathbf{V} = 0$$

$$\rho\left(\frac{\partial V_x}{\partial t} + \frac{\partial (V_x V_x)}{\partial x} + \frac{\partial (V_y V_x)}{\partial y} + \frac{\partial (V_z V_x)}{\partial z}\right) = -\frac{\partial p}{\partial x} + \eta \nabla^2 V_x$$
$$+ \left(C_{66}\frac{\partial^2}{\partial y^2} + C_{44}\frac{\partial^2}{\partial z^2}\right)u_x$$
$$+ (C_{12} + C_{66})\frac{\partial^2 u_y}{\partial x \partial y}$$

$$
\begin{aligned}
&\hspace{3cm}+(C_{13}+C_{44}-C_{11})\frac{\partial^2 u_z}{\partial x\partial z}+(C_{11}-B)\frac{\partial}{\partial x}\nabla\cdot\mathbf{u} \\
&\hspace{3cm}+R\frac{\partial}{\partial x}\nabla_1\cdot\mathbf{w}-R\frac{\partial}{\partial y}\left(\frac{\partial w_x}{\partial y}-\frac{\partial w_y}{\partial x}\right)-(A-B)\frac{1}{\rho_0}\frac{\partial\delta\rho}{\partial x} \\[4pt]
&\rho\left(\frac{\partial V_y}{\partial t}+\frac{\partial(V_x V_y)}{\partial x}+\frac{\partial(V_y V_y)}{\partial y}+\frac{\partial(V_z V_y)}{\partial z}\right)=-\frac{\partial p}{\partial y}+\eta\nabla^2 V_y \\
&\hspace{3cm}+(C_{12}+C_{66})\frac{\partial^2 u_x}{\partial x\partial y}+\left(C_{66}\frac{\partial^2}{\partial x^2}+C_{11}\frac{\partial^2}{\partial y^2}+C_{44}\frac{\partial^2}{\partial z^2}\right)u_y \\
&\hspace{3cm}+(C_{13}+C_{44})\frac{\partial^2 u_z}{\partial y\partial z}+(C_{11}-B)\frac{\partial}{\partial y}\nabla\cdot\mathbf{u}-R\frac{\partial}{\partial x}\left(\frac{\partial w_x}{\partial y}-\frac{\partial w_y}{\partial x}\right) \\
&\hspace{3cm}-R\frac{\partial}{\partial y}\nabla_1\cdot\mathbf{w}-(A-B)\frac{1}{\rho_0}\frac{\partial\delta\rho}{\partial y} \\[4pt]
&\rho\left(\frac{\partial V_z}{\partial t}+\frac{\partial(V_x V_z)}{\partial x}+\frac{\partial(V_y V_z)}{\partial y}+\frac{\partial(V_z V_z)}{\partial z}\right)=-\frac{\partial p}{\partial z}+\eta\nabla^2 V_z \\
&\hspace{3cm}+\left(C_{44}\frac{\partial^2}{\partial x^2}+C_{44}\frac{\partial^2}{\partial y^2}+(C_{33}-C_{13}-C_{44})\frac{\partial^2}{\partial z^2}\right)u_z \\
&\hspace{3cm}+(C_{13}+C_{44}-B)\frac{\partial}{\partial z}\nabla\cdot\mathbf{u}-(A-B)\frac{1}{\rho_0}\frac{\partial\delta\rho}{\partial z} \\[6pt]
&\frac{\partial u_x}{\partial t}+V_x\frac{\partial u_x}{\partial x}+V_y\frac{\partial u_x}{\partial y}+V_z\frac{\partial u_x}{\partial z}=V_x+\Gamma_{\mathbf{u}} \\
&\hspace{2cm}\left[\begin{array}{l}\left(C_{11}\frac{\partial^2}{\partial x^2}+C_{66}\frac{\partial^2}{\partial y^2}+C_{44}\frac{\partial^2}{\partial z^2}\right)u_x+(C_{12}+C_{66})\frac{\partial^2 u_y}{\partial x\partial y} \\[6pt] +(C_{13}+C_{44})\frac{\partial^2 u_z}{\partial x\partial z}+R\frac{\partial}{\partial x}\nabla_1\cdot\mathbf{w}-R\frac{\partial}{\partial y}\left(\frac{\partial w_x}{\partial y}-\frac{\partial w_y}{\partial x}\right)\end{array}\right] \\[6pt]
&\frac{\partial u_y}{\partial t}+V_x\frac{\partial u_y}{\partial x}+V_y\frac{\partial u_y}{\partial y}+V_z\frac{\partial u_y}{\partial z}=V_y \\
&\hspace{1cm}+\Gamma_{\mathbf{u}}\left[\begin{array}{l}(C_{12}+C_{66})\frac{\partial^2 u_x}{\partial x\partial y}+\left(C_{66}\frac{\partial^2}{\partial x^2}+C_{11}\frac{\partial^2}{\partial y^2}+C_{44}\frac{\partial^2}{\partial z^2}\right)u_y \\[6pt] +(C_{13}+C_{44})\frac{\partial^2 u_z}{\partial y\partial z}-R\frac{\partial}{\partial x}\left(\frac{\partial w_x}{\partial y}-\frac{\partial w_y}{\partial x}\right)-R\frac{\partial}{\partial y}\nabla_1\cdot\mathbf{w}\end{array}\right] \\[6pt]
&\frac{\partial u_z}{\partial t}+V_x\frac{\partial u_z}{\partial x}+V_y\frac{\partial u_z}{\partial y}+V_z\frac{\partial u_z}{\partial z}=V_z \\
&\hspace{1cm}+\Gamma_{\mathbf{u}}\left[\begin{array}{l}(C_{13}+C_{44})\left(\frac{\partial^2 u_x}{\partial x\partial z}+\frac{\partial^2 u_y}{\partial y\partial z}\right) \\[6pt] +\left(C_{44}\frac{\partial^2}{\partial x^2}+C_{44}\frac{\partial^2}{\partial y^2}+C_{33}\frac{\partial^2}{\partial z^2}\right)u_z\end{array}\right] \\[6pt]
&\frac{\partial w_x}{\partial t}+V_x\frac{\partial w_x}{\partial x}+V_y\frac{\partial w_x}{\partial y}+V_z\frac{\partial w_x}{\partial z}=\Gamma_{\mathbf{w}}\left[\begin{array}{l}K_1\nabla_1^2 w_x+K_4\frac{\partial^2 w_x}{\partial z^2}+K_2\frac{\partial}{\partial y}\left(\frac{\partial}{\partial x}-\frac{\partial}{\partial y}\right)w_y \\[6pt] +R\frac{\partial}{\partial x}\left(\frac{\partial u_x}{\partial x}-\frac{\partial u_y}{\partial y}\right)-R\frac{\partial}{\partial y}\left(\frac{\partial u_x}{\partial y}+\frac{\partial u_y}{\partial x}\right)\end{array}\right] \\[6pt]
&\frac{\partial w_y}{\partial t}+V_x\frac{\partial w_y}{\partial x}+V_y\frac{\partial w_y}{\partial y}+V_z\frac{\partial w_y}{\partial z}=\Gamma_{\mathbf{w}}\left[\begin{array}{l}K_1\nabla_1^2 w_y+K_4\frac{\partial^2 w_y}{\partial z^2}+R\frac{\partial}{\partial x}\left(\frac{\partial u_x}{\partial y}+\frac{\partial u_y}{\partial x}\right) \\[6pt] +R\frac{\partial}{\partial y}\left(\frac{\partial u_x}{\partial x}-\frac{\partial u_y}{\partial y}\right)\end{array}\right]
\end{aligned}
$$

$$\tag{8.7.1}$$

in which the equation of state has been excluded already, the number of field variables is reduced to 9, and the number for governing equations is also 9, and the initial- and boundary-value conditions are similar to those for the Eq. (8.6.2) so that solving for the initial- and boundary-value problems of Eq. (8.7.1) will be simpler. Some examples of the computations can be referred to in Chaps. 7 and 9 because there are similarities between these systems in the solving processes.

8.8 Conclusion and Discussion

A complete equation set of generalized dynamics of a two-dimensional field of soft-matter decagonal quasicrystals is derived, in which the new equation of state is included. The complete equation set of generalized dynamics is the basis of the analysis.

The computation is stable and shows the solvability of the equations and the well-conditionality of the proposed initial-boundary-value problem of the equations. All field variables through the specimen are determined numerically, including the important hydrodynamic variable: the mass density ρ which shows the material is compressible.

This specimen is quite simple and can be easily tested experimentally, at present lack of the material constants is a great difficulty, although we referred some work, e.g. Refs. [12–14] and so on, but the experiments are developed recently refer to the new advances reported in the preface of this edition.

It is evident that the present study is a continuation and development of hydro-dynamics for solid quasicrystals, for which Fan [1] gave a detailed description. In our work, we pay attention to collaboration among hydrodynamics, thermody-namics, mathematical physics, and computational physics, which helps us to deter-mine observable physical quantities quantitatively, such as displacements, velocities, and stresses in the time-spatial domain. The new equation set of motion including the new equation of state and developed method appears to be very important for soft-matter quasicrystals. Of course, the verification experimentally of the equation of state is one of our attempts. At last, the three-dimensional hydrodynamics of point group 10 mm soft-matter quasi crystals is also discussed.

The incompressible model of the quasicrystals with 10-fold symmetry is newly included in this chapter, but the study of which is simpler than the model introduced in Sects. 8.1–8.6.

References

1. Fan, T.Y.: Equation system of generalized hydrodynamics of soft-matter quasicrystals. Appl. Math. Mech. **37**(4), 331–347 (2016). in Chinese
2. Fan, T.Y.: Generalized hydrodynamics generalized of second kind of two-dimensional soft-matter quasicrystals. Appl. Math. Mech. **38**, 189–199 (2017). in Chinese
3. Fan, T., Yand Tang, Z.Y.: Three-dimensional generalized dynamics of soft-matter quasicrystals. Appl. Math. Mech. **38**, 1195–1027 (2017) (in Chinese); Adv. Mater. Sci. Eng. Article ID 4875854 (2020)
4. Fan, T.: Mathematical Theory of Elasticity of Quasicrystals and Its Applications. Science Press, Beijing/Springer, Heidelberg 1st edn. (2010), 2nd edn. (2016)
5. Li, X.F., Duan, X.Y., Fan, T.Y., Sun, Y.F.: Elastic field for a straight dislocation in a decagonal quasicrystal. J. Phys. Conden. Matter, **11**, 703–711 (1999)
6. Li, X.F., Fan, T.Y.: New method for solving plane elasticity of planar quasicrystals and solution. Chin Phys Lett **15**, 278–280 (1998)

7. Ding, D.H., Wang, R.H., Yang We, G., Hu, C.Z.: General expressions for the elastic displacement fields induced by dislocation in quasicrystals. J. Phys. Conden. Matter **7**(28), 5423–5436 (1995)
8. de Gennes, P.G., Prost, J.: The Physics of Liquid Crystals, 2nd edn. Clarendon, Oxford (1993)
9. Cheng, H., Fan, T.Y., Wei, H.: Characters of deformation and motion of soft-matter quasicrystals with 5-and 10-fold symmetries (2016) (unpublished work)
10. KeithJ, M., King, J.A., Miskioglu, I., Roache, S.C.: Tensile modulus modeling of carbon-filled liquid crystal polymer composites. Polym. Compos. **30**, 1166–1174 (2009)
11. Hu, C.Z., Wang, R.H., Ding, D.H.: Symmetry groups, physical property tensors, elasticity and dislocations in quasicrystals. Rep. Prog. Phys. **63**, 1–39 (2000)
12. Dotera, T.: Quasicrystals in soft matter. Isr. J. Chem. **51**, 1197–1205 (2011)
13. Soon, C.F., Youseffib, M., Blagden, N., Denyera, M.C.T.: Influence of time dependent factors to the phases and Poisson's ratio of Cholesteryl Ester liquid crystals. J. Sci. Technol. **3**, 1–14 (2011)
14. Witten, T.A.: Insight from soft condensed matter. Rev. Mod. Phys. **71**, 367–373 (1999)

Chapter 9
Dynamics of 8-Fold Symmetric Soft-Matter Quasicrystal Models

Apart from the observed 12-, 18-, and 10-fold symmetric soft-matter quasicrystals, the 8-fold symmetric soft-matter quasicrystals are plausible to be observed soon. With the consideration of the angles and symmetry, the 8-fold symmetric quasicrystals exhibit similarities with their 5-, 10-, and 12-fold symmetric equivalents. The 8-fold symmetric soft-matter quasicrystals are as important as their solid equivalents. They are expected to exhibit a strong phonon-phason coupling that leads to interesting physicomechanical properties and mathematical solutions.

9.1 Basic Equations of 8-Fold Symmetric Soft-Matter Quasicrystal Models

The concrete constitutive laws for phonons and phasons can be obtained from the respective symmetry groups. For 8-fold symmetric models of quasicrystals, one can use the 8 mm point group. If one then considers a plane filled in the xy-plane, and 8-fold symmetric z-axis, then the elastic constitutive law has the following form:

$$
\left.
\begin{aligned}
\sigma_{xx} &= L(\varepsilon_{xx} + \varepsilon_{yy}) + 2M\varepsilon_{xx} + R(w_{xx} + w_{yy}) \\
\sigma_{yy} &= L(\varepsilon_{xx} + \varepsilon_{yy}) + 2M\varepsilon_{yy} - R(w_{xx} + w_{yy}) \\
\sigma_{xy} &= \sigma_{yx} = 2M\varepsilon_{xy} + R(w_{yx} - w_{xy}) \\
H_{xx} &= K_1 w_{xx} + K_2 w_{yy} + R(\varepsilon_{xx} - \varepsilon_{yy}) \\
H_{yy} &= K_1 w_{yy} + K_2 w_{xx} + R(\varepsilon_{xx} - \varepsilon_{yy}) \\
H_{xy} &= (K_1 + K_2 + K_3)w_{xy} + K_3 w_{yx} - 2R\varepsilon_{xy} \\
H_{yx} &= (K_1 + K_2 + K_3)w_{yx} + K_3 w_{xy} + 2R\varepsilon_{xy}
\end{aligned}
\right\}
\qquad (9.1.1a)
$$

in addition, the fluid constitutive law:

T.-Y. Fan et al., *Generalized Dynamics of Soft-Matter Quasicrystals*, Springer Series in Materials Science 260, https://doi.org/10.1007/978-981-16-6628-5_9

$$\left.\begin{aligned}
p_{xx} &= -p + \sigma'_{xx} = -p + 2\eta(\dot{\xi}_{xx} - \tfrac{1}{3}\dot{\xi}_{kk}) \\
p_{yy} &= -p + \sigma'_{yy} = -p + 2\eta(\dot{\xi}_{yy} - \tfrac{1}{3}\dot{\xi}_{kk}) \\
p_{xy} &= p_{yx} = \sigma'_{xy} = \sigma'_{yx} = 2\eta\dot{\xi}_{xy} \\
\dot{\xi}_{kk} &= \dot{\xi}_{xx} + \dot{\xi}_{yy}
\end{aligned}\right\} \qquad (9.1.1b)$$

Using specific derivations of the Poisson bracket method (see Chap. 5) on the constitutive law (9.1.1a, 9.1.1b) [1], one can derive the final governing equation system of the generalized dynamics:

$$\left.\begin{aligned}
&\frac{\partial \rho}{\partial t} + \nabla \cdot (\rho \mathbf{V}) = 0 \\[4pt]
&\frac{\partial(\rho V_x)}{\partial t} + \frac{\partial(V_x \rho V_x)}{\partial x} + \frac{\partial(V_y \rho V_x)}{\partial y} = -\frac{\partial p}{\partial x} + \eta \nabla^2(\rho V_x) + \frac{1}{3}\eta\frac{\partial}{\partial x}\nabla \cdot \mathbf{V} + M\nabla^2 u_x + (L + M - B)\frac{\partial}{\partial x}\nabla \cdot \mathbf{u} \\
&\quad + R\left(\frac{\partial^2 w_x}{\partial x^2} + 2\frac{\partial^2 w_y}{\partial x \partial y} - \frac{\partial^2 w_x}{\partial y^2}\right) - (A - B)\frac{1}{\rho_0}\frac{\partial \delta\rho}{\partial x} \\[4pt]
&\frac{\partial(\rho V_y)}{\partial t} + \frac{\partial(V_x \rho V_y)}{\partial x} + \frac{\partial(V_y \rho V_y)}{\partial y} = -\frac{\partial p}{\partial y} + \eta \nabla^2(\rho V_y) + \frac{1}{3}\eta\frac{\partial}{\partial y}\nabla \cdot \mathbf{V} + M\nabla^2 u_y + (L + M - B)\frac{\partial}{\partial y}\nabla \cdot \mathbf{u} \\
&\quad + R\left(\frac{\partial^2 w_y}{\partial x^2} - 2\frac{\partial^2 w_x}{\partial x \partial y} - \frac{\partial^2 w_y}{\partial y^2}\right) - (A - B)\frac{1}{\rho_0}\frac{\partial \delta\rho}{\partial y} \\[4pt]
&\frac{\partial u_x}{\partial t} + V_x\frac{\partial u_x}{\partial x} + V_y\frac{\partial u_x}{\partial y} = V_x + \Gamma_{\mathbf{u}}\left[M\nabla^2 u_x + (L + M)\frac{\partial}{\partial x}\nabla \cdot \mathbf{u}\right. \\
&\quad \left. + R\left(\frac{\partial^2 w_x}{\partial x^2} + 2\frac{\partial^2 w_y}{\partial x \partial y} - \frac{\partial w_x}{\partial y^2}\right)\right] \\[4pt]
&\frac{\partial u_y}{\partial t} + V_x\frac{\partial u_y}{\partial x} + V_y\frac{\partial u_y}{\partial y} = V_y + \Gamma_{\mathbf{u}}\left[M\nabla^2 u_y + (L + M)\frac{\partial}{\partial y}\nabla \cdot \mathbf{u}\right. \\
&\quad \left. + R_1\left(\frac{\partial^2 w_y}{\partial x^2} - 2\frac{\partial^2 w_x}{\partial x \partial y} - \frac{\partial^2 w_y}{\partial y^2}\right)\right] \\[4pt]
&\frac{\partial w_x}{\partial t} + V_x\frac{\partial w_x}{\partial x} + V_y\frac{\partial w_x}{\partial y} = \Gamma_{\mathbf{w}}\left[K_1\nabla^2 w_x + (K_2 + K_3)\frac{\partial^2 w_x}{\partial y^2} + R\left(\frac{\partial^2 u_x}{\partial x^2} - 2\frac{\partial^2 u_y}{\partial x \partial y} - \frac{\partial^2 u_x}{\partial y^2}\right)\right] \\[4pt]
&\frac{\partial w_y}{\partial t} + V_x\frac{\partial w_y}{\partial x} + V_y\frac{\partial w_y}{\partial y} = \Gamma_{\mathbf{w}}\left[K_1\nabla_1^2 w_y + (K_2 + K_3)\frac{\partial^2 w_y}{\partial x^2} + R\left(\frac{\partial^2 u_y}{\partial x^2} + 2\frac{\partial^2 u_x}{\partial x \partial y} - \frac{\partial^2 u_y}{\partial y^2}\right)\right] \\[4pt]
&p = f(\rho)
\end{aligned}\right\}$$

$$(9.1.2)$$

However, the equation of state in (9.1.2) is not derived by applying the Poisson brackets but belongs to the result of thermodynamics.

9.2 Dislocation in 8-Fold Symmetric Soft-Matter Quasicrystals

9.2.1 Elastic Static Solution

If one neglects the fluid effect, then the final governing equation of elasticity of two-dimensional quasicrystals with eight-fold symmetry can be expressed [2] as:

$$(\nabla^2\nabla^2\nabla^2\nabla^2 - 4\varepsilon\nabla^2\nabla^2\Lambda^2\Lambda^2 + 4\varepsilon\Lambda^2\Lambda^2\Lambda^2\Lambda^2)F = 0 \tag{9.2.1}$$

where:

$$\left.\begin{array}{c}
\nabla^2 = \dfrac{\partial^2}{\partial x^2} + \dfrac{\partial^2}{\partial y^2}, \quad \Lambda^2 = \dfrac{\partial^2}{\partial x^2} - \dfrac{\partial^2}{\partial y^2} \\[3mm]
\varepsilon = \dfrac{R^2(L+M)(K_2+K_3)}{[M(K_1+K_2+K_3)-R^2][(L+2M)K_1-R^2]}
\end{array}\right\} \tag{9.2.2}$$

in which $F(x, y)$ is a displacement potential function (see Ref. [2] for detail). Equation (9.2.1) is more complicated than those of (7.2.2), (7.2.3), and (8.3.2), and thus its solution is also more complicated. This chapter discusses the most relevant results, while the reader is encouraged to refer to the primary literature for detailed derivation.

With consideration of the dislocation problem $b^{\parallel}\oplus b^{\perp} = (b_1^{\parallel}, 0, b_1^{\perp}, 0, 0)$, one can try to determine the displacement field under the boundary constraints, that is, the condition of semi-planarity ($y > 0$, or $y < 0$):

$$\left.\begin{array}{c}
\sigma_{ij}(x, y) \to 0, \; H_{ij}(x, y) \to 0 \quad (\sqrt{x^2+y^2} \to \infty) \\[2mm]
\sigma_{yy}(x, 0) = 0, \; H_{yy}(x, 0) = 0 \\[2mm]
\displaystyle\int_{\Gamma} du_x = b_1^{\parallel} \;, \qquad \int_{\Gamma} dw_x = b_1^{\perp}
\end{array}\right\} \tag{9.2.3}$$

Fourier transform reduces (9.2.1) to:

$$\left[\left(\frac{d^2}{dy^2} - \xi^2\right)^4 - 4\varepsilon\left(\frac{d^2}{dy^2} - \xi^2\right)^2 + 4\varepsilon\left(\frac{d^2}{dy^2} + \xi^2\right)^4 \hat{F} = 0\right] \tag{9.2.4}$$

The eigenroots of Eq. (9.2.4) depend on the value of the parameter ε. Zhou provided a detailed discussion for the solutions corresponding to two cases: (1): $0 < \varepsilon < 1$; (2): $\varepsilon < 0$; however, his calculation is very complex and laborious (see reference for [3] more details). For case (1) the obtained solution is:

$$u_x(x, y) = \frac{1}{2\pi}\left\{\frac{b_1^{\parallel}}{2}\left[\arctan\left(\frac{\lambda_1^2 + \lambda_2^2}{\lambda_1}\frac{y}{x} + \frac{\lambda_2}{\lambda_1}\right) + \arctan\left(\frac{\lambda_1^2 + \lambda_2^2}{\lambda_1}\frac{y}{x} - \frac{\lambda_2}{\lambda_1}\right)\right]\right.$$

$$\left. + (F_3 C + F_4 D)\left[\arctan\frac{2\lambda_3 xy}{x^2 - (\lambda_3^2 + \lambda_4^2)y^2} - \arctan\frac{2\lambda_1 xy}{x^2 - (\lambda_1^2 + \lambda_2^2)y^2}\right]\right\}$$

$$ + \frac{1}{4\pi}\left[F_5 \ln\frac{x^2 + 2\lambda_2 xy + (\lambda_1^2 + \lambda_2^2)y^2}{x^2 - 2\lambda_2 xy + (\lambda_1^2 + \lambda_2^2)y^2} + F_6 \ln\frac{x^2 + 2\lambda_4 xy + (\lambda_3^2 + \lambda_4^2)y^2}{x^2 - 2\lambda_4 xy + (\lambda_3^2 + \lambda_4^2)y^2}\right]$$

$$u_y = \frac{1}{2\pi}\left\{H_1\left[\arctan\frac{2\lambda_1\lambda_2 y^2}{x^2 + (\lambda_1^2 - \lambda_2^2)y^2} - 2\arctan\frac{\lambda_2}{\lambda_1}\right]\right.$$

$$\left. + H_2\left[\arctan\frac{2\lambda_3\lambda_4 y^2}{x^2 + (\lambda_3^2 - \lambda_4^2)y^2} - 2\arctan\frac{\lambda_4}{\lambda_3}\right]\right\}$$

$$ + \frac{1}{4\pi}\left\{H_3 \ln\left[1 + \frac{x^4 + 2(\lambda_1^2 - \lambda_2^2)x^2 y^2}{(\lambda_1^2 + \lambda_2^2)^2 y^4}\right]\right\}$$

$$ + \left\{H_4 \ln\left[1 + \frac{x^4 + 2(\lambda_3^2 - \lambda_4^2)x^2 y^2}{(\lambda_3^2 + \lambda_4^2)^2 y^4}\right]\right\}$$

$$w_x(x, y) = \frac{1}{2\pi}\left\{\frac{b_1^{\perp}}{2}\left[\arctan\left(\frac{\lambda_1^2 + \lambda_2^2}{\lambda_1}\frac{y}{x} + \frac{\lambda_2}{\lambda_1}\right) + \arctan\left(\frac{\lambda_1^2 + \lambda_2^2}{\lambda_1}\frac{y}{x} - \frac{\lambda_2}{\lambda_1}\right)\right]\right.$$

$$\left. + (G_3 C + G_4 D) \times \left[\arctan\frac{2\lambda_3 xy}{x^2 - (\lambda_3^2 + \lambda_4^2)y^2} - \arctan\frac{2\lambda_1 xy}{x^2 - (\lambda_1^2 + \lambda_2^2)y^2}\right]\right\}$$

$$ + \frac{1}{4\pi}\left[G_5 \ln\frac{x^2 + 2\lambda_2 xy + (\lambda_1^2 + \lambda_2^2)y^2}{x^2 - 2\lambda_2 xy + (\lambda_1^2 + \lambda_2^2)y^2} + G_6 \ln\frac{x^2 + 2\lambda_4 xy + (\lambda_3^2 + \lambda_4^2)y^2}{x^2 - 2\lambda_4 xy + (\lambda_3^2 + \lambda_4^2)y^2}\right]$$

$$w_y = \frac{1}{2\pi}\left\{I_1\left[\arctan\frac{2\lambda_1\lambda_2 y^2}{x^2 + (\lambda_1^2 - \lambda_2^2)y^2} - 2\arctan\frac{\lambda_2}{\lambda_1}\right]\right.$$

$$\left. + I_2\left[\arctan\frac{2\lambda_3\lambda_4 y^2}{x^2 + (\lambda_3^2 - \lambda_4^2)y^2} - 2\arctan\frac{\lambda_4}{\lambda_3}\right]\right\}$$

$$ + \frac{1}{4\pi}\left\{I_3 \ln\left[1 + \frac{x^4 + 2(\lambda_1^2 - \lambda_2^2)x^2 y^2}{(\lambda_1^2 + \lambda_2^2)^2 y^4}\right]\right.$$

$$+ I_4 \ln\left[1 + \frac{x^4 + 2(\lambda_3^2 - \lambda_4^2)x^2 y^2}{(\lambda_3^2 + \lambda_4^2)^2 y^4}\right]\right\} \tag{9.2.5}$$

in which $F_1, \ldots, F_6, G_1, \ldots, G_6, H_1, \ldots, H_4$, and $I_1, \ldots, I_4$, are functions of $\lambda_1, \lambda_2, \lambda_3$, and λ_4, where the latter are constants deriving from the original material constants M, L, K_1, K_2, K_3, and R. Owing to the lengthiness of these expressions, they have not been included here, but the reader is encouraged to check references [1, 3].

Similarly, a solution for case (2) can be obtained by solving the final governing Eq. (9.2.1). This can be difficult and tedious, and for the sake of simplicity, the derivation has been omitted here.

9.2.2 Modification with Consideration of the Fluid Effect

The pure elastic solution of dislocation in 8-fold symmetric soft-matter quasicrystals represents a zero-order approximation of the practical solution. Strict analytic solution concerning the elasticity and fluidity effects remains currently unknown. The modification of the fluid effect for 8-fold symmetric quasicrystals in a similar way as already showcased in Sect. 8.4 will be far more challenging to implement than for the 12-, 5-, and 10-fold symmetric equivalents. Therefore, such implementation is not discussed here.

9.3 Transient Dynamics Analysis

9.3.1 Specimen

Equations (9.1.2) are nonlinear dynamic equations and the transient problems in these equations can be solved only numerically. Wang et al. [4] took the finite difference method and obtained the detailed solution on the transient dynamics of the specimen, as shown in Fig. 9.1. The configuration of the transient dynamics of a specimen is similar to that in Fig. 9.1.

Wang et al. [4] used a more precise numerical grid than the one introduced in Sect. 8.5. They aimed to explore the spatial distribution of the field variables in the transient dynamic process. The numerical results and analysis of this study are available for further reading under reference [4].

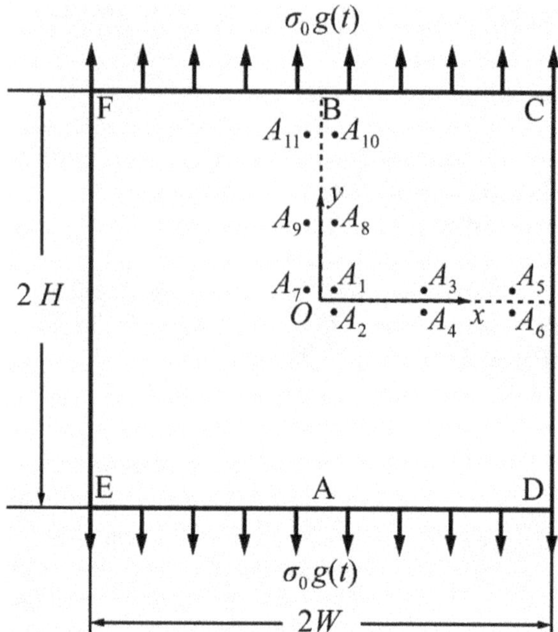

Fig. 9.1 Specimen under impact tension and computed points

9.4   Flow Past a Circular Cylinder

Similarly to Sect. 7.6, we carried out a numerical analysis to model the flow of 8-fold symmetric soft-matter quasicrystals past a cylinder or other obstacle. We used the same configuration for the flow past a cylinder, as shown in Figs. 7.1 and 7.2. The boundary conditions were taken the same as those given in Eqs. (7.6.1) and (7.6.2), respectively. After applying Oseen's modification to Eqs. (9.1.2) and taking the form of the equations in the polar coordinate system, we have solved the boundary-value problem by finite difference method similar to that displayed in Sect. 7.6. We obtained numerical results that illustrate the phason stresses as shown in Figs. 9.2, 9.3, 9.4 and 9.5. The material constants adopted in the calculation are:

$L = 10\,\text{MPa}, M = 4\,\text{MPa}, K_1 = 0.5\,\text{L}, K_2 = -0.1\,\text{L}, K_3 = 0.05\,\text{L}, R = 0.04\,M$, and the data of geometry and external flow field are the same as in Sect. 7.6.

There is a strong phason-phonon coupling in the 8-fold symmetric soft-matter quasicrystals, which is responsive on the phason field under an external flow field (see Figs. 9.2, 9.3, 9.4 and 9.5). This is qualitatively different than that in the 12-fold symmetric equivalents, which do not exhibit phason-phonon coupling (see Sect. 7.6). Although not shown here, the results on the phonon and the fluid phonon fields are similar for the soft-matter quasicrystals of different symmetries. An in-depth analysis of the effects of the interactions among different phasons is needed to continue our discussion.

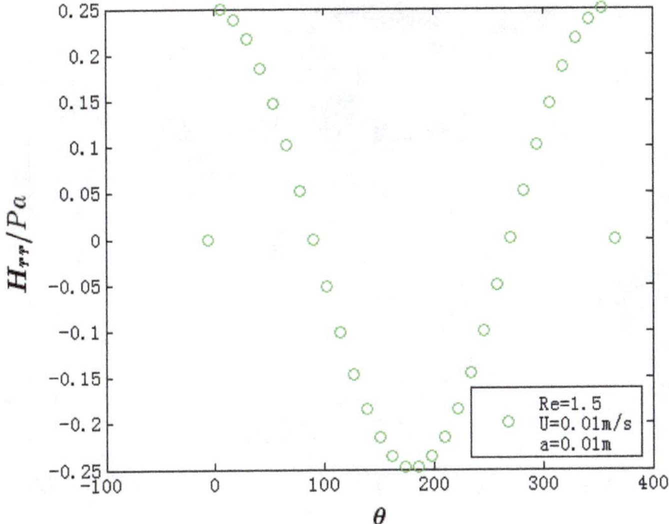

Fig. 9.2 Angular distribution of radical normal stress of phason filed at $r = 1.55a$

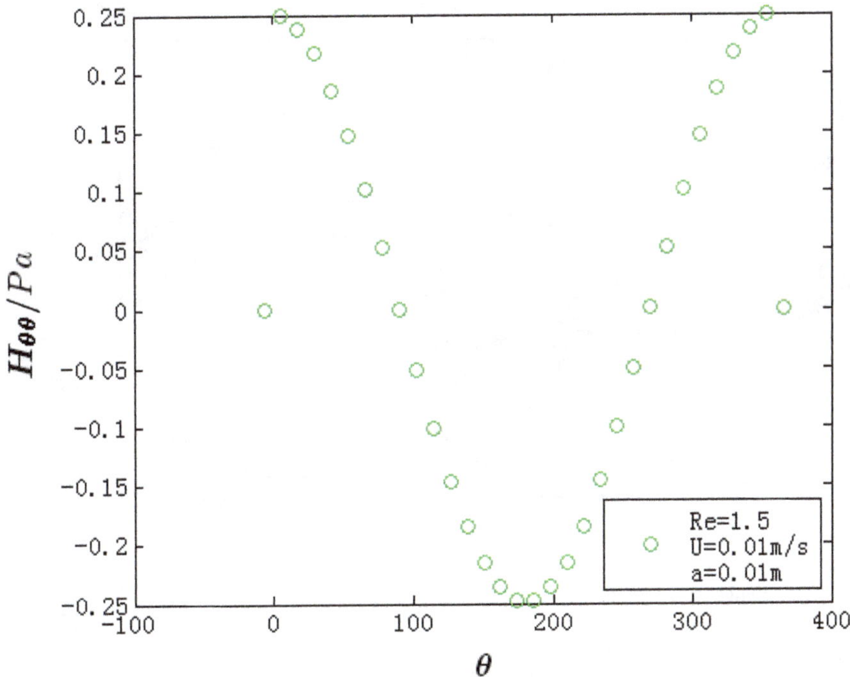

Fig. 9.3 Angular distribution of circumferential normal stress of phason filed at $r = 1.55a$

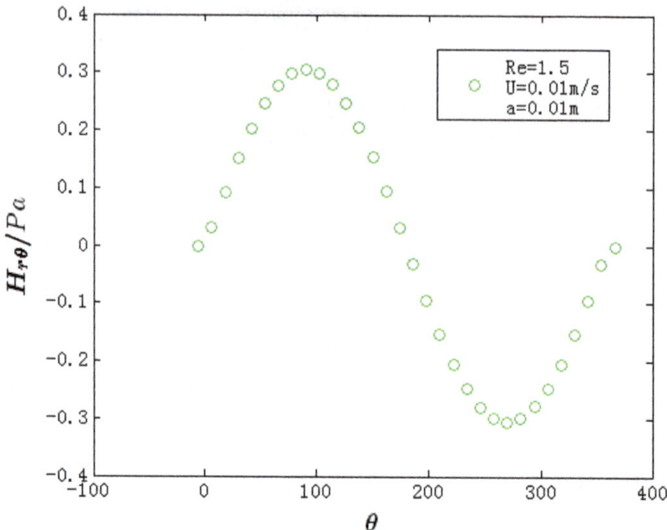

Fig. 9.4 Angular distribution of shear stress $H_{r\theta}$ of phason filed at $r = 1.55a$

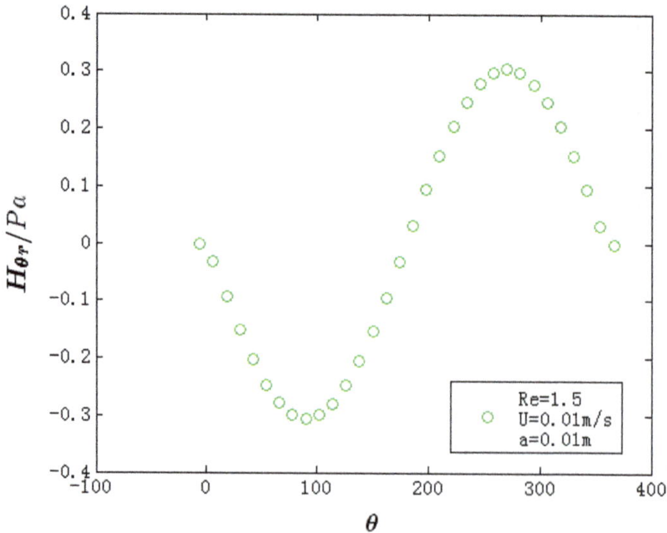

Fig. 9.5 Angular distribution of shear stress $H_{\theta r}$ of phason filed at $r = 1.55a$

9.5 Three-Dimensional Systems with 8-Fold Symmetric Soft-Matter Quasicrystals

The previous descriptions showcased the behavior of the 8-fold symmetric soft-matter quasicrystals and the importance of the strong phonon-phason couplings at the two-dimensional, quasiperiodic limit (i.e., the xy-plane). Herein, we expand to the three-dimensional dynamics, where the z-axis is the 8-fold symmetric axis. Following similar derivations as shown previously, the final governing equations of the generalized dynamics follow the constitutive law [1, 5] where:

$$
\left.
\begin{aligned}
\sigma_{xx} &= C_{11}\varepsilon_{xx} + C_{12}\varepsilon_{yy} + C_{13}\varepsilon_{zz} + R\left(w_{xx} + w_{yy}\right) \\
\sigma_{yy} &= C_{12}\varepsilon_{xx} + C_{11}\varepsilon_{yy} + C_{13}\varepsilon_{zz} - R\left(w_{xx} + w_{yy}\right) \\
\sigma_{zz} &= C_{13}\varepsilon_{xx} + C_{13}\varepsilon_{yy} + C_{33}\varepsilon_{zz} \\
\sigma_{yz} &= \sigma_{zy} = 2C_{44}\varepsilon_{yz} \\
\sigma_{zx} &= \sigma_{xz} = 2C_{44}\varepsilon_{zx} \\
\sigma_{xy} &= \sigma_{yx} = 2C_{66}\varepsilon_{xy} - Rw_{xy} + Rw_{yx}
\end{aligned}
\right\}
\tag{9.5.1a}
$$

$$
\left.
\begin{aligned}
H_{xx} &= K_1 w_{xx} + K_2 w_{yy} + R\left(\varepsilon_{xx} - \varepsilon_{yy}\right) \\
H_{yy} &= K_2 w_{xx} + K_1 w_{yy} + R\left(\varepsilon_{xx} - \varepsilon_{yy}\right) \\
H_{yz} &= K_4 w_{yz} \\
H_{xy} &= (K_1 + K_2 + K_3)w_{xy} + K_2 w_{yz} - 2R\varepsilon_{xy} \\
H_{xz} &= K_4 w_{xz} \\
H_{yx} &= K_3 w_{xy} + (K_1 + K_2 + K_3)w_{yx} + 2R\varepsilon_{xy}
\end{aligned}
\right\}
\tag{9.5.1b}
$$

$$
\left.
\begin{aligned}
p_{xx} &= -p + 2\eta\dot{\xi}_{xx} - \frac{2}{3}\eta\dot{\xi}_{kk} \\
p_{yy} &= -p + 2\eta\dot{\xi}_{yy} - \frac{2}{3}\eta\dot{\xi}_{kk} \\
p_{zz} &= -p + 2\eta\dot{\xi}_{zz} - \frac{2}{3}\eta\dot{\xi}_{kk} \\
p_{yz} &= 2\eta\dot{\xi}_{yz}, \; p_{zx} = 2\eta\dot{\xi}_{zx}, \; p_{xy} = 2\eta\dot{\xi}_{xy} \\
\dot{\xi}_{kk} &= \dot{\xi}_{xx} + \dot{\xi}_{yy} + \dot{\xi}_{zz}
\end{aligned}
\right\}
\tag{9.5.1c}
$$

With these basic relations, the governing equations of the three-dimensional dynamics of 8-fold symmetric soft-matter quasicrystals are given as follows [6]:

$$\frac{\partial \rho}{\partial t} + \nabla \cdot (\rho \mathbf{V}) = 0$$

$$\frac{\partial(\rho V_x)}{\partial t} + \frac{\partial(V_x \rho V_x)}{\partial x} + \frac{\partial(V_y \rho V_x)}{\partial y} + \frac{\partial(V_z \rho V_x)}{\partial z}$$

$$= -\frac{\partial p}{\partial x} + \eta \nabla^2 (\rho V_x) + \frac{1}{3}\eta \frac{\partial}{\partial x} \nabla \cdot \mathbf{V}$$

$$+ \left(C_{66}\frac{\partial^2}{\partial y^2} + C_{44}\frac{\partial^2}{\partial z^2} \right) u_x + (C_{12} + C_{66})\frac{\partial^2 u_y}{\partial x \partial y}$$

$$+ (C_{13} + C_{44} - C_{11})\frac{\partial^2 u_z}{\partial x \partial z} + (C_{11} - B)\frac{\partial}{\partial x}\nabla \cdot \mathbf{u} + R\frac{\partial}{\partial x}\nabla_1 \cdot \mathbf{w}$$

$$- R\frac{\partial}{\partial y}\left(\frac{\partial w_x}{\partial y} - \frac{\partial w_y}{\partial x} \right) - (A - B)\frac{1}{\rho_0}\frac{\partial \delta \rho}{\partial x}$$

$$\frac{\partial(\rho V_y)}{\partial t} + \frac{\partial(V_x \rho V_y)}{\partial x} + \frac{\partial(V_y \rho V_y)}{\partial y} + \frac{\partial(V_z \rho V_y)}{\partial z}$$

$$= -\frac{\partial p}{\partial y} + \eta \nabla^2 (\rho V_y) + \frac{1}{3}\eta \frac{\partial}{\partial y} \nabla \cdot \mathbf{V} + (C_{12} + C_{66})\frac{\partial^2 u_x}{\partial x \partial y}$$

$$+ \left(C_{66}\frac{\partial^2}{\partial x^2} + C_{11}\frac{\partial^2}{\partial y^2} + C_{44}\frac{\partial^2}{\partial z^2} \right) u_y + (C_{13} + C_{44})\frac{\partial^2 u_z}{\partial y \partial z}$$

$$+ (C_{11} - B)\frac{\partial}{\partial y}\nabla \cdot \mathbf{u}$$

$$- R\frac{\partial}{\partial x}\left(\frac{\partial w_x}{\partial y} - \frac{\partial w_y}{\partial x} \right) - R\frac{\partial}{\partial y}\nabla_1 \cdot \mathbf{w} - (A - B)\frac{1}{\rho_0}\frac{\partial \delta \rho}{\partial y}$$

$$\frac{\partial(\rho V_z)}{\partial t} + \frac{\partial(V_x \rho V_z)}{\partial x} + \frac{\partial(V_y \rho V_z)}{\partial y} + \frac{\partial(V_z \rho V_z)}{\partial z}$$

$$= -\frac{\partial p}{\partial z} + \eta \nabla^2 (\rho V_z) + \frac{1}{3}\eta \frac{\partial}{\partial z} \nabla \cdot \mathbf{V}$$

$$+ \left(C_{44}\frac{\partial^2}{\partial x^2} + C_{44}\frac{\partial^2}{\partial y^2} + (C_{33} - C_{13} - C_{44})\frac{\partial^2}{\partial z^2} \right) u_z$$

$$+ (C_{13} + C_{44} - B)\frac{\partial}{\partial z}\nabla \cdot \mathbf{u} - (A - B)\frac{1}{\rho_0}\frac{\partial \delta \rho}{\partial z}$$

$$\frac{\partial u_x}{\partial t} + V_x \frac{\partial u_x}{\partial x} + V_y \frac{\partial u_x}{\partial y} + V_z \frac{\partial u_x}{\partial z} = V_x$$

$$
\left.
\begin{aligned}
&+ \Gamma_{\mathbf{u}}
\begin{bmatrix}
\left(C_{11}\dfrac{\partial^2}{\partial x^2} + C_{66}\dfrac{\partial^2}{\partial y^2} + C_{44}\dfrac{\partial^2}{\partial z^2} \right) u_x \\[6pt]
+ (C_{12} + C_{66})\dfrac{\partial^2 u_y}{\partial x \partial y} \\[6pt]
+ (C_{13} + C_{44})\dfrac{\partial^2 u_z}{\partial x \partial z} + R\dfrac{\partial}{\partial x}\nabla_1 \cdot \mathbf{w} \\[6pt]
- R\dfrac{\partial}{\partial y}\left(\dfrac{\partial w_x}{\partial y} - \dfrac{\partial w_y}{\partial x} \right)
\end{bmatrix} \\[10pt]
&\dfrac{\partial u_y}{\partial t} + V_x\dfrac{\partial u_y}{\partial x} + V_y\dfrac{\partial u_y}{\partial y} + V_z\dfrac{\partial u_y}{\partial z} = V_y \\[10pt]
&+ \Gamma_{\mathbf{u}}
\begin{bmatrix}
(C_{12} + C_{66})\dfrac{\partial^2 u_x}{\partial x \partial y} \\[6pt]
+ \left(C_{66}\dfrac{\partial^2}{\partial x^2} + C_{11}\dfrac{\partial^2}{\partial y^2} + C_{44}\dfrac{\partial^2}{\partial z^2} \right) u_y \\[6pt]
+ (C_{13} + C_{44})\dfrac{\partial^2 u_z}{\partial y \partial z} \\[6pt]
- R\dfrac{\partial}{\partial x}\left(\dfrac{\partial w_x}{\partial y} - \dfrac{\partial w_y}{\partial x} \right) - R\dfrac{\partial}{\partial y}\nabla_1 \cdot \mathbf{w}
\end{bmatrix} \\[10pt]
&\dfrac{\partial u_z}{\partial t} + V_x\dfrac{\partial u_z}{\partial x} + V_y\dfrac{\partial u_z}{\partial y} + V_z\dfrac{\partial u_z}{\partial z} = V_z \\[10pt]
&+ \Gamma_{\mathbf{u}}
\begin{bmatrix}
(C_{13} + C_{44})\left(\dfrac{\partial^2 u_x}{\partial x \partial z} + \dfrac{\partial^2 u_y}{\partial y \partial z} \right) \\[6pt]
+ \left(C_{44}\dfrac{\partial^2}{\partial x^2} + C_{44}\dfrac{\partial^2}{\partial y^2} + C_{33}\dfrac{\partial^2}{\partial z^2} \right) u_z
\end{bmatrix} \\[10pt]
&\dfrac{\partial w_x}{\partial t} + V_x\dfrac{\partial w_x}{\partial x} + V_y\dfrac{\partial w_x}{\partial y} + V_z\dfrac{\partial w_x}{\partial z} \\[10pt]
&= \Gamma_{\mathbf{w}}
\begin{bmatrix}
K_1\nabla_1^2 w_x + (K_2 + K_3)\dfrac{\partial^2 w_x}{\partial y^2} + K_4\dfrac{\partial^2 w_x}{\partial z^2} + K_2\dfrac{\partial}{\partial y}\left(\dfrac{\partial}{\partial x} + \dfrac{\partial}{\partial z} \right) w_y \\[6pt]
+ R\dfrac{\partial}{\partial x}\left(\dfrac{\partial u_x}{\partial x} - \dfrac{\partial u_y}{\partial y} \right) - R\dfrac{\partial}{\partial y}\left(\dfrac{\partial u_x}{\partial y} + \dfrac{\partial u_y}{\partial x} \right)
\end{bmatrix} \\[10pt]
&\dfrac{\partial w_y}{\partial t} + V_x\dfrac{\partial w_y}{\partial x} + V_y\dfrac{\partial w_y}{\partial y} V_z\dfrac{\partial w_y}{\partial z} \\[10pt]
&= \Gamma_{\mathbf{w}}
\begin{bmatrix}
(K_2 + K_3)\dfrac{\partial^2 w_x}{\partial x \partial y} + K_3\dfrac{\partial^2 w_x}{\partial y \partial z} + K_1\nabla_1^2 w_y \\[6pt]
+ (K_2 + K_3)\dfrac{\partial^2 w_y}{\partial x^2} + (K_1 + K_2 + K_3)\dfrac{\partial^2 w_y}{\partial x \partial z} \\[6pt]
+ R\dfrac{\partial}{\partial x}\left(\dfrac{\partial u_x}{\partial y} + \dfrac{\partial u_y}{\partial x} \right) + R\dfrac{\partial}{\partial y}\left(\dfrac{\partial u_x}{\partial x} - \dfrac{\partial u_y}{\partial y} \right)
\end{bmatrix} \\[10pt]
&p = f(\rho)
\end{aligned}
\right\}
\tag{9.5.2}
$$

in which $\nabla^2 = \frac{\partial^2}{\partial x^2} + \frac{\partial^2}{\partial y^2} + \frac{\partial^2}{\partial z^2}$, $\nabla_1^2 = \frac{\partial^2}{\partial x^2} + \frac{\partial^2}{\partial y^2}$, $\nabla = \mathbf{i}\frac{\partial}{\partial x} + \mathbf{j}\frac{\partial}{\partial y} + \mathbf{k}\frac{\partial}{\partial z}$, $\nabla_1 = \mathbf{i}\frac{\partial}{\partial x} + \mathbf{j}\frac{\partial}{\partial y}$, $\mathbf{V} = \mathbf{i}V_x + \mathbf{j}V_y + \mathbf{k}V_z$, $\mathbf{u} = \mathbf{i}u_x + \mathbf{j}u_y + \mathbf{k}u_z$, $\mathbf{w} = \mathbf{i}w_x + \mathbf{j}w_y$, and $C_{11}, C_{12}, C_{13}, C_{33}, C_{44}, C_{66} = (C_{11} - C_{12})/2$ are the phonon elastic constants,

K_1, K_2, K_3, K_4 are the phason elastic constants, R is the phonon-phason coupling constant, η the fluid dynamic viscosity, Γ_u and Γ_w the phonon and phason dissipation coefficients, A and B are the material constants due to variation of mass density and are named as LRT constants.

Equation (9.5.2) are the final governing equations of dynamics of the 8-fold symmetric soft-matter quasicrystals in a three-dimensional case with field variables u_x, u_y, u_z, w_x, w_y, V_x, V_y, V_z, ρ, and p. The number of field variables is ten and equals the number of field equations. Among the equations, (9.5.2a) is the mass conservation equation, (9.5.2b–d) are the momentum conservation equations or the generalized Navier-Stikes equations, (9.5.2e–g) are the equations of motion of phonons due to the symmetry breaking, (9.5.2h, i) are the phason dissipation equations, and (9.5.2j) is the equation of state. The equations are mathematically solvable. The equation of state is essential because, in its absence, the equation system would not have been closed, and thus it would not be mathematically or physically meaningful.

From the viewpoint of hydrodynamics, the governing equations reveal (a) the nature of wave propagation of fields $\mathbf{u}$ and $\mathbf{V}$ with phonon wave speeds $c_1 = \sqrt{A + C_{11} - 2B/\rho}$, $c_2 = c_3 = \sqrt{C_{11} - C_{12}/2\rho}$; (b) the fluid phonon wave speed $c_4 = \sqrt{(\partial p/\partial \rho)_s}$; (c) the nature of the diffusion of the field $\mathbf{w}$ with major diffusive coefficient $D_1 = \Gamma_w K_1$; (d) other less significant diffusive coefficients $D_2 = \Gamma_w K_2$, etc.

9.6 Incompressible Model of the 8-Fold Symmetric Soft-Matter Quasicrystals

The incompressible model of soft-matter quasicrystals simplifies the mathematical treatment and enables approximate solutions.

$$\nabla \cdot \mathbf{V} = 0$$

$$\rho\left(\frac{\partial V_x}{\partial t} + \frac{\partial(V_x V_x)}{\partial x} + \frac{\partial(V_y V_x)}{\partial y} + \frac{\partial(V_z V_x)}{\partial z}\right) = -\frac{\partial p}{\partial x} + \eta\rho\nabla^2 V_x$$

$$+ \left(C_{66}\frac{\partial^2}{\partial y^2} + C_{44}\frac{\partial^2}{\partial z^2}\right)u_x + (C_{12} + C_{66})\frac{\partial^2 u_y}{\partial x \partial y} + (C_{13} + C_{44} - C_{11})\frac{\partial^2 u_z}{\partial x \partial z}$$

$$+ (C_{11} - B)\frac{\partial}{\partial x}\nabla \cdot \mathbf{u} + R\frac{\partial}{\partial x}\nabla_1 \cdot \mathbf{w} - R\frac{\partial}{\partial y}\left(\frac{\partial w_x}{\partial y} - \frac{\partial w_y}{\partial x}\right)$$

$$\rho\left(\frac{\partial V_y}{\partial t} + \frac{\partial(V_x V_y)}{\partial x} + \frac{\partial(V_y V_y)}{\partial y} + \frac{\partial(V_z V_y)}{\partial z}\right) = -\frac{\partial p}{\partial y} + \eta\rho\nabla^2 V_y$$

$$+ (C_{12} + C_{66})\frac{\partial^2 u_x}{\partial x \partial y} + \left(C_{66}\frac{\partial^2}{\partial x^2} + C_{11}\frac{\partial^2}{\partial y^2} + C_{44}\frac{\partial^2}{\partial z^2}\right)u_y + (C_{13} + C_{44})\frac{\partial^2 u_z}{\partial y \partial z}$$

$$+ (C_{11} - B)\frac{\partial}{\partial y}\nabla \cdot \mathbf{u} - R\frac{\partial}{\partial x}\left(\frac{\partial w_x}{\partial y} - \frac{\partial w_y}{\partial x}\right) - R\frac{\partial}{\partial y}\nabla_1 \cdot \mathbf{w}$$

$$\rho\left(\frac{\partial V_z}{\partial t} + \frac{\partial(V_x V_z)}{\partial x} + \frac{\partial(V_y V_z)}{\partial y} + \frac{\partial(V_z V_z)}{\partial z}\right) = -\frac{\partial p}{\partial z} + \eta\rho\nabla^2 V_z$$

$$+\left(C_{44}\frac{\partial^2}{\partial x^2} + C_{44}\frac{\partial^2}{\partial y^2} + (C_{33} - C_{13} - C_{44})\frac{\partial^2}{\partial z^2}\right)u_z + (C_{13} + C_{44} - B)\frac{\partial}{\partial z}\nabla\cdot\mathbf{u}$$

$$\frac{\partial u_x}{\partial t} + V_x\frac{\partial u_x}{\partial x} + V_y\frac{\partial u_x}{\partial y} + V_z\frac{\partial u_x}{\partial z} = V_x$$

$$+\Gamma_{\mathbf{u}}\left[\begin{array}{l}\left(C_{11}\dfrac{\partial^2}{\partial x^2} + C_{66}\dfrac{\partial^2}{\partial y^2} + C_{44}\dfrac{\partial^2}{\partial z^2}\right)u_x + (C_{12} + C_{66})\dfrac{\partial^2 u_y}{\partial x\partial y} \\[2ex] +(C_{13} + C_{44})\dfrac{\partial^2 u_z}{\partial x\partial z} + R\dfrac{\partial}{\partial x}\nabla_1\cdot\mathbf{w} - R\dfrac{\partial}{\partial y}\left(\dfrac{\partial w_x}{\partial y} - \dfrac{\partial w_y}{\partial x}\right)\end{array}\right]$$

$$\frac{\partial u_y}{\partial t} + V_x\frac{\partial u_y}{\partial x} + V_y\frac{\partial u_y}{\partial y} + V_z\frac{\partial u_y}{\partial z} = V_y$$

$$+\Gamma_{\mathbf{u}}\left[\begin{array}{l}(C_{12} + C_{66})\dfrac{\partial^2 u_x}{\partial x\partial y} + \left(C_{66}\dfrac{\partial^2}{\partial x^2} + C_{11}\dfrac{\partial^2}{\partial y^2} + C_{44}\dfrac{\partial^2}{\partial z^2}\right)u_y \\[2ex] +(C_{13} + C_{44})\dfrac{\partial^2 u_z}{\partial y\partial z} - R\dfrac{\partial}{\partial x}\left(\dfrac{\partial w_x}{\partial y} - \dfrac{\partial w_y}{\partial x}\right) - R\dfrac{\partial}{\partial y}\nabla_1\cdot\mathbf{w}\end{array}\right]$$

$$\frac{\partial u_z}{\partial t} + V_x\frac{\partial u_z}{\partial x} + V_y\frac{\partial u_z}{\partial y} + V_z\frac{\partial u_z}{\partial z} = V_z$$

$$+\Gamma_{\mathbf{u}}\left[(C_{13} + C_{44})\left(\frac{\partial^2 u_x}{\partial x\partial z} + \frac{\partial^2 u_y}{\partial y\partial z}\right) + \left(C_{44}\frac{\partial^2}{\partial x^2} + C_{44}\frac{\partial^2}{\partial y^2} + C_{33}\frac{\partial^2}{\partial z^2}\right)u_z\right]$$

$$\frac{\partial w_x}{\partial t} + V_x\frac{\partial w_x}{\partial x} + V_y\frac{\partial w_x}{\partial y} + V_z\frac{\partial w_x}{\partial z}$$

$$= \Gamma_{\mathbf{w}}\left[\begin{array}{l}K_1\nabla_1^2 w_x + (K_2 + K_3)\dfrac{\partial^2 w_x}{\partial y^2} + K_4\dfrac{\partial^2 w_x}{\partial z^2} + K_2\dfrac{\partial}{\partial y}\left(\dfrac{\partial}{\partial x} + \dfrac{\partial}{\partial z}\right)w_y \\[2ex] +R\dfrac{\partial}{\partial x}\left(\dfrac{\partial u_x}{\partial x} - \dfrac{\partial u_y}{\partial y}\right) - R\dfrac{\partial}{\partial y}\left(\dfrac{\partial u_x}{\partial y} + \dfrac{\partial u_y}{\partial x}\right)\end{array}\right]$$

$$\frac{\partial w_y}{\partial t} + V_x\frac{\partial w_y}{\partial x} + V_y\frac{\partial w_y}{\partial y} V_z\frac{\partial w_y}{\partial z}$$

$$= \Gamma_{\mathbf{w}}\left[\begin{array}{l}(K_2 + K_3)\dfrac{\partial^2 w_x}{\partial x\partial y} + K_3\dfrac{\partial^2 w_x}{\partial y\partial z} + K_1\nabla_1^2 w_y + (K_2 + K_3)\dfrac{\partial^2 w_y}{\partial x^2} + (K_1 + K_2 + K_3)\dfrac{\partial^2 w_y}{\partial x\partial z} \\[2ex] +R\dfrac{\partial}{\partial x}\left(\dfrac{\partial u_x}{\partial y} + \dfrac{\partial u_y}{\partial x}\right) + R\dfrac{\partial}{\partial y}\left(\dfrac{\partial u_x}{\partial x} - \dfrac{\partial u_y}{\partial y}\right)\end{array}\right]$$

$$\tag{9.6.1}$$

In these equations, the mass density is taken as a constant. Consequently, the equation of state is not needed, and thus the numbers of field variables and field equations are reduced to nine. The initial and boundary conditions are similar to those associated with Eqs. (9.5.2). The solution to the initial- and the boundary-value problems of these equations is relatively more straightforward than that associated with Eqs. (9.5.2). Still, because of this approximation, some physically relevant information is also lost.

9.7 Solution Example of an Incompressible Model

Following the theory on the incompressible model of the 8-fold symmetric soft-matter quasicrystals, we develop a model for the two-dimensional case. The equations for this type of quasicrystals are:

$$
\left.
\begin{aligned}
&\nabla \cdot (\mathbf{V}) = 0 \\[4pt]
&\rho\left(\frac{\partial(V_x)}{\partial t} + \frac{\partial(V_x V_x)}{\partial x} + \frac{\partial(V_y V_x)}{\partial y}\right) \\[4pt]
&= -\frac{\partial p}{\partial x} + \rho\eta\nabla^2(V_x) + M\nabla^2 u_x + (L+M)\frac{\partial}{\partial x}\nabla \cdot \mathbf{u} \\[4pt]
&\rho\left(\frac{\partial(V_y)}{\partial t} + \frac{\partial(V_x V_y)}{\partial x} + \frac{\partial(V_y V_y)}{\partial y}\right) \\[4pt]
&= -\frac{\partial p}{\partial y} + \rho\eta\nabla^2(V_y) + M\nabla^2 u_y + (L+M)\frac{\partial}{\partial y}\nabla \cdot \mathbf{u} \\[4pt]
&\quad + R\left(\frac{\partial^2 w_y}{\partial x^2} - 2\frac{\partial^2 w_x}{\partial x \partial y} - \frac{\partial^2 w_y}{\partial y^2}\right) \\[4pt]
&\frac{\partial u_x}{\partial t} + V_x\frac{\partial u_x}{\partial x} + V_y\frac{\partial u_x}{\partial y} \\[4pt]
&= V_x + \Gamma_{\mathbf{u}}\left[M\nabla^2 u_x + (L+M)\frac{\partial}{\partial x}\nabla \cdot \mathbf{u}\right. \\[4pt]
&\quad \left. + R\left(\frac{\partial^2 w_x}{\partial x^2} + 2\frac{\partial^2 w_y}{\partial x \partial y} - \frac{\partial w_x}{\partial y^2}\right)\right. \\[4pt]
&\frac{\partial u_y}{\partial t} + V_x\frac{\partial u_y}{\partial x} + V_y\frac{\partial u_y}{\partial y} \\[4pt]
&= V_y + \Gamma_{\mathbf{u}}\left[M\nabla^2 u_y + (L+M)\frac{\partial}{\partial y}\nabla \cdot \mathbf{u}\right. \\[4pt]
&\quad \left. R_1\left(\frac{\partial^2 w_y}{\partial x^2} - 2\frac{\partial^2 w_x}{\partial x \partial y} - \frac{\partial^2 w_y}{\partial y^2}\right)\right] \\[4pt]
&\frac{\partial w_x}{\partial t} + V_x\frac{\partial w_x}{\partial x} + V_y\frac{\partial w_x}{\partial y} \\[4pt]
&= \Gamma_{\mathbf{w}}\left[K_1\nabla^2 w_x + (K_2+K_3)\frac{\partial^2 w_x}{\partial y^2}\right. \\[4pt]
&\quad \left. + R\left(\frac{\partial^2 u_x}{\partial x^2} - 2\frac{\partial^2 u_y}{\partial x \partial y} - \frac{\partial^2 u_x}{\partial y^2}\right)\right. \\[4pt]
&\frac{\partial w_y}{\partial t} + V_x\frac{\partial w_y}{\partial x} + V_y\frac{\partial w_y}{\partial y} = \Gamma_{\mathbf{w}} \\[4pt]
&\quad \left[K_1\nabla_1^2 w_y + (K_2+K_3)\frac{\partial^2 w_y}{\partial x^2}\right. \\[4pt]
&\quad \left. + R\left(\frac{\partial^2 u_y}{\partial x^2} + 2\frac{\partial^2 u_x}{\partial x \partial y} - \frac{\partial^2 u_y}{\partial y^2}\right)\right]
\end{aligned}
\right\}. \qquad (9.7.1)
$$

The finite difference method and iterative procedure are used for solving the specimen with height 2H and width 2W under impact loading of Heaviside-type as shown in Fig. 8.1. The use of an approximation makes the numerical computation simpler than that in Sect. 9.3. Simultaneously, the results on time and space evolution of mass density ρ, mass density variation $\delta\rho/\rho_0$ and $c_4 = \sqrt{\left(\frac{\partial p}{\delta\rho}\right)_s}$ cannot be obtained. This is because when approximating the effective free energy as a function of the mass density ρ which is assumed to be constant (i.e., $\rho = $ const) then the free energy loses its meaning. This reduces the quality of the physical insights and information and represents a shortcoming on the side of the incompressible model.

9.8   Conclusion and Discussion

In this chapter, the dislocation solution (Sect. 9.2), the solution of transient dynamics as a function of the impact stress (Sect. 9.3), and other solutions on flow past an obstacle similar to that in (Sect. 7.6) were introduced. The flow of 8-fold symmetric soft-matter quasicrystals past a circular cylinder was discussed. It was indicated that Eqs. (9.1.2) should be modified by considering the Oseen modification as implemented in (7.4.1). The results on the phason stresses (Figs. 9.2, 9.3, 9.4 and 9.5) show an interesting behavior due to the strong phason-phonon coupling, which is very different from that outlined in Sect. 7.6.

Finally, the three-dimensional dynamics of point group $8mm$ soft-matter quasicrystals have been discussed, but the solutions for the three-dimensional problems have not been carried out in this chapter. Readers can refer to Chap. 13 for the thermodynamic stability. The incompressible model has been introduced, although the computational result did not provide novel insights.

References

1. Fan, T.Y.: Mathematical Theory of Elasticity of Quasicrystals and Its Applications, Beijing, 1st edn. (2010), 2nd edn. (2016). Science Press/Heidelberg, Springer
2. Li, X.F., Fan, T.Y.: New method for solving plane elasticity of planar quasicrystals and solution. Chin. Phys. Lett. **15**, 278–280 (1998)
3. Zhou, W. M.: Dislocation, crack and contact problems in two- and three-dimensional quasicrystals. Dissertation (in Chinese), Beijing Institute of Technology (2000); Zhou, W.M., Fan, T.Y.: Plane elasticity of octagonal quasicrystals and solutions. Chin. Phys. **10**, 743–747 (2000)
4. Wang, F., Chen, H., Fan, T.Y., Hu, H.Y.: A stress analysis of some fundamental specimens of soft matter quasicrystals with 8-fold symmetry based on generalized dynamics. Adv. Mater. Sci. Eng. Article 1D 8789151 (2019)
5. Fan, T.Y., Tang, Z.Y.: Three-dimensional generalized dynamics of soft matter quasicrystals. Appl. Math. Mech. **38**, 1095–1207 (in Chinese) (2017); Adv. Mater. Sci. Eng. Article 1D 4875854 (2020)
6. Fan, T.Y.: Equation system of generalized hydrodynamics of soft matter quasicrystals. Appl. Math. Mech. **37**, 331–347 (2016) (in Chinese). arXiv:1908.06425], Oct 15, 2019

Chapter 10
Dynamics of 18-Fold Symmetric Soft-Matter Quasicrystals

The discovery of 18-fold symmetric quasicrystals in colloids by Fischer et al. [1] raised broad fundamental importance. They are topologically different from the previous reports on pentagonal, decagonal, octagonal, and dodecagonal solid quasicrystals and the dodecagonal and decagonal soft-matter quasicrystals. This discovery is an important milestone not only in materials science but also in symmetry theory and algebra (group theory).

10.1 Six-Dimensional Embedded Space

The previous reported two-dimensional quasiperiodic solid and soft-matter structures present some common features:

(1) Four rationally independent reciprocal basis vectors need to be set to index the diffraction pattern with integers;
(2) The basis vectors can be considered as a projection from a 4-dimensional embedding space (V) upon to the 2-dimensional physical space (V_E);
(3) Space V is the direct sum of V_E and V_I where V_I is the orthogonal complement of space;
(4) The four hydrodynamic degrees of freedom in phases can be parametrized by a two-dimensional vector field. One of them is the phonon field (denoted by **u**), and the other is the phason field (denoted by **w**).

The study of the symmetry operations for quasiperiodic structures by Janssen [2] suggests that all two-dimensional quasicrystal structures with four-dimensional embedding space have been discovered. The noncrystallographic orientational symmetries in such materials are only decagonal (or pentagonal), octagonal, and dodecagonal. As pointed out in Chap. 2, these quasicrystals can be classified as the first kind of two-dimensional quasicrystals. The next two-dimensional quasicrystalline structures, with reservations to their formation, may have a six-dimensional

© The Author(s), under exclusive license to Springer Nature Singapore Pte Ltd. 2022
T.-Y. Fan et al., *Generalized Dynamics of Soft-Matter Quasicrystals*, Springer Series
in Materials Science 260, https://doi.org/10.1007/978-981-16-6628-5_10

embedding space. The symmetries should be seven-, nine-, fourteen-, and 18-fold. These types of structures constitute the second kind of two-dimensional quasicrystals.

The first and second kinds of two-dimensional quasicrystals differ from one another in terms of structure. The first kind of two-dimensional quasicrystals of solid and soft matter have been known for a longer time, while the second one has been more recently discovered. Based on group representation theory, Hu et al. [3] predicted the second kind of two-dimensional quasicrystals: a six-dimensional embedding space is needed to describe the quasiperiodic plane symmetry, which consists of parallel space $E_\parallel^2$ and two perpendicular spaces $E_{\perp1}^2$ and $E_{\perp2}^2$, i.e.,

$$E^6 = E_\parallel^2 \oplus E_{\perp1}^2 \oplus E_{\perp2}^2 \tag{10.1.1}$$

We call $E_{\perp1}^2$ as the first and $E_{\perp2}^2$ the second perpendicular space. Based on the Landau-Anderson expansion, this concept can be further expanded:

$$\rho(\mathbf{r}) = \sum_{\mathbf{G}\in L_R} \rho_\mathbf{G} \exp\{i\mathbf{G}\cdot\mathbf{r}\} = \sum_{\mathbf{G}\in L_R} |\rho_\mathbf{G}| \exp\{-i\Phi_\mathbf{G} + i\mathbf{G}\cdot\mathbf{r}\} \tag{10.1.2}$$

with the extended angular phase

$$\Phi_n = \mathbf{G}_n^\parallel \cdot \mathbf{u} + \mathbf{G}_n^{\perp^1} \cdot \mathbf{v} + \mathbf{G}_n^{\perp^2} \cdot \mathbf{w} \tag{10.1.3}$$

in which $\mathbf{G}_n^\parallel$ represents reciprocal lattice vector in parallel space $E_\parallel^2$, $\mathbf{G}_n^{\perp^1}$, and $\mathbf{G}_n^{\perp^2}$ the reciprocal lattice vector in the first and second perpendicular space $E_{\perp1}^2$ and $E_{\perp2}^2$, $\mathbf{u}$ the phonon displacement field in parallel space, $\mathbf{v}$ and $\mathbf{w}$ the first and second phason displacement field in the $E_{\perp1}^2$ and $E_{\perp2}^2$ perpendicular space, respectively.

10.2 Elasticity of the Possible 18-Fold Symmetric Solid Quasicrystals

The 18-fold symmetric colloidal quasicrystals were first reported experimentally in 2011 [1], as we introduced in Chap. 2. The diffraction pattern is the only experimental evidence available to date (see Fig. 2.7). Hu et al. [3] predicted the existence of an 18-fold symmetry structure and assigned it with the point group 18 mm symmetry. The description by Hu et al. refers to solid instead of soft-matter quasicrystals. Although 18-fold symmetric solid quasicrystals have not been experimentally reported, this does not render their predictions irrelevant. Since 2004, there were proposals for the synthesis of the first kind of soft-matter quasicrystals, not a prediction on the formation of 18-fold symmetric soft-matter quasicrystals (i.e., the second kind ones) till 2011. One may question whether the diffraction pattern of Fig. 2.7 can be assigned to point group 18 mm uniquely, considering other possible point groups with 18-fold symmetry as well (see Chap. 2). For simplicity, in this chapter, we will discuss 18 mm point group only.

For the two-dimensional 18-fold symmetric quasicrystals, we assign the z- axis along the 18-fold rotation axis, and the two-dimensional displacement fields in the quasiperiodic plane are $u = (u_x, u_y), v = (v_x, v_y), w = (w_x, w_y)$. The corresponding strain fields are:

$$\varepsilon_{ij} = \frac{1}{2}\left(\frac{\partial u_i}{\partial x_j} + \frac{\partial u_j}{\partial x_i}\right), v_{ij} = \frac{\partial v_i}{\partial x_j}, w_{ij} = \frac{\partial w_i}{\partial x_j} \tag{10.2.1}$$

and the corresponding generalized Hooke's law is:

$$\left.\begin{aligned}
\sigma_{ij} &= \frac{\partial f_{\text{def}}}{\partial \varepsilon_{ij}} = C_{ijkl}\varepsilon_{kl} + r_{ijkl}v_{kl} + R_{ijkl}w_{kl} \\
\tau_{ij} &= \frac{\partial f_{\text{def}}}{\partial v_{ij}} = T_{ijkl}v_{kl} + r_{klij}\varepsilon_{kl} + G_{ijkl}w_{kl} \\
H_{ij} &= \frac{\partial f_{\text{def}}}{\partial w_{ij}} = K_{ijkl}w_{kl} + R_{klij}\varepsilon_{kl} + G_{klij}v_{kl}
\end{aligned}\right\} \tag{10.2.2}$$

in which $f_{\text{def}} = f_{\text{def}}(u, v, w)$ denotes the elastic deformation energy density (or strain energy density) of the system, given by (10.3.11) (please refer to the next section). The σ_{ij} and C_{ijkl} are same as those defined in previous chapters, r_{ijkl} are the phonon-first phason coupling elastic constants (i.e., the $u - v$ coupling), R_{ijkl} are the phonon-second phason coupling (i.e., the $u - w$ coupling) elastic constants, τ_{ij} is the stress tensor associated with the phason strain tensor v_{ij}, T_{ijkl} is the phason elastic constants corresponding to $\tau_{ij} - v_{ij}$. The meanings of H_{ij} and K_{ijkl} are the same as introduced in the previous chapters, and they correspond to field w, but it is the second phason field, in this case, G_{ijkl} are the elastic constants of coupling (i.e., the $v - w$ coupling) between first–second phason fields.

According to group representation theory, the phonons have only two independent nonzero elastic constants L, and M, where:

$$\begin{aligned}
C_{ijkl} &= L\delta_{ij}\delta_{kl} + M(\delta_{ik}\delta_{jl} + \delta_{il}\delta_{jk}) \quad (i, j, k, l = 1, 2) \\
L &= C_{12}, M = (C_{11} - C_{12})/2 = C_{66}
\end{aligned} \tag{10.2.3}$$

This is identical to the two-dimensional solid quasicrystals of 5-, 8-, 10-, and 12-fold symmetries (see Table 10.1).

where $x = x_1, y = x_2$.

There is no coupling term between phonon and either first or second phason, i.e.,

$$r_{ijkl} = 0 \quad (i, j, k, l = 1, 2) \tag{10.2.4}$$

$$R_{ijkl} = 0 \quad (i, j, k, l = 1, 2) \tag{10.2.5}$$

Table 10.1 Phonon elastic constants of 18-fold symmetric quasicrystals

	11	22	12	21
11	C_{11}	C_{12}	0	0
22	C_{12}	C_{11}	0	0
12	0	0	C_{66}	C_{66}
21	0	0	C_{66}	C_{66}

There are two independent elastic constants of second phasons T_{ijkl}

$$T_{ijkl} = T_1 \delta_{ik}\delta_{jl} + T_2(\delta_{ij}\delta_{kl} - \delta_{il}\delta_{jk}) \quad (i, j, k, l = 1, 2)$$
$$T_{1111} = T_{2222} = T_{2121} = T_1$$
$$T_{1122} = T_{2211} = -T_{2112} = -T_{1221} = T_2 \tag{10.2.6}$$

which are given in Table 10.2.

The coupling elastic constants between the first and the second phasons are

$$G_{ijkl} = G(\delta_{i1} - \delta_{i2})(\delta_{ij}\delta_{kl} - \delta_{ik}\delta_{jl} + \delta_{il}\delta_{jk}) \quad (i, j, k, l = 1, 2) \tag{10.2.7}$$

as listed in Table 10.3.

Table 10.2 Elastic constants of the second phasons

	11	22	12	21
11	T_1	T_2	0	0
22	T_2	T_1	0	0
12	0	0	T_1	$-T_2$
21	0	0	$-T_2$	T_1

Table 10.3 The coupling elastic constants between the first and second phason fields

$v_{ij} \backslash w_{ij}$	11	22	12	21
11	G	G	0	0
22	$-G$	$-G$	0	0
12	0	0	$-G$	G
21	0	0	$-G$	G

10.3 Dynamics of 18-Fold Symmetric Quasicrystals with 18 mm Point Group

The dynamics of 18-fold symmetric quasicrystals are described by (i) the elastic displacement fields $u = (u_x, u_y)$, $v = (v_x, v_y)$, $w = (w_x, w_y)$; (ii) the fluid velocity field $V = (V_x, V_y)$; (iii) the independent field variables mass density ρ and fluid pressure p. The elastic constitutive law has been listed by (10.2.2), the fluid constitutive law is

$$p_{ij} = -p\delta_{ij} + \sigma'_{ij} = -p\delta_{ij} + \eta_{ijkl}\dot{\xi}_{kl} \tag{10.3.1}$$

where

$$\dot{\xi}_{ij} = \frac{1}{2}\left(\frac{\partial V_i}{\partial x_j} + \frac{\partial V_j}{\partial x_i}\right) \tag{10.3.2}$$

represents the fluid deformation rate tensor.

The generalized dynamics equations are: the mass conservation equation

$$\frac{\partial \rho(r,t)}{\partial t} = -\nabla_i(r)(\rho V_i) \tag{10.3.3}$$

the momentum conservation equations (or the generalized Navier–Stokes equations)

$$\begin{aligned}
\frac{\partial g_i(r,t)}{\partial t} = &-\nabla_k(r)(V_k g_i) + \nabla_j(r)\left(-p\delta_{ij} + \eta_{ijkl}\nabla_k(r)V_l\right) \\
&- (\delta_{ij} - \nabla_i(r)u_j)\frac{\delta H}{\delta u_j} + (\nabla_i v_j)\frac{\delta H}{\delta v_j(r,t)} + (\nabla_i(r)w_j)\frac{\delta H}{\delta w_j} \\
&- \rho\nabla_i(r)\frac{\delta H}{\delta \rho}, \quad g_j = \rho V_j
\end{aligned} \tag{10.3.4}$$

the equations of motion of phonons due to the symmetry breaking

$$\frac{\partial u_i(r, t)}{\partial t} = -V_j \nabla_j(r) u_i + \Gamma_u \frac{\delta H}{\delta u_i(r, t)} + V_i \tag{10.3.5}$$

the first phason dissipation equations

$$\frac{\partial v_i(\mathbf{r}, t)}{\partial t} = -V_j \nabla_j(\mathbf{r}) v_i + \Gamma_v \frac{\delta H}{\delta v_i(\mathbf{r}, t)} \tag{10.3.6}$$

the second phason dissipation equations

$$\frac{\partial w_i(r, t)}{\partial t} = -V_j \nabla_j(r) w_i + \Gamma_w \frac{\delta H}{\delta w_i(r, t)} \tag{10.3.7}$$

and the equation of state, for example,

$$p = f(\rho) \tag{10.3.8}$$

where H denotes the Hamiltonian of the system

$$H = H[\Psi(r, t)] = \int \frac{g^2}{2\rho} d^d r + F_{el}, \quad F_{el}$$
$$= F_u + F_v + F_w + F_{uv} + F_{uw} + F_{vw}, \quad g = \rho V \tag{10.3.9}$$

in which the last six terms on the right-hand side of (10.3.9) represent free energies of phonons, first phasons and second phasons, phonon-phason couplings, first–second phason coupling, respectively:

$$F_u = \int \frac{1}{2} C_{ijkl}\varepsilon_{ij}\varepsilon_{kl} d^d r$$

$$F_v = \int \frac{1}{2} T_{ijkl} v_{ij} v_{kl} d^d r$$

$$F_w = \int \frac{1}{2} K_{ijkl} w_{ij} w_{kl} d^d r$$

$$F_{uv} = \int \left(r_{ijkl}\varepsilon_{ij}v_{kl} + r_{klij}v_{ij}\varepsilon_{kl} \right) d^d r$$

$$F_{uw} = \int \left(R_{ijkl}\varepsilon_{ij}w_{kl} + R_{klij}w_{ij}\varepsilon_{kl} \right) d^d r$$

$$F_{vw} = \int \left(G_{ijkl}v_{ij}w_{kl} + G_{klij}w_{ij}v_{kl} \right) d^d r \tag{10.3.10}$$

for the present case, then F_u, F_w, F_{uw} in Chap. 5 are replaced by $F_u, F_v, F_w, F_{uv}, F_{uw}, F_{vw}$, respectively.

The deformation energy density for defining constitutive equation in (10.2.2) is given for the following

$$
\begin{aligned}
f_{\text{def}}(u, v, w) = f_{\text{def}}(\varepsilon_{ij}, v_{ij}, w_{ij}) =\ & \frac{1}{2}L(\nabla \cdot u)^2 + M\varepsilon_{ij}\varepsilon_{ij} \\
& + T_1\big[(v_{11} + v_{22})^2 + (v_{21} - v_{12})^2\big] \\
& + T_2\big[(v_{11} - v_{22})^2 + (v_{21} + v_{12})^2\big] \\
& + K_1\big[(w_{11} + w_{22})^2 + (w_{21} - w_{12})^2\big] \\
& + T_2\big[(w_{11} - w_{22})^2 + (w_{21} + w_{12})^2\big] \\
& + G[(v_{11} - v_{22})(w_{11} - w_{22}) + (v_{21} + v_{12})(w_{21} + w_{12})] \\
=\ & F_u + F_v + F_w + F_{vw},\ (x = x_1,\, y = x_2,\, i = 1, 2,\, j = 1, 2)
\end{aligned}
$$

$$(10.3.11)$$

then the elastic constitutive equations so the whole constitutive equations are determined at last [4]

$$
\left.
\begin{aligned}
\sigma_{xx} &= (L + 2M)\varepsilon_{xx} + L\varepsilon_{yy} \\
\sigma_{yy} &= L\varepsilon_{xx} + (L + 2M)\varepsilon_{yy} \\
\sigma_{xy} &= \sigma_{yx} = 2M\varepsilon_{xy} \\
\tau_{xx} &= T_1 v_{xx} + T_2 v_{yy} + G\big(w_{xx} - w_{yy}\big) \\
\tau_{yy} &= T_2 v_{xx} + T_1 v_{yy} + G\big(w_{xx} - w_{yy}\big) \\
\tau_{xy} &= T_1 v_{xy} - T_2 v_{yx} - G\big(w_{yx} + w_{xy}\big) \\
\tau_{yx} &= -T_2 v_{xy} + T_1 v_{yx} + G\big(w_{yx} + w_{xy}\big) \\
H_{xx} &= K_1 w_{xx} + K_2 w_{yy} + G\big(v_{xx} + v_{yy}\big) \\
H_{yy} &= K_2 w_{xx} + K_1 w_{yy} - G\big(v_{xx} + v_{yy}\big) \\
H_{xy} &= K_1 w_{xy} - K_2 w_{yx} - G\big(v_{xy} + v_{yx}\big) \\
H_{yx} &= K_1 w_{yx} - K_2 w_{xy} + G\big(v_{xy} - v_{yx}\big) \\
p_{xx} &= -p + 2\eta\left(\dot{\xi}_{xx} - \frac{1}{3}\dot{\xi}_{kk}\right) \\
p_{yy} &= -p + 2\eta\left(\dot{\xi}_{yy} - \frac{1}{3}\dot{\xi}_{kk}\right) \\
p_{xy} &= p_{yx} = 2\eta\dot{\xi}_{xy} \\
\dot{\xi}_{kk} &= \dot{\xi}_{xx} + \dot{\xi}_{yy}
\end{aligned}
\right\}
\qquad (10.3.12)
$$

The equations of motion, including the equation of state, can also be obtained

$$\frac{\partial \rho}{\partial t} + \nabla \cdot (\rho \mathbf{V}) = 0$$

$$\frac{\partial(\rho V_x)}{\partial t} + \frac{\partial(V_x \rho V_x)}{\partial x} + \frac{\partial(V_y \rho V_x)}{\partial y} = -\frac{\partial p}{\partial x} + \eta \nabla^2 V_x + \frac{1}{3}\eta\frac{\partial}{\partial x}\nabla \cdot \mathbf{V} + M\nabla^2 u_x$$

$$+(L+M-B)\frac{\partial}{\partial x}\nabla \cdot \mathbf{u} - (A-B)\frac{1}{\rho_0}\frac{\partial \delta\rho}{\partial x}$$

$$\frac{\partial(\rho V_y)}{\partial t} + \frac{\partial(V_x \rho V_y)}{\partial x} + \frac{\partial(V_y \rho V_y)}{\partial y} = -\frac{\partial p}{\partial y} + \eta \nabla^2 V_y + \frac{1}{3}\eta\frac{\partial}{\partial y}\nabla \cdot \mathbf{V} + M\nabla^2 u_y$$

$$+(L+M-B)\frac{\partial}{\partial y}\nabla \cdot \mathbf{u} - (A-B)\frac{1}{\rho_0}\frac{\partial \delta\rho}{\partial y}$$

$$\frac{\partial u_x}{\partial t} + V_x\frac{\partial u_x}{\partial x} + V_y\frac{\partial u_x}{\partial y} = V_x + \Gamma_{\mathbf{u}}\left[M\nabla^2 u_x + (L+M)\frac{\partial}{\partial x}\nabla \cdot \mathbf{u}\right]$$

$$\frac{\partial u_y}{\partial t} + V_x\frac{\partial u_y}{\partial x} + V_y\frac{\partial u_y}{\partial y} = V_y + \Gamma_{\mathbf{u}}\left[M\nabla^2 u_y + (L+M)\frac{\partial}{\partial y}\nabla \cdot \mathbf{u}\right]$$

$$\frac{\partial v_x}{\partial t} + V_x\frac{\partial v_x}{\partial x} + V_y\frac{\partial v_x}{\partial y} = \Gamma_v\left[T_1\nabla^2 v_x + G\left(\frac{\partial^2 w_x}{\partial x^2} - \frac{\partial^2 w_x}{\partial y^2}\right) - 2G\frac{\partial^2 w_y}{\partial x \partial y}\right]$$

$$\frac{\partial v_y}{\partial t} + V_x\frac{\partial v_y}{\partial x} + V_y\frac{\partial v_y}{\partial y} = \Gamma_v\left[T_1\nabla^2 v_y + 2G\frac{\partial^2 w_x}{\partial x \partial y} + G\left(\frac{\partial^2 w_y}{\partial x^2} - \frac{\partial^2 w_y}{\partial y^2}\right)\right]$$

$$\frac{\partial w_x}{\partial t} + V_x\frac{\partial w_x}{\partial x} + V_y\frac{\partial w_x}{\partial y} = \Gamma_{\mathbf{w}}\left[K_1\nabla^2 w_x + G\left(\frac{\partial^2 v_x}{\partial x^2} - \frac{\partial^2 v_x}{\partial y^2}\right) + 2G\frac{\partial^2 v_y}{\partial x \partial y}\right]$$

$$\frac{\partial w_y}{\partial t} + V_x\frac{\partial w_y}{\partial x} + V_y\frac{\partial w_y}{\partial y} = \Gamma_{\mathbf{w}}\left[K_1\nabla^2 w_y - 2G\frac{\partial^2 v_x}{\partial x \partial y} + G\left(\frac{\partial^2 v_y}{\partial x^2} - \frac{\partial^2 v_y}{\partial y^2}\right)\right]$$

$$p = f(\rho)$$

$$(10.3.13)$$

in which $\nabla \cdot = \mathbf{i}\frac{\partial}{\partial x} + \mathbf{j}\frac{\partial}{\partial y}$, $\mathbf{V} = \mathbf{i}V_x + \mathbf{j}V_y$, $\mathbf{u} = \mathbf{i}u_x + \mathbf{j}u_y$, and $L = C_{12}$, $M = (C_{11}-C_{12})/2$ the phonon elastic constants, and T_1, K_1 are the first and second phason elastic constants, G is the coupling elastic constant between the first and second phasons, η is the fluid dynamic viscosity (for simplicity only the scalar quantity form of η_{ijkl} is considered here), and Γ_u, Γ_v, and Γ_w the phonon, first and second phason dissipation coefficients, A and B the material constants (LRT constants) due to variation of mass density, respectively.

The phonon fields of displacement, stress, and strain, and elastic constants are defined in Sect. 10.2. In that section definitions of the first and second phason fields and their stresses and strains are given; the corresponding elastic constants refer to Tables 10.1, 10.2 and 10.3, respectively.

In (10.3.13) there are ten equations with ten field variables. The variables are: $u_x, u_y, v_x, v_y w_x, w_x, V_x, V_y, \rho$, and p. Among the equations, (10.3.13a) is the mass conservation equation, (10.3.13b, c) are the momentum conservation equations or the generalized Navier–Stokes equations, (10.3.13d, e) are the equations of motion of phonons due to symmetry breaking, (10.3.13f, g) are the first phason dissipation equations, (10.3.13h, i) are the second phason dissipation equations, and (10.3.13j) is the equation of state. The equations are mathematically solvable if the initial and the boundary conditions are well set. The equation of state has been discussed in Chap. 4 and reference [4]. Here we take a general form as:

$$p = f(\rho)$$

For the detail on equations of states, see (4.2.2) and (4.2.3) in Chap. 4.

10.4　The Steady Dynamic and the Static Case of the First and the Second Phason Fields

Similarly, as shown in the previous chapters, the hydrodynamics described by Eqs. (10.3.13) can be reduced to the hydrostatic case and the steady hydrodynamics of the phason fields:

$$
\begin{aligned}
V_x \frac{\partial v_x}{\partial x} + V_y \frac{\partial v_x}{\partial y} &= \Gamma_v \left[T_1 \nabla^2 v_x + G \left(\frac{\partial^2 w_x}{\partial x^2} - \frac{\partial^2 w_x}{\partial y^2} \right) - 2G \frac{\partial^2 w_y}{\partial x \partial y} \right] \\
V_x \frac{\partial v_y}{\partial x} + V_y \frac{\partial v_y}{\partial y} &= \Gamma_v \left[T_1 \nabla^2 v_y + 2G \frac{\partial^2 w_x}{\partial x \partial y} + G \left(\frac{\partial^2 w_y}{\partial x^2} - \frac{\partial^2 w_y}{\partial y^2} \right) \right] \\
V_x \frac{\partial w_x}{\partial x} + V_y \frac{\partial w_x}{\partial y} &= \Gamma_w \left[K_1 \nabla^2 w_x + G \left(\frac{\partial^2 v_x}{\partial x^2} - \frac{\partial^2 v_x}{\partial y^2} \right) + 2G \frac{\partial^2 v_y}{\partial x \partial y} \right] \\
V_x \frac{\partial w_y}{\partial x} + V_y \frac{\partial w_y}{\partial y} &= \Gamma_w \left[K_1 \nabla^2 w_y - 2G \frac{\partial^2 v_x}{\partial x \partial y} + G \left(\frac{\partial^2 v_y}{\partial x^2} - \frac{\partial^2 v_y}{\partial y^2} \right) \right]
\end{aligned}
\tag{10.4.1}
$$

In the hydrostatic case, the first five equations are the same with (10.2.1) and (10.2.2), and the remaining one. The sixth to the ninth equations are reduced to:

$$
\begin{aligned}
T_1 \nabla^2 v_x + G \left(\frac{\partial^2 w_x}{\partial x^2} - \frac{\partial^2 w_x}{\partial y^2} \right) - 2G \frac{\partial^2 w_y}{\partial x \partial y} &= 0, \\
T_1 \nabla^2 v_y + 2G \frac{\partial^2 w_1}{\partial x \partial y} + G \left(\frac{\partial^2 w_y}{\partial x^2} - \frac{\partial^2 w_y}{\partial y^2} \right) &= 0, \\
K_1 \nabla^2 w_x + G \left(\frac{\partial^2 v_x}{\partial x^2} - \frac{\partial^2 v_x}{\partial y^2} \right) + 2G \frac{\partial^2 v_y}{\partial x \partial y} &= 0, \\
K_1 \nabla^2 w_y - 2G \frac{\partial^2 v_x}{\partial x \partial y} + G \left(\frac{\partial^2 v_y}{\partial x^2} - \frac{\partial^2 v_y}{\partial y^2} \right) &= 0,
\end{aligned}
\tag{10.4.2}
$$

These equations are decoupled with both fluid fields as well as phonon fields. Li et al. [6] solved them using displacement potential functions such as:

$$
v_x = -G \left(\frac{\partial^2 F_1}{\partial x^2} - \frac{\partial^2 F_1}{\partial y^2} - 2 \frac{\partial^2 F_2}{\partial x \partial y} \right)
\tag{10.4.3}
$$

$$
v_y = -G \left(\frac{\partial^2 F_2}{\partial x^2} - \frac{\partial^2 F_2}{\partial y^2} + 2 \frac{\partial^2 F_1}{\partial x \partial y} \right)
\tag{10.4.4}
$$

$$
w_x = T_1 \left(\frac{\partial^2 F_1}{\partial x^2} + \frac{\partial^2 F_1}{\partial y^2} \right)
\tag{10.4.5}
$$

$$
w_y = T_1 \left(\frac{\partial^2 F_2}{\partial x^2} + \frac{\partial^2 F_2}{\partial y^2} \right)
\tag{10.4.6}
$$

or

$$v_x = K_1 \left(\frac{\partial^2 F_1}{\partial x^2} + \frac{\partial^2 F_1}{\partial y^2} \right) \tag{10.4.7}$$

$$v_y = K_1 \left(\frac{\partial^2 F_2}{\partial x^2} + \frac{\partial^2 F_2}{\partial y^2} \right) \tag{10.4.8}$$

$$w_x = -G \left(\frac{\partial^2 F_1}{\partial x^2} - \frac{\partial^2 F_1}{\partial y^2} + 2 \frac{\partial^2 F_2}{\partial x \partial y} \right) \tag{10.4.9}$$

$$w_y = G \left(2 \frac{\partial^2 F_1}{\partial x \partial y} - \frac{\partial^2 F_2}{\partial x^2} + \frac{\partial^2 F_2}{\partial y^2} \right) \tag{10.4.10}$$

where $F_j (j = 1, 2)$ satisfy biharmonic equations

$$\nabla^2 \nabla^2 F_j = 0 \tag{10.4.11}$$

These biharmonic equations can be solved through complex analysis or Fourier transforms. This is a common approach for solid matter quasicrystals [8], soft-matter quasicrystals (Chaps. 7–11), and for smectic A liquid crystals (Chap. 16).

10.5   Dislocations and Solutions

10.5.1   The Zero-Order Approximate Solution for Dislocations in 18-Fold Symmetric Soft-Matter Quasicrystals

By approximating a static case, the fluid velocities are omitted, and the phonons become decoupled from the fluid phonons and phasons. Consequently, one obtains the zero-order approximate solution of dislocation for the phonon field (7.3.1).

To derive an analytical expression for the displacement and the stress components induced by a dislocation onto an 18-fold symmetric quasicrystal, one should consider a dislocation with a core at the origin with the Burgers vector $\mathbf{b} = \mathbf{b}^{\parallel} \oplus \mathbf{b}_1^{\perp} \oplus \mathbf{b}_2^{\perp} = \left(b_1^{\parallel}, b_2^{\parallel}, b_{11}^{\perp}, b_{12}^{\perp}, b_{21}^{\perp}, b_{22}^{\perp} \right)$, where

$$\int_{\Gamma} du_j = b_j, \qquad \int_{\Gamma} dv_j = b_{1j}^{\perp}, \qquad \int_{\Gamma} dw_j = b_{2j}^{\perp}, \tag{10.5.1}$$

in which the integral path should be taken together with the Burgers circuit surrounding the dislocation core in the physical space. It is sufficient to solve the phason fields induced by the vector components $b_{ij}^{\perp} (i, j = 1, 2)$. The phonon fields induced by $b_j^{\parallel}(j = 1, 2)$ have been evaluated by (7.2.4), which hold for the present quasicrystal system.

Li et al. provided the complex representation of solution (10.4.11) from which the zero-approximate solution of the dislocation for phason field can be obtained [6]. Admissible solution for the dislocation has been further developed [7, 8], and one has the form:

$$F_j = \left(x^2 + y^2\right)\arctan\frac{y}{x}, \quad j = 1, 2 \tag{10.5.2}$$

For the Burgers vector components, $\mathbf{b}^{\perp} = \left(0, 0, b_{11}^{\perp}, b_{12}^{\perp}, b_{21}^{\perp}, b_{22}^{\perp}\right)$ Li and Fan [9] obtained the displacements induced by the dislocation as follows

$$v_x(x, y) = \frac{b_{11}^{\perp}}{2\pi}\arctan\left(\frac{y}{x}\right) + \frac{b_{21}^{\perp}}{2\pi}\frac{G}{T_1}\frac{xy}{x^2 + y^2} + \frac{b_{22}^{\perp}}{2\pi}\frac{G}{T_1}\frac{x^2 - y^2}{2\left(x^2 + y^2\right)}, \tag{10.5.3}$$

$$v_y(x, y) = \frac{b_{12}^{\perp}}{2\pi}\arctan\left(\frac{y}{x}\right) - \frac{b_{21}^{\perp}}{2\pi}\frac{G}{T_1}\frac{x^2 - y^2}{2\left(x^2 + y^2\right)} + \frac{b_{22}^{\perp}}{2\pi}\frac{G}{T_1}\frac{xy}{x^2 + y^2}, \tag{10.5.4}$$

$$w_x(x, y) = \frac{b_{11}^{\perp}}{2\pi}\frac{G}{K_1}\frac{xy}{x^2 + y^2} - \frac{b_{12}^{\perp}}{2\pi}\frac{G}{K_1}\frac{x^2 - y^2}{2\left(x^2 + y^2\right)} + \frac{b_{21}^{\perp}}{2\pi}\arctan\left(\frac{y}{x}\right), \tag{10.5.5}$$

$$w_y(x, y) = \frac{b_{11}^{\perp}}{2\pi}\frac{G}{K_1}\frac{x^2 - y^2}{2\left(x^2 + y^2\right)} + \frac{b_{12}^{\perp}}{2\pi}\frac{G}{K_1}\frac{xy}{x^2 + y^2} + \frac{b_{22}^{\perp}}{2\pi}\arctan\left(\frac{y}{x}\right). \tag{10.5.6}$$

The previous displacement fields do not have a logarithmic singularity. With consideration of the displacement fields and the constitutive equations follows

$$\tau_{xx} = T_1\frac{\partial v_x}{\partial x} + T_2\frac{\partial v_y}{\partial y} + G\left(\frac{\partial w_x}{\partial x} - \frac{\partial w_y}{\partial y}\right), \tag{10.5.7}$$

$$\tau_{yy} = T_2\frac{\partial v_x}{\partial x} + T_1\frac{\partial v_y}{\partial y} + G\left(\frac{\partial w_x}{\partial x} - \frac{\partial w_y}{\partial y}\right), \tag{10.5.8}$$

$$\tau_{xy} = T_1\frac{\partial v_y}{\partial x} - T_2\frac{\partial v_x}{\partial y} + G\left(\frac{\partial w_y}{\partial x} + \frac{\partial w_x}{\partial y}\right), \tag{10.5.9}$$

$$\tau_{yx} = -T_2\frac{\partial v_x}{\partial y} + T_1\frac{\partial v_y}{\partial x} + G\left(\frac{\partial w_y}{\partial x} + \frac{\partial w_x}{\partial y}\right), \tag{10.5.10}$$

$$H_{xx} = K_1\frac{\partial w_x}{\partial x} + K_2\frac{\partial w_y}{\partial y} + G\left(\frac{\partial v_x}{\partial x} + \frac{\partial v_y}{\partial y}\right), \tag{10.5.11}$$

$$H_{yy} = K_2\frac{\partial w_x}{\partial x} + K_1\frac{\partial w_y}{\partial y} - G\left(\frac{\partial v_x}{\partial x} + \frac{\partial v_y}{\partial y}\right), \tag{10.5.12}$$

$$H_{xy} = K_1 \frac{\partial w_x}{\partial y} - K_2 \frac{\partial w_y}{\partial x} + G\left(\frac{\partial v_x}{\partial y} - \frac{\partial v_y}{\partial x}\right), \qquad (10.5.13)$$

$$H_{yx} = -K_2 \frac{\partial w_x}{\partial y} + K_1 \frac{\partial w_y}{\partial x} + G\left(\frac{\partial v_x}{\partial y} - \frac{\partial v_y}{\partial x}\right). \qquad (10.5.14)$$

then one can obtain the stress expressions

$$
\begin{aligned}
\tau_{xx} = &-\frac{b_{11}^{\perp}\left(K_1 T_1 - G^2\right)}{2\pi K_1}\frac{y}{x^2 + y^2} + \frac{b_{12}^{\perp}\left(K_1 T_2 - G^2\right)}{2\pi K_1}\frac{x}{x^2 + y^2} \\
&-\frac{b_{21}^{\perp}(T_1 - T_2)G}{\pi T_1}\frac{x^2 y}{\left(x^2 + y^2\right)^2} - \frac{b_{22}^{\perp}(T_1 - T_2)G}{2\pi T_1}\frac{x\left(x^2 - y^2\right)}{\left(x^2 + y^2\right)^2},
\end{aligned} \qquad (10.5.15)
$$

$$
\begin{aligned}
\tau_{yy} = &-\frac{b_{11}^{\perp}\left(K_1 T_2 - G^2\right)}{2\pi K_1}\frac{y}{x^2 + y^2} + \frac{b_{12}^{\perp}\left(K_1 T_1 - G^2\right)}{2\pi K_1}\frac{x}{x^2 + y^2} \\
&+\frac{b_{21}^{\perp}(T_1 - T_2)G}{2\pi T_1}\frac{y\left(x^2 - y^2\right)}{\left(x^2 + y^2\right)^2} - \frac{b_{22}^{\perp}(T_1 - T_2)G}{\pi T_1}\frac{xy^2}{\left(x^2 + y^2\right)^2},
\end{aligned} \qquad (10.5.16)
$$

$$
\begin{aligned}
\tau_{xy} = &\frac{b_{11}^{\perp}\left(K_1 T_1 - G^2\right)}{2\pi K_1}\frac{x}{x^2 + y^2} + \frac{b_{12}^{\perp}\left(K_1 T_2 - G^2\right)}{2\pi K_1}\frac{y}{x^2 + y^2} \\
&-\frac{b_{21}^{\perp}(T_1 - T_2)G}{\pi T_1}\frac{xy^2}{\left(x^2 + y^2\right)^2} - \frac{b_{22}^{\perp}(T_1 - T_2)G}{2\pi T_1}\frac{y\left(x^2 - y^2\right)}{\left(x^2 + y^2\right)^2},
\end{aligned} \qquad (10.5.17)
$$

$$
\begin{aligned}
\tau_{yx} = &-\frac{b_{11}^{\perp}\left(K_1 T_2 - G^2\right)}{2\pi K_1}\frac{x}{x^2 + y^2} - \frac{b_{12}^{\perp}\left(K_1 T_1 - G^2\right)}{2\pi K_1}\frac{y}{x^2 + y^2} \\
&+\frac{b_{21}^{\perp}(T_1 - T_2)G}{2\pi T_1}\frac{x\left(x^2 - y^2\right)}{\left(x^2 + y^2\right)^2} - \frac{b_{22}^{\perp}(T_1 - T_2)G}{\pi T_1}\frac{x^2 y}{\left(x^2 + y^2\right)^2},
\end{aligned} \qquad (10.5.18)
$$

$$
\begin{aligned}
H_{xx} = &-\frac{b_{11}^{\perp}(K_1 + K_2)G}{\pi K_1}\frac{x^2 y}{\left(x^2 + y^2\right)^2} + \frac{b_{12}^{\perp}(K_1 + K_2)G}{2\pi K_1}\frac{x\left(x^2 - y^2\right)}{\left(x^2 + y^2\right)^2} \\
&-\frac{b_{21}^{\perp}\left(K_1 T_1 - G^2\right)}{2\pi T_1}\frac{y}{x^2 + y^2} + \frac{b_{22}^{\perp}\left(K_2 T_1 + G^2\right)}{2\pi T_1}\frac{x}{x^2 + y^2},
\end{aligned} \qquad (10.5.19)
$$

$$
\begin{aligned}
H_{yy} = &-\frac{b_{11}^{\perp}(K_1 + K_2)G}{2\pi K_1}\frac{y\left(x^2 - y^2\right)}{\left(x^2 + y^2\right)^2} - \frac{b_{12}^{\perp}(K_1 + K_2)G}{\pi K_1}\frac{xy^2}{\left(x^2 + y^2\right)^2} \\
&-\frac{b_{21}^{\perp}\left(K_2 T_1 + G^2\right)}{2\pi T_1}\frac{y}{x^2 + y^2} + \frac{b_{22}^{\perp}\left(K_1 T_1 - G^2\right)}{2\pi T_1}\frac{x}{x^2 + y^2},
\end{aligned} \qquad (10.5.20)
$$

$$
H_{xy} = -\frac{b_{11}^{\perp}(K_1 + K_2)G}{\pi K_1}\frac{xy^2}{\left(x^2 + y^2\right)^2} + \frac{b_{12}^{\perp}(K_1 + K_2)G}{2\pi K_1}\frac{y\left(x^2 - y^2\right)}{\left(x^2 + y^2\right)^2}
$$

$$+ \frac{b_{21}^{\perp}\left(K_1 T_1 - G^2\right)}{2\pi T_1} \frac{x}{x^2 + y^2} + \frac{b_{22}^{\perp}\left(K_2 T_1 + G^2\right)}{2\pi T_1} \frac{y}{x^2 + y^2}, \qquad (10.5.21)$$

$$H_{yx} = -\frac{b_{11}^{\perp}(K_1 + K_2)G}{2\pi K_1} \frac{x\left(x^2 - y^2\right)}{\left(x^2 + y^2\right)^2} - \frac{b_{12}^{\perp}(K_1 + K_2)G}{\pi K_1} \frac{x^2 y}{\left(x^2 + y^2\right)^2}$$

$$- \frac{b_{21}^{\perp}\left(K_2 T_1 + G^2\right)}{2\pi T_1} \frac{x}{x^2 + y^2} - \frac{b_{22}^{\perp}\left(K_1 T_1 - G^2\right)}{2\pi T_1} \frac{y}{x^2 + y^2}. \qquad (10.5.22)$$

The above results indicate that both phason stresses exhibit a r^{-1} singularity near the dislocation core. Some errors appearing in [10] have been corrected. After the substitution of the above phason, it is readily checked that the equilibrium equations are identically fulfilled.

For 18-fold symmetric quasicrystals, the dislocation solution is simpler than those of 5-, 10-, and 8-fold symmetry quasicrystals (see Chaps. 8 and 9). The reason is that the phonon and phason fields are decoupled, although, between the first and the second phason fields, there is a coupling. By omitting the fluid effect, the above solution is a zero-order approximate solution.

10.5.2 *Modification to the Solution (10.5.3) to (10.5.6) Considering the Fluid Effect*

In soft-matter quasicrystals, the existence of fluid phonon requires consideration of the fluid effect on the dislocations; however, once considered, it makes the derivation of the pure analytic solution virtually impossible. More clarification is presented in Sect. 8.4.

10.6 Discussion on Transient Dynamics Analysis

Using a similar argumentation as in Sects. 8.5 and 9.3, Gao et al. [10] analyzed the specimen with the same configuration as shown in Fig. 8.1, but using 12- or 18-fold symmetric quasicrystals of soft matter. In this work, Gao et al., based on the computational results, claimed that the phason field w in 12-fold and phason fields v, w in 18-fold symmetric quasicrystals vanish. The reason for this is not merely the $u-w$ decoupling for 12-fold symmetry quasicrystals or the $u-v$, $u-w$ decoupling for the 18-fold symmetry quasicrystals. The authors have tried to add phason boundary stress conditions $H_{yy} = H_0 f(t), y = \pm H, |x| < W$ for the 12-fold symmetric quasicrystals and to place first and second phason stress boundary conditions $\tau_{yy} = \tau_0 f(t), H_{yy} = H_0 f(t), y = \pm H, |x| < W$ for 18-fold symmetry quasicrystals of the impact specimens. Following this arrangement, they have obtained a nonzero computational solution for w of the 12-fold symmetry quasicrystals and for v and w

of the 18-fold symmetric quasicrystals. Their results are very limiting. For instance, one may question how to determine realistic boundary conditions of the loading, considering that currently, there is a lack of experimental approaches to solving such a problem. There is a serious criticism of this work.

10.7 Three-Dimensional Equations of Generalized Dynamics of 18-Fold Symmetric Soft-Matter Quasicrystals

10.7.1 Introduction

The two-dimensional form of the governing generalized dynamics equations for the 18-fold symmetric soft-matter quasicrystals is outlined in reference [4]. The monograph in reference [2] introduced some solutions to those equations. Fan and Tang [3] obtained the three-dimensional equations of generalized dynamics of the first kind of soft-matter quasicrystals. The theory is partially based on the six-dimensional embedding space concept of Hu et al. [4]. The latter concept and the group representation theory provided the elastic constitutive law of the two-dimensional 18-fold symmetric structure. Hu et al. expected that their theory could be used to describe the possible, solid quasicrystals, although this type of quasicrystals in solids has not been reported so far. It is fascinating that after 17 years of the original proposal, Fischer et al. discovered such quasicrystals in colloids [5]. This highlights the power of group and group representation. After the experimental report, Fan and his group carry out a series of works to study the quasicrystals [2]. They suggested the concept of the first kind and second kind of soft-matter quasicrystals (see above). For the second kind of soft-matter quasicrystals, the studies remain in the domain of the two-dimension.

For fundamental and applied understanding, a three-dimensional analysis is needed. Herein, we show how the group theory representation can be applied, while the results can be readily used to study the stability of the 18-fold symmetric soft-matter quasicrystals [6] and Chap. 13 in this book. This problem shows a debate that remains open and has to do with the different ideas on the formation mechanism of soft-matter and solid quasicrystals.

10.7.2 Some Basic Relations

According to the hypothesis of Hu et al. [4] on the six-dimensional embedding space for the 7-, 9-, 14-, and 18-fold symmetric quasicrystals, there are elementary excitations phonons, first and second phasons, whose fields are u, v, and w respectively. Fan [1, 2] showed a need to introduce another elementary excitation—the fluid phonon and its corresponding field with the fluid velocity V for soft-matter quasicrystals. With the purpose to describe the deformation and motion of matter, the tensors of phonon strain, first phason strain, second phason strain, and fluid phonon deformation rate are given as follows:

$$\varepsilon_{ij} = \frac{1}{2}\left(\frac{\partial u_i}{\partial x_j} + \frac{\partial u_j}{\partial x_i}\right), v_{ij} = \frac{\partial v_i}{\partial x_j}, w_{ij} = \frac{\partial w_i}{\partial x_j}, \dot{\xi}_{ij} = \frac{1}{2}\left(\frac{\partial V_i}{\partial x_j} + \frac{\partial V_j}{\partial x_i}\right) \quad (10.7.1)$$

in which $x = x_1, y = x_2, z = x_3, \quad i, j = 1, 2, 3$, and the corresponding constitutive law [1, 2]

$$\left.\begin{aligned}
\sigma_{ij} &= C_{ijkl}\varepsilon_{kl} + r_{ijkl}v_{kl} + R_{ijkl}w_{kl} \\
\tau_{ij} &= T_{ijkl}v_{kl} + r_{klij}\varepsilon_{kl} + G_{ijkl}w_{kl} \\
H_{ij} &= K_{ijkl}w_{kl} + R_{klij}\varepsilon_{kl} + G_{klij}v_{kl} \\
p_{ij} &= -p\delta_{ij} + \sigma'_{ij} = -p\delta_{ij} + \eta_{ijkl}\dot{\xi}_{kl}
\end{aligned}\right\} \quad i, j, k, l = 1, 2, 3 \qquad (10.7.2)$$

where σ_{ij} denotes the phonon stress tensor associated with the phonon strain tensor ε_{ij}, C_{ijkl} is the phonon elastic constants, τ_{ij} is the first phason stress tensor associated to the first phason strain tensor v_{ij}, T_{ijkl} is the first phason elastic constants, H_{ij} is the second phason stress tensor associated with the second phason strain tensor w_{ij}, K_{ijkl} is the second phason elastic constants, r_{ijkl}, r_{klij} are the phonon-first phason coupling elastic constants(i.e., $u - v$ coupling), R_{ijkl}, R_{klij} are the phonon-second phason coupling (i.e., $u - w$ coupling) elastic constants, G_{ijkl}, G_{klij} are the first–second phason coupling (i.e., $v - w$ coupling) elastic constants, p_{ij} is the fluid stress tensor, p is the fluid pressure, δ_{ij} is the unit tensor, σ'_{ij} is the fluid viscous stress tensor, η_{ijkl} is the fluid viscous coefficient constants, respectively.

By using the group representation theory developed by Hu et al. [7], all of the independent nonzero elastic constants for 18-fold symmetric soft-matter quasicrystals with 18 mm point group symmetry in a three-dimensional case were found:

$$\left.\begin{aligned}
&C_{1111} = C_{11}, C_{1122} = C_{12}, C_{3333} = C_{33}, \\
&C_{1133} = C_{13}, C_{2323} = C_{44}, C_{1212} = C_{66}, \\
&(C_{11} - C_{12})/2 = C_{66}, \\
&T_{1111} = T_{2222} = T_{2121} = T_{1212} = T_1, \\
&T_{1122} = T_{2211} = -T_{2112} = -T_{1221} = T_2, \\
&T_{2323} = T_{1313} = T_3 \\
&K_{1111} = K_{2222} = K_{2121} = K_{1212} = K_1, \\
&K_{1122} = K_{2211} = -K_{2112} = -K_{1221} = K_2, \\
&K_{2323} = K_{1313} = K_3 \\
&G_{1111} = -G_{1122} = G_{2211} = -G_{2222} \\
&\quad = -G_{1212} = -G_{1221} = G_{2112} = G_{2121} = G
\end{aligned}\right\} \qquad (10.7.3)$$

This provides the concrete form of the constitutive laws for the phonons

$$
\left.
\begin{aligned}
\sigma_{xx} &= C_{11}\varepsilon_{xx} + C_{12}\varepsilon_{yy} + C_{13}\varepsilon_{zz} \\
\sigma_{yy} &= C_{12}\varepsilon_{xx} + C_{11}\varepsilon_{yy} + C_{13}\varepsilon_{zz} \\
\sigma_{zz} &= C_{13}\varepsilon_{xx} + C_{13}\varepsilon_{yy} + C_{33}\varepsilon_{zz} \\
\sigma_{yz} &= \sigma_{zy} = 2C_{44}\varepsilon_{yz} \\
\sigma_{zx} &= \sigma_{xz} = 2C_{44}\varepsilon_{zx} \\
\sigma_{xy} &= \sigma_{yx} = 2C_{66}\varepsilon_{xy}
\end{aligned}
\right\}
\tag{10.7.4}
$$

for the first phasons

$$
\left.
\begin{aligned}
\tau_{xx} &= T_1 v_{xx} + T_2 v_{yy} + G\left(w_{xx} - w_{yy}\right) \\
\tau_{yy} &= T_2 v_{xx} + T_1 v_{yy} + G\left(w_{xx} - w_{yy}\right) \\
\tau_{xy} &= T_1 v_{xy} - T_2 v_{yx} - G\left(w_{xy} + w_{yx}\right) \\
\tau_{yx} &= T_1 v_{yx} - T_2 v_{xy} + G\left(w_{xy} + w_{yx}\right) \\
\tau_{xz} &= T_3 v_{xz}, \qquad\qquad \tau_{yz} = T_3 v_{yz}
\end{aligned}
\right\}
\tag{10.7.5}
$$

for the second phasons

$$
\left.
\begin{aligned}
H_{xx} &= K_1 w_{xx} + K_2 w_{yy} + G\left(v_{xx} + v_{yy}\right) \\
H_{yy} &= K_2 w_{xx} + K_1 w_{yy} - G\left(v_{xx} + v_{yy}\right) \\
H_{xy} &= K_1 w_{xy} - K_2 w_{yx} - G\left(v_{xy} - v_{yx}\right) \\
H_{yx} &= K_1 w_{yx} - K_2 w_{xy} - G\left(v_{xy} - v_{yx}\right) \\
H_{xz} &= K_3 w_{xz}, \qquad\qquad H_{yz} = K_3 w_{yz}
\end{aligned}
\right\}
\tag{10.7.6}
$$

In addition, there is the constitutive law for the fluid phonon

$$
\left.
\begin{aligned}
p_{xx} &= -p + 2\eta\dot{\xi}_{xx} - \frac{2}{3}\eta\dot{\xi}_{kk} \\
p_{yy} &= -p + 2\eta\dot{\xi}_{yy} - \frac{2}{3}\eta\dot{\xi}_{kk} \\
p_{zz} &= -p + 2\eta\dot{\xi}_{zz} - \frac{2}{3}\eta\dot{\xi}_{kk} \\
p_{yz} &= p_{zy} = 2\eta\dot{\xi}_{yz} \\
p_{zx} &= p_{xz} = 2\eta\dot{\xi}_{zx} \\
p_{xy} &= p_{yx} = 2\eta\dot{\xi}_{xy} \\
\dot{\xi}_{kk} &= \dot{\xi}_{xx} + \dot{\xi}_{yy} + \dot{\xi}_{zz}
\end{aligned}
\right\}
\tag{10.7.7}
$$

The detail on the treatment with group theory representation have been omitted, but the readers are encouraged to check the primary literature [11].

10.7.3 Three-Dimensional Equations of Generalized Dynamics of Point Group 18 mm Soft-Matter Quasicrystals

The generalized dynamics of soft-matter quasicrystals [1, 2] have been developed and based on Sect. 10.2. This defines the Hamiltonian as

$$
\left.
\begin{aligned}
H &= H[\Psi(r, t)] \\
&= \int \frac{g^2}{2\rho} d^d r + \int \left[\frac{1}{2} A \left(\frac{\delta\rho}{\rho_0} \right)^2 + B \left(\frac{\delta\rho}{\rho_0} \right) \nabla \cdot u \right] d^d r + F_{\mathrm{el}} \\
&= H_{\mathrm{kin}} + H_{\mathrm{density}} + F_{\mathrm{el}} \\
F_{\mathrm{el}} &= F_u + F_v + F_w + F_{uv} + F_{uw} + F_{vw}, \quad g = \rho V
\end{aligned}
\right\}
\tag{10.7.8}
$$

where $\mathbf{V}$ represents the fluid velocity field mentioned above, A, B are the constants describing mass density variation named as LRT constants. The last term of (10.7.8) represents the elastic energies, which consists of phonons, phasons, the phonon-phason coupling, and the phason-phason coupling parts, respectively:

$$
\left.
\begin{aligned}
F_u &= \int \frac{1}{2} C_{ijkl} \varepsilon_{ij} \varepsilon_{kl} d^d r \\
F_v &= \int \frac{1}{2} T_{ijkl} v_{ij} v_{kl} d^d r \\
F_w &= \int \frac{1}{2} K_{ijkl} w_{ij} w_{kl} d^d r \\
F_{uv} &= \int \left(r_{ijkl} \varepsilon_{ij} v_{kl} + r_{klij} v_{ij} \varepsilon_{kl} \right) d^d r \\
F_{uw} &= \int \left(R_{ijkl} \varepsilon_{ij} w_{kl} + R_{klij} w_{ij} \varepsilon_{kl} \right) d^d r \\
F_{vw} &= \int \left(G_{ijkl} v_{ij} w_{kl} + G_{klij} w_{ij} v_{kl} \right) d^d r
\end{aligned}
\right\}
\tag{10.7.9}
$$

The fundamental laws of the mass conservation

$$
\frac{\partial \rho}{\partial t} + \nabla_k (\rho V_k) = 0
\tag{10.7.10}
$$

The momentum of conservation or the generalized Navier–Stokes equation is

$$\left.\begin{aligned}
\frac{\partial g_i(r,t)}{\partial t} &= -\nabla_k(r)(V_k g_i) + \nabla_j(r)\left(-p\delta_{ij} + \eta_{ijkl}\nabla_k(r)V_l\right) \\
&\quad -\left(\delta_{ij} - \nabla_i u_j\right)\frac{\delta H}{\delta u_j(r,t)} + \left(\nabla_i v_j\right)\frac{\delta H}{\delta v_j(r,t)} + \left(\nabla_i w_j\right)\frac{\delta H}{\delta w_j(r,t)} \\
&\quad -\rho\nabla_i(r)\frac{\delta H}{\delta\rho(r,t)}, \quad g_j = \rho V_j
\end{aligned}\right\} \qquad (10.7.11)$$

The symmetry breaking rule on the motion of phonons is

$$\frac{\partial u_i(r,t)}{\partial t} = -V_j\nabla_j(r)u_i - \Gamma_u\frac{\delta H}{\delta u_i(r,t)} + V_i \qquad (10.7.12)$$

in which Γ_u represents phonon dissipation coefficient, and the symmetry breaking rules on the motion of the first phasons

$$\frac{\partial v_i(r,t)}{\partial t} = -V_j\nabla_j(r)v_i - \Gamma_v\frac{\delta H}{\delta v_i(r,t)} \qquad (10.7.13)$$

and the second phasons

$$\frac{\partial w_i(r,t)}{\partial t} = -V_j\nabla_j(r)w_i - \Gamma_w\frac{\delta H}{\delta w_i(r,t)} \qquad (10.7.14)$$

in which Γ_v and Γ_w represent the first and second phason dissipation coefficients. However, the equation set is not closed because the number of field variables is greater than the number of field equations. One must provide the equation of state which stipulates the relation between fluid pressure and mass density:

$$p = f(\rho) \qquad (10.7.15)$$

This is a complex topic in soft matter.

After some probes [1] the equation of state can be taken as (4.2.2) and (4.2.3) given in Chap. 4, where, ρ_0 is the initial value of the mass density, or the rest-mass density.

Substituting (10.7.9) into (10.7.8), then into (10.7.11)–(10.7.14) and finally combining (10.7.10) and (10.7.15) but by omitting the higher terms $\left(\nabla_i u_j\right)\frac{\delta H}{\delta u_j(r,t)}$, $\left(\nabla_i v_j\right)\frac{\delta H}{\delta v_j(r,t)}$ and $\left(\nabla_i w_j\right)\frac{\delta H}{\delta w_j(r,t)}$ in (10.7.11) we obtain the final governing equations of three-dimensional dynamics of 18-fold symmetric soft-matter quasicrystals as follows [12]

$$\frac{\partial\rho}{\partial t} + \nabla\cdot(\rho\mathbf{V}) = 0 \qquad (10.7.16a)$$

$$\frac{\partial(\rho V_x)}{\partial t} + \frac{\partial(V_x\rho V_x)}{\partial x} + \frac{\partial(V_y\rho V_x)}{\partial y} + \frac{\partial(V_z\rho V_x)}{\partial z} = -\frac{\partial p}{\partial x} + \eta\nabla^2 V_x$$

$$+ \frac{1}{3}\eta\frac{\partial}{\partial x}\nabla\cdot\mathbf{V} + \left(C_{11}\frac{\partial^2}{\partial x^2} + C_{66}\frac{\partial^2}{\partial y^2} + C_{44}\frac{\partial^2}{\partial z^2}\right)u_x$$

$$+ (C_{12} + C_{66})\frac{\partial^2 u_y}{\partial x\partial y} + (C_{13} + C_{44})\frac{\partial^2 u_z}{\partial x\partial z}$$

$$- B\frac{\partial}{\partial x}\nabla\cdot\mathbf{u} - (A - B)\frac{1}{\rho_0}\frac{\partial\delta\rho}{\partial x} \tag{10.7.16b}$$

$$\frac{\partial(\rho V_y)}{\partial t} + \frac{\partial(V_x\rho V_y)}{\partial x} + \frac{\partial(V_y\rho V_y)}{\partial y} + \frac{\partial(V_z\rho V_y)}{\partial z} = -\frac{\partial p}{\partial y} + \eta\nabla^2 V_y$$

$$+ \frac{1}{3}\eta\frac{\partial}{\partial y}\nabla\cdot\mathbf{V} + (C_{12} + C_{66})\frac{\partial^2 u_x}{\partial x\partial y} + \left(C_{66}\frac{\partial^2}{\partial x^2} + C_{11}\frac{\partial^2}{\partial y^2} + C_{44}\frac{\partial^2}{\partial z^2}\right)u_y$$

$$+ (C_{13} + C_{44})\frac{\partial^2 u_z}{\partial y\partial z} - B\frac{\partial}{\partial y}\nabla\cdot\mathbf{u} - (A - B)\frac{1}{\rho_0}\frac{\partial\delta\rho}{\partial y} \tag{10.7.16c}$$

$$\frac{\partial(\rho V_z)}{\partial t} + \frac{\partial(V_x\rho V_z)}{\partial x} + \frac{\partial(V_y\rho V_z)}{\partial y} + \frac{\partial(V_z\rho V_z)}{\partial z} = -\frac{\partial p}{\partial z} + \eta\nabla^2 V_z$$

$$+ \frac{1}{3}\eta\frac{\partial}{\partial z}\nabla\cdot\mathbf{V} + (C_{13} + C_{44})\left(\frac{\partial^2 u_x}{\partial x\partial z} + \frac{\partial^2 u_y}{\partial y\partial z}\right)$$

$$+ \left(C_{44}\frac{\partial^2}{\partial x^2} + C_{44}\frac{\partial^2}{\partial y^2} + C_{33}\frac{\partial^2}{\partial z^2}\right)u_z$$

$$- B\frac{\partial}{\partial z}\nabla\cdot\mathbf{u} - (A - B)\frac{1}{\rho_0}\frac{\partial\delta\rho}{\partial z} \tag{10.7.16d}$$

$$\frac{\partial u_x}{\partial t} + V_x\frac{\partial u_x}{\partial x} + V_y\frac{\partial u_x}{\partial y} + V_z\frac{\partial u_x}{\partial z} = V_x$$

$$+ \Gamma_{\mathbf{u}}\left[\left(C_{11}\frac{\partial^2}{\partial x^2} + C_{66}\frac{\partial^2}{\partial y^2} + C_{44}\frac{\partial^2}{\partial z^2}\right)u_x\right.$$

$$\left. + (C_{12} + C_{66})\frac{\partial^2 u_y}{\partial x\partial y} + (C_{13} + C_{44})\frac{\partial^2 u_z}{\partial x\partial z}\right] \tag{10.7.16e}$$

$$\frac{\partial u_y}{\partial t} + V_x\frac{\partial u_y}{\partial x} + V_y\frac{\partial u_y}{\partial y} + V_z\frac{\partial u_y}{\partial z} = V_y$$

$$+ \Gamma_{\mathbf{u}}\left[(C_{12} + C_{66})\frac{\partial^2 u_x}{\partial x\partial y} + \left(C_{66}\frac{\partial^2}{\partial x^2} + C_{11}\frac{\partial^2}{\partial y^2} + C_{44}\frac{\partial^2}{\partial z^2}\right)u_y\right.$$

$$\left. + (C_{13} + C_{44})\frac{\partial^2 u_z}{\partial y\partial z}\right] \tag{10.17.16f}$$

$$\frac{\partial u_z}{\partial t} + V_x\frac{\partial u_z}{\partial x} + V_y\frac{\partial u_z}{\partial y} + V_z\frac{\partial u_z}{\partial z} = V_z$$

$$+ \Gamma_{\mathbf{u}}\left[(C_{13} + C_{44})\left(\frac{\partial^2 u_x}{\partial x\partial z} + \frac{\partial^2 u_y}{\partial y\partial z}\right) + \left(C_{44}\frac{\partial^2}{\partial x^2} + C_{44}\frac{\partial^2}{\partial y^2} + C_{33}\frac{\partial^2}{\partial z^2}\right)u_z\right]$$

$$\tag{10.7.16g}$$

$$\frac{\partial v_x}{\partial t} + V_x \frac{\partial v_x}{\partial x} + V_y \frac{\partial v_x}{\partial y} + V_z \frac{\partial v_x}{\partial z} =$$

$$\Gamma_v \left[\left(T_1 \frac{\partial^2}{\partial x^2} + T_1 \frac{\partial^2}{\partial y^2} + T_3 \frac{\partial^2}{\partial z^2} \right) v_x + G \left(\frac{\partial^2}{\partial x^2} - \frac{\partial^2}{\partial y^2} \right) w_x - 2G \frac{\partial^2 w_y}{\partial x \partial y} \right]$$

$$(10.7.16h)$$

$$\frac{\partial v_y}{\partial t} + V_x \frac{\partial v_y}{\partial x} + V_y \frac{\partial v_y}{\partial y} + V_z \frac{\partial v_y}{\partial z} =$$

$$\Gamma_v \left[\left(T_1 \frac{\partial^2}{\partial x^2} + T_1 \frac{\partial^2}{\partial y^2} + T_3 \frac{\partial^2}{\partial z^2} \right) v_y + 2G \frac{\partial^2 w_x}{\partial x \partial y} + G \left(\frac{\partial^2}{\partial x^2} - \frac{\partial^2}{\partial y^2} \right) w_y \right]$$

$$(10.7.16i)$$

$$\frac{\partial w_x}{\partial t} + V_x \frac{\partial w_x}{\partial x} + V_y \frac{\partial w_x}{\partial y} + V_z \frac{\partial w_x}{\partial z} =$$

$$\Gamma_w \left[\left(K_1 \frac{\partial^2}{\partial x^2} + K_1 \frac{\partial^2}{\partial y^2} + K_3 \frac{\partial^2}{\partial z^2} \right) w_x + G \left(\frac{\partial^2}{\partial x^2} - \frac{\partial^2}{\partial y^2} \right) v_x + 2G \frac{\partial^2 v_y}{\partial x \partial y} \right]$$

$$(10.7.16j)$$

$$\frac{\partial w_y}{\partial t} + V_x \frac{\partial w_y}{\partial x} + V_y \frac{\partial w_y}{\partial y} + V_z \frac{\partial w_y}{\partial z} =$$

$$\Gamma_w \left[\left(K_1 \frac{\partial^2}{\partial x^2} + K_1 \frac{\partial^2}{\partial y^2} + K_3 \frac{\partial^2}{\partial z^2} \right) w_y - 2G \frac{\partial^2 v_x}{\partial x \partial y} + G \left(\frac{\partial^2}{\partial x^2} - \frac{\partial^2}{\partial y^2} \right) v_y \right]$$

$$(10.7.16k)$$

$$p = f(\rho) \qquad (10.7.16l)$$

These are in total twelve equations with twelve field variables. Equation (10.7.16a) is the mass conservation equation, (10.7.16b–d) are the momentum conservation equations or generalized Navier–Stokes equations, (10.7.16e–g) are the equations of motion of phonons due to the symmetry breaking, (10.7.16h, i) are the dissipation equations of the first phasons, (10.7.16j, k) are the dissipation equations of the second phasons, (10.7.16l) is the equation of state. The field variables are: $\rho, p, V_x, V_y, V_z, u_x, u_y, u_z, v_x, v_y, w_x, w_y$. The number of field variables and field equations needs to be equal for the governing equations to be consistent and to be mathematically solvable. In the absence of the equation of state, the equation set (10.7.16) would not be closed, and thus, it would not be able to be solved. Consequently, it would not hold any physical or mathematical meaning. This highlights the importance of the equation of state.

10.8 Incompressible Generalized Dynamics of 18-Fold Symmetric Soft-Matter Quasicrystals

The incompressible approximation treatment applies to the 18-fold symmetric quasicrystals. The following governing equations are in place

$$\nabla \cdot \mathbf{V} = 0 \tag{10.8.1a}$$

$$\rho\left(\frac{\partial V_x}{\partial t} + \frac{\partial(V_x V_x)}{\partial x} + \frac{\partial(V_y V_x)}{\partial y} + \frac{\partial(V_z V_x)}{\partial z}\right) = -\frac{\partial p}{\partial x} + \eta\nabla^2 V_x$$

$$+ \left(C_{11}\frac{\partial^2}{\partial x^2} + C_{66}\frac{\partial^2}{\partial y^2} + C_{44}\frac{\partial^2}{\partial z^2}\right)u_x + (C_{12} + C_{66})\frac{\partial^2 u_y}{\partial x \partial y}$$

$$+ (C_{13} + C_{44})\frac{\partial^2 u_z}{\partial x \partial z} - B\frac{\partial}{\partial x}\nabla \cdot \mathbf{u} \tag{10.8.1b}$$

$$\rho\left(\frac{\partial V_y}{\partial t} + \frac{\partial(V_x V_y)}{\partial x} + \frac{\partial(V_y V_y)}{\partial y} + \frac{\partial(V_z V_y)}{\partial z}\right) = -\frac{\partial p}{\partial y} + \eta\nabla^2 V_y$$

$$+ (C_{12} + C_{66})\frac{\partial^2 u_x}{\partial x \partial y} + \left(C_{66}\frac{\partial^2}{\partial x^2} + C_{11}\frac{\partial^2}{\partial y^2} + C_{44}\frac{\partial^2}{\partial z^2}\right)u_y$$

$$+ (C_{13} + C_{44})\frac{\partial^2 u_z}{\partial y \partial z} - B\frac{\partial}{\partial y}\nabla \cdot \mathbf{u} \tag{10.8.1c}$$

$$\rho\left(\frac{\partial V_z}{\partial t} + \frac{\partial(V_x V_z)}{\partial x} + \frac{\partial(V_y V_z)}{\partial y} + \frac{\partial(V_z V_z)}{\partial z}\right) = -\frac{\partial p}{\partial z} + \eta\nabla^2 V_z$$

$$+ (C_{13} + C_{44})\left(\frac{\partial^2 u_x}{\partial x \partial z} + \frac{\partial^2 u_y}{\partial y \partial z}\right) + \left(C_{44}\frac{\partial^2}{\partial x^2} + C_{44}\frac{\partial^2}{\partial y^2} + C_{33}\frac{\partial^2}{\partial z^2}\right)u_z$$

$$- B\frac{\partial}{\partial z}\nabla \cdot \mathbf{u} \tag{10.8.1d}$$

$$\frac{\partial u_x}{\partial t} + V_x\frac{\partial u_x}{\partial x} + V_y\frac{\partial u_x}{\partial y} + V_z\frac{\partial u_x}{\partial z} = V_x$$

$$+ \Gamma_{\mathbf{u}}\left[\left(C_{11}\frac{\partial^2}{\partial x^2} + C_{66}\frac{\partial^2}{\partial y^2} + C_{44}\frac{\partial^2}{\partial z^2}\right)u_x\right.$$

$$\left. + (C_{12} + C_{66})\frac{\partial^2 u_y}{\partial x \partial y} + (C_{13} + C_{44})\frac{\partial^2 u_z}{\partial x \partial z}\right] \tag{10.8.1.e}$$

$$\frac{\partial u_y}{\partial t} + V_x\frac{\partial u_y}{\partial x} + V_y\frac{\partial u_y}{\partial y} + V_z\frac{\partial u_y}{\partial z} = V_y$$

$$+ \Gamma_{\mathbf{u}}\left[(C_{12} + C_{66})\frac{\partial^2 u_x}{\partial x \partial y} + \left(C_{66}\frac{\partial^2}{\partial x^2} + C_{11}\frac{\partial^2}{\partial y^2} + C_{44}\frac{\partial^2}{\partial z^2}\right)u_y\right.$$

$$+(C_{13} + C_{44})\frac{\partial^2 u_z}{\partial y \partial z}\Bigg] \tag{10.8.1f}$$

$$\frac{\partial u_z}{\partial t} + V_x \frac{\partial u_z}{\partial x} + V_y \frac{\partial u_z}{\partial y} + V_z \frac{\partial u_z}{\partial z} = V_z$$

$$+ \Gamma_{\mathbf{u}}\Bigg[(C_{13} + C_{44})\left(\frac{\partial^2 u_x}{\partial x \partial z} + \frac{\partial^2 u_y}{\partial y \partial z} \right) + \left(C_{44}\frac{\partial^2}{\partial x^2} + C_{44}\frac{\partial^2}{\partial y^2} + C_{33}\frac{\partial^2}{\partial z^2} \right) u_z \Bigg] \tag{10.8.1g}$$

$$\frac{\partial v_x}{\partial t} + V_x \frac{\partial v_x}{\partial x} + V_y \frac{\partial v_x}{\partial y} + V_z \frac{\partial v_x}{\partial z} =$$

$$\Gamma_v\Bigg[\left(T_1\frac{\partial^2}{\partial x^2} + T_1\frac{\partial^2}{\partial y^2} + T_3\frac{\partial^2}{\partial z^2} \right) v_x + G\left(\frac{\partial^2}{\partial x^2} - \frac{\partial^2}{\partial y^2} \right) w_x - 2G\frac{\partial^2 w_y}{\partial x \partial y} \Bigg] \tag{10.8.1h}$$

$$\frac{\partial v_y}{\partial t} + V_x \frac{\partial v_y}{\partial x} + V_y \frac{\partial v_y}{\partial y} + V_z \frac{\partial v_y}{\partial z} =$$

$$\Gamma_v\Bigg[\left(T_1\frac{\partial^2}{\partial x^2} + T_1\frac{\partial^2}{\partial y^2} + T_3\frac{\partial^2}{\partial z^2} \right) v_y + 2G\frac{\partial^2 w_x}{\partial x \partial y} + G\left(\frac{\partial^2}{\partial x^2} - \frac{\partial^2}{\partial y^2} \right) w_y \Bigg] \tag{10.8.1i}$$

$$\frac{\partial w_x}{\partial t} + V_x \frac{\partial w_x}{\partial x} + V_y \frac{\partial w_x}{\partial y} + V_z \frac{\partial w_x}{\partial z} =$$

$$\Gamma_{\mathbf{w}}\Bigg[\left(K_1\frac{\partial^2}{\partial x^2} + K_1\frac{\partial^2}{\partial y^2} + K_3\frac{\partial^2}{\partial z^2} \right) w_x + G\left(\frac{\partial^2}{\partial x^2} - \frac{\partial^2}{\partial y^2} \right) v_x + 2G\frac{\partial^2 v_y}{\partial x \partial y} \Bigg] \tag{10.8.1j}$$

$$\frac{\partial w_y}{\partial t} + V_x \frac{\partial w_y}{\partial x} + V_y \frac{\partial w_y}{\partial y} + V_z \frac{\partial w_y}{\partial z} =$$

$$\Gamma_{\mathbf{w}}\Bigg[\left(K_1\frac{\partial^2}{\partial x^2} + K_1\frac{\partial^2}{\partial y^2} + K_3\frac{\partial^2}{\partial z^2} \right) w_y - 2G\frac{\partial^2 v_x}{\partial x \partial y} + G\left(\frac{\partial^2}{\partial x^2} - \frac{\partial^2}{\partial y^2} \right) v_y \Bigg] \tag{10.8.1k}$$

In these equations, the mass density is a constant, while the number of field variables is reduced to eleven. This is because the equation of state has been excluded.

10.9   Other Solutions and Applications

Besides dislocation and transient dynamics, other solutions have been carried out. For example, the flow of 18-fold symmetric soft-matter quasicrystals past an obstacle can be carried similarly as that in Sects. 7.6 and in Chap. 9. Due to the phonon-phason decoupling, the results only exhibit new features describing the first phason field and coupling between the first and second phason fields in addition to those summarized in Chap. 7, which partly can be reported in Chap. 11 on solution of soft-matter quasicrystals of 14-fold symmetry.

This chapter provided only a general overview of the 18-fold symmetric soft-matter quasicrystals. A complete description also touches on details from symmetry groups and group representation, which were not outlined here, but readers can check reference [11] for more details.

References

1. Fischer, S., Exner, A., Zielske, K., Perlich, J., Deloudi, S., Steuer, W., Linder, P., Foestor, S.: Colloidal quasicrystals with 12-fold and 18-fold symmetry. Proc. Nat. Acad. Sci. **108**, 1810–1814 (2011)
2. Janssen, T.: The symmetry operations for N-dimensional periodic and quasiperiodic structures. Z. Kristallogr. **198**, 17–32 (1992)
3. Hu, C.Z., Ding, D.H., Yang, W.G., Wang, R.H.: Possible two-dimensional quasicystal structure. Phys. Rev. B **49**, 9423–9427 (1994)
4. Fan, T.Y.: Equation system of generalized hydrodynamics of soft-matter quasicrystals. Appl. Math. Mech. **37**, 331–347 (in Chinese (2016)); arXiv:1908235[cond.Mat-soft] Oct 15, (2019)
5. Fan, T,Y,: Generalized hydrodynamics of soft-matter second kind of two-dimensional quasicrystals. Appl. Math. Mech. **38**, 189–199 (in Chinese (2017)); arXiv:1908240[cond.Mat-soft] Oct 15, (2019)
6. Li, X.F., Xie, L.Y., Fan, T.Y.: Elasticity and dislocations in quasicrystals with 18-fold symmetry. Phys. Lett. A **377**, 2810–2814 (2013)
7. Li, X.F., Fan, T.Y.: New method for solving plane elasticity of planar quasicrystals and solution. Chin. Phys. Lett. **1998**(15), 278–280 (1998)
8. Fan, T.Y.: Mathematical Theory of Elasticity of Quascrystals and Its Applications. Science Press, Beijing/Springer, Heidelberg, 1st edition (2010); 2nd edition (2016)
9. Li, X.F., Fan, T.Y.: Dislocations in the second kind two-dimensional quasicrystals of soft matter. Physica B **52**, 175–180 (2016)
10. Gao, H., Fan, T.Y., Cheng, H.: Importance of coupling between phonons and phasons in soft-matter quasicrystals (unpublished work) (2018)
11. Tang, Z.Y., Fan, T.Y.: Point groups and group representation theory of second kind of two-dimensional quasicrystals (unpublished work) (2017)
12. Tang, Z.Y., Fan, T.Y.: Three-dimensional equations of 18-fold symmetry quasicrystals of soft matter. Mod. Phys. Lett. B **34**, 2050109 (2020)

Chapter 11
The Possible 7-, 9-, and 14-fold Symmetry Quasicrystals in Soft Matter

The possible 7-, 9-, and 14-fold symmetry quasicrystals are similar to those of 18-fold symmetry, and belong to the second kind of two-dimensional quasicrystals, in which the possible 7- and 14-fold symmetry quasicrystals are more interesting because the phonons and second phasons are coupled apart from the coupling between the first and second phasons. In this chapter, some mathematical presentations on the generalized dynamics of the soft-matter quasicrystals are introduced, which is suggested by Fan [1, 2]. Due to a lack of any experimental data, the solutions are very limited, which are also discussed.

11.1 The Possible 7-fold Symmetry Quasicrystals with Point Group $7m$ of Soft Matter and the Dynamic Theory

In Chaps. 2 and 10 we pointed out that the possible 7-, 9-, and 14-fold symmetrical quasicrystals present similar symmetry to 18-fold symmetry quasicrystals, and are different from those of 5-, 8-, 10-, and 12-fold symmetry quasicrystals. For this reason, we must introduce the so-called six-dimensional embedding space that we have been introduced in Chap. 10 in detail, so the discussion here can be omitted.

According to the hypothesis of the six-dimensional embedding space, there is a phonon displacement field $\mathbf{u}$ in parallel space, and the first and second phason displacement fields $\mathbf{v}$ and $\mathbf{w}$ in the two perpendicular spaces, respectively.

The quasicrystals of 7-fold symmetry have not been observed that far yet, so there is no diffraction pattern of the matter like that was shown by Fig. 2.7. Hu et al. [3] predicted the existence of a 7-fold symmetry structure. They gave a point group $7m$ to describe the possible solid quasicrystals, but the solid quasicrystals have not been observed too. Although there is a lack of experimental results, the prediction of Hu et al. is meaningful. Based on the point group $7m$ we can discuss the phonon and phason constitutive laws. We should point out that except point group $7m$ there might be other point groups of 7-fold symmetry quasicrystals refer to Chap. 2, but the discussion here is only for point group $7m$ quasicrystals.

If we assume the z—direction is the direction of the 7-fold rotation axis, the displacements are $\boldsymbol{u} = (u_x, u_y)$, $\boldsymbol{v} = (v_x, v_y)$, $\boldsymbol{w} = (w_x, w_y)$, which are two-dimensional displacement fields. The corresponding strain fields are

$$\varepsilon_{ij} = \frac{1}{2}\left(\frac{\partial u_i}{\partial x_j} + \frac{\partial u_j}{\partial x_i}\right), v_{ij} = \frac{\partial v_i}{\partial x_j}, w_{ij} = \frac{\partial w_i}{\partial x_j}$$

Based on the group representation theory that there are two couplings: between phonons and second phasons, and first and second phasons. Accordingly, the corresponding free energy densities defined by

$$f_u = \frac{1}{2}L\varepsilon_{ii}\varepsilon_{ii} + M\varepsilon_{ij}\varepsilon_{ij}$$

$$f_v = \frac{1}{2}T_1 v_{ij} v_{ij} + T_2\left(v_{xx} v_{yy} - v_{xy} v_{xy}\right)$$

$$f_w = \frac{1}{2}K_1 w_{ij} w_{ij} + K_2\left(w_{xx} w_{yy} - w_{xy} w_{xy}\right)$$

$$f_{uw} = R\left[\left(\varepsilon_{xx} - \varepsilon_{yy}\right)\left(w_{xx} + w_{yy}\right) + 2\varepsilon_{xy}\left(w_{yx} - w_{xy}\right)\right]$$

$$f_{vw} = G[(v_{11} - v_{22})(w_{11} - w_{22}) + (v_{21} + v_{12})(w_{21} + w_{12})] \qquad (11.1.1)$$

the total deformation energy density

$$f_{\text{def}} = f_u + f_v + f_w + f_{uw} + f_{vw}$$

can be obtained. The generalized Hooke's law is defined by

$$\sigma_{ij} = \frac{\partial f_{\text{def}}}{\partial \varepsilon_{ij}} = C_{ijkl}\varepsilon_{kl} + r_{ijkl}v_{kl} + R_{ijkl}w_{kl}$$

$$\tau_{ij} = \frac{\partial f_{\text{def}}}{\partial v_{ij}} = T_{ijkl}v_{kl} + r_{klij}\varepsilon_{kl} + G_{ijkl}w_{kl}$$

$$H_{ij} = \frac{\partial f_{\text{def}}}{\partial w_{ij}} = K_{ijkl}w_{kl} + R_{klij}\varepsilon_{kl} + G_{klij}v_{kl}$$

In addition, we have the fluid constitutive law is

$$p_{ij} = -p\delta_{ij} + \sigma'_{ij} = -p\delta_{ij} + 2\eta\dot{\xi}_{ij} \tag{11.1.2}$$

where

$$\dot{\xi}_{ij} = \frac{1}{2}\left(\frac{\partial V_i}{\partial x_j} + \frac{\partial V_j}{\partial x_i}\right)$$

represents the fluid deformation rate tensor and η the viscous coefficient. Furthermore, the concrete version of the constitutive law

$$\left.\begin{aligned}
\sigma_{xx} &= (L + 2M)\varepsilon_{xx} + L\varepsilon_{yy} + R(w_{xx} + w_{yy}) \\
\sigma_{yy} &= L\varepsilon_{xx} + (L + 2M)\varepsilon_{yy} - R(w_{xx} + w_{yy}) \\
\sigma_{xy} &= \sigma_{yx} = 2M\varepsilon_{xy} + R(w_{yx} - w_{xy}) \\
\tau_{xx} &= T_1 v_{xx} + T_2 v_{yy} + G(w_{xx} - w_{yy}) \\
\tau_{yy} &= T_2 v_{xx} + T_1 v_{yy} - G(w_{xx} - w_{yy}) \\
\tau_{xy} &= T_1 v_{xy} - T_2 v_{yx} + G(w_{yx} + w_{xy}) \\
\tau_{yx} &= -T_2 v_{xy} + T_1 v_{yx} + G(w_{yx} + w_{xy}) \\
H_{xx} &= K_1 w_{xx} + K_2 w_{yy} + R(\varepsilon_{xx} - \varepsilon_{yy}) + G(v_{xx} - v_{yy}) \\
H_{yy} &= K_2 w_{xx} + K_1 w_{yy} + R(\varepsilon_{xx} - \varepsilon_{yy}) - G(v_{xx} - v_{yy}) \\
H_{xy} &= K_1 w_{xy} - K_2 w_{yx} - 2R\varepsilon_{xy} + G(v_{xy} + v_{yx}) \\
H_{yx} &= K_1 w_{yx} - K_2 w_{xy} + 2R\varepsilon_{xy} + G(v_{xy} + v_{yx}) \\
p_{xx} &= -p + 2\eta\left(\dot{\xi}_{xx} - \frac{1}{3}\dot{\xi}_{kk}\right) \\
p_{yy} &= -p + 2\eta\left(\dot{\xi}_{yy} - \frac{1}{3}\dot{\xi}_{kk}\right) \\
p_{xy} &= p_{yx} = 2\eta\dot{\xi}_{xy} \\
\dot{\xi}_{kk} &= \dot{\xi}_{xx} + \dot{\xi}_{yy}
\end{aligned}\right\} \tag{11.1.3}$$

According to a similar derivation in Chap. 10 we can obtain the equation system of generalized dynamics of possible 7-fold symmetry soft-matter quasicrystals as below [2]

$$\frac{\partial \rho}{\partial t} + \nabla \cdot (\rho \mathbf{V}) = 0$$

$$\frac{\partial(\rho V_x)}{\partial t} + \frac{\partial(V_x \rho V_x)}{\partial x} + \frac{\partial(V_y \rho V_x)}{\partial y} = -\frac{\partial p}{\partial x} + \eta \nabla^2 V_x + \frac{1}{3}\eta \frac{\partial}{\partial x} \nabla \cdot \mathbf{V} + M \nabla^2 u_x$$
$$+ (L + M - B)\frac{\partial}{\partial x}\nabla \cdot \mathbf{u} - (A - B)\frac{1}{\rho_0}\frac{\partial \delta \rho}{\partial x}$$

$$\frac{\partial(\rho V_y)}{\partial t} + \frac{\partial(V_x \rho V_y)}{\partial x} + \frac{\partial(V_y \rho V_y)}{\partial y} = -\frac{\partial p}{\partial y} + \eta \nabla^2 V_y + \frac{1}{3}\eta \frac{\partial}{\partial y} \nabla \cdot \mathbf{V} + M \nabla^2 u_y$$
$$+ (L + M - B)\frac{\partial}{\partial y}\nabla \cdot \mathbf{u} - (A - B)\frac{1}{\rho_0}\frac{\partial \delta \rho}{\partial y}$$

$$\frac{\partial u_x}{\partial t} + V_x \frac{\partial u_x}{\partial x} + V_y \frac{\partial u_x}{\partial y} = V_x + \Gamma_{\mathbf{u}}\left[M\nabla^2 u_x + (L + M)\frac{\partial}{\partial x}\nabla \cdot \mathbf{u}\right.$$
$$\left. + R\left(\frac{\partial^2 w_x}{\partial x^2} + 2\frac{\partial^2 w_y}{\partial x \partial y} - \frac{\partial w_x}{\partial y^2}\right)\right]$$

$$\frac{\partial u_y}{\partial t} + V_x \frac{\partial u_y}{\partial x} + V_y \frac{\partial u_y}{\partial y} = V_y + \Gamma_{\mathbf{u}}\left[M\nabla^2 u_y + (L + M)\frac{\partial}{\partial y}\nabla\rangle\mathbf{u}\right.$$
$$\left. + R\left(\frac{\partial^2 w_y}{\partial x^2} - 2\frac{\partial^2 w_x}{\partial x \partial y} - \frac{\partial^2 w_y}{\partial y^2}\right)\right]$$

$$\frac{\partial v_x}{\partial t} + V_x \frac{\partial v_x}{\partial x} + V_y \frac{\partial v_x}{\partial y} = \Gamma_v\left[T_1 \nabla^2 v_x + G\nabla^2 w_x\right]$$

$$\frac{\partial v_y}{\partial t} + V_x \frac{\partial v_y}{\partial x} + V_y \frac{\partial v_y}{\partial y} = \Gamma_v\left[T_1 \nabla^2 v_y + G\nabla^2 w_y\right]$$

$$\frac{\partial w_x}{\partial t} + V_x \frac{\partial w_x}{\partial x} + V_y \frac{\partial w_x}{\partial y} = \Gamma_{\mathbf{w}}\left[K_1 \nabla^2 w_x + R\left(\frac{\partial^2 u_x}{\partial x^2} - 2\frac{\partial^2 u_y}{\partial x \partial y} - \frac{\partial^2 u_x}{\partial y^2}\right) + G\nabla^2 v_x\right]$$

$$\frac{\partial w_y}{\partial t} + V_x \frac{\partial w_y}{\partial x} + V_y \frac{\partial w_y}{\partial y} = \Gamma_{\mathbf{w}}\left[K_1 \nabla^2 w_y + R\left(\frac{\partial^2 u_y}{\partial x^2} + 2\frac{\partial^2 u_x}{\partial x \partial y} - \frac{\partial^2 u_y}{\partial y^2}\right) + G\nabla^2 v_y\right]$$

$$p = f(\rho)$$

$$(11.1.4)$$

in which $\mathbf{V} = \mathbf{i}V_x + \mathbf{j}V_y$, $\mathbf{u} = \mathbf{i}u_x + \mathbf{j}u_y$, $\nabla = \mathbf{i}\frac{\partial}{\partial x} + \mathbf{j}\frac{\partial}{\partial y}$, $\nabla^2 = \frac{\partial^2}{\partial x^2} + \frac{\partial^2}{\partial y^2}$, and $L = C_{12}$, $M = (C_{11} - C_{12})/2$ are phonon elastic constants, T_1, K_1 the elastic constants of the first and second phasons, R, G the coupling elastic constants between the phonons and second phasons and between first and second phasons, η the fluid viscosity, Γ_u, Γ_v, and Γ_w the phonon, first phason and second phason dissipation coefficients, A and B the LRT constants, respectively.

Equations (11.1.4) consist of 10 field variables, i.e., phonon field $u = (u_x, u_y)$, first phason field $v = (v_x, v_y)$ and second phason field $w = (w_x, w_y)$, fluid velocity field $V = (V_x, V_y)$, mass density ρ, and fluid pressure p, respectively. The amount of the equations is also 10: (11.1.4a) is the mass conservation equation, (11.1.4b) and (11.1.4c) the momentum conservation equations or the generalized Navier–Stokes equations, (11.1.4d) and (11.1.4e) the equations of motion of phonons due to symmetry breaking, (11.1.4f) and (11.1.4g) the first phason dissipation equations, (11.1.4h) and (11.1.4i) the second phason dissipation equations, and (11.1.4j) the equation of state, respectively. If there is no equation of state, the equation system is not closed, and the importance of the equation is evident, for which one can refer to Chap. 4, where further verification by experiments must be done.

In the present case, Eqs. (11.1.4) are consistent mathematically and solvable.

The kind of quasicrystals with both couplings between phonons and second phasons and between first and second phasons is the very interesting ones and will be observed soon.

11.2 The Possible 9-fold Symmetrical Quasicrystals with Point Group *9m* of Soft Matter and Their Dynamics

The quasicrystals of 9-fold symmetry have not been observed yet, so no diffraction pattern of the matter like that was shown by Fig. 2.7. Hu et al. [3] predicted the existence of a 9-fold symmetry structure in solid. They gave a point group *9m* to describe the possible solid quasicrystals, but the solid quasicrystals have not been observed, either. Although there is a lack of experimental results, the prediction of Hu et al. is meaningful. We should point out that except point group *9m*, there might be other point groups of 9-fold symmetry quasicrystals (please refer to Chap. 2). But the discussion here is only on point group *9m* quasicrystals for simplicity purposes.

Based on the point group *9m* we can discuss the phonon and phason constitutive laws.

According to the theory of group representation, the possible 9-fold symmetrical quasicrystals of soft matter have similar constitutive equations to those of 18-fold symmetrical quasicrystals. The concrete results are as follows

$$
\begin{cases}
f_{\text{def}}(u, v, w) = f_{\text{def}}(\varepsilon_{ij}, v_{ij}, w_{ij}) = f_u + f_v + f_w + f_{vw} \\
f_u = \frac{1}{2}L(\nabla \cdot u)^2 + M\varepsilon_{ij}\varepsilon_{ij} \\
f_v = T_1\left[(v_{xx} + v_{yy})^2 + (v_{yx} - v_{xy})^2\right] + T_2\left[(v_{xx} - v_{yy})^2 + (v_{yx} + v_{xy})^2\right] \\
f_w = K_1\left[(w_{xx} + w_{yy})^2 + (w_{yx} - w_{xy})^2\right] + K_2\Big[(w_{xx} - w_{yy})^2 \\
\quad + (w_{yx} + w_{xy})^2\Big] \\
f_{vw} = G\left[(v_{xx} - v_{yy})(w_{xx} + w_{yy}) + (v_{yx} + v_{xy})(w_{yx} - w_{xy})\right] \\
(x = x_1, y = x_2, i = 1, 2, j = 1, 2)
\end{cases}
$$

$$(11.2.1)$$

for 9-fold symmetry quasicrystals with point group *9m*, and substituting the elastic deformation energy density (11.2.1) into the general relationship

$$
\sigma_{ij} = \frac{\partial f_{\text{def}}}{\partial \varepsilon_{ij}} = C_{ijkl}\varepsilon_{kl} + r_{ijkl}v_{kl} + R_{ijkl}w_{kl}
$$

$$
\tau_{ij} = \frac{\partial f_{\text{def}}}{\partial v_{ij}} = T_{ijkl}v_{kl} + r_{klij}\varepsilon_{kl} + G_{ijkl}w_{kl}
$$

$$
H_{ij} = \frac{\partial f_{\text{def}}}{\partial w_{ij}} = K_{ijkl}w_{kl} + R_{klij}\varepsilon_{kl} + G_{klij}v_{kl}
$$

one can obtain the elastic constitutive law

$$\left.\begin{aligned}
\sigma_{xx} &= (L + 2M)\varepsilon_{xx} + L\varepsilon_{yy} \\
\sigma_{yy} &= L\varepsilon_{xx} + (L + 2M)\varepsilon_{yy} \\
\sigma_{xy} &= \sigma_{yx} = 2M\varepsilon_{xy} \\
\tau_{xx} &= T_1 v_{xx} + T_2 v_{yy} + G(w_{xx} + w_{yy}) \\
\tau_{yy} &= T_2 v_{xx} + T_1 v_{yy} - G(w_{xx} + w_{yy}) \\
\tau_{xy} &= T_1 v_{xy} - T_2 v_{yx} + G(w_{yx} - w_{xy}) \\
\tau_{yx} &= -T_2 v_{xy} + T_1 v_{yx} + G(w_{yx} - w_{xy}) \\
H_{xx} &= K_1 w_{xx} + K_2 w_{yy} + G(v_{xx} - v_{yy}) \\
H_{yy} &= K_2 w_{xx} + K_1 w_{yy} + G(v_{xx} - v_{yy}) \\
H_{xy} &= K_1 w_{xy} - K_2 w_{yx} - G(v_{xy} + v_{yx}) \\
H_{yx} &= K_1 w_{yx} - K_2 w_{xy} + G(v_{xy} + v_{yx})
\end{aligned}\right\} \tag{11.2.2}$$

for 9-fold symmetry quasicrystals and collaborating the fluid constitutive law

$$\left.\begin{aligned}
p_{xx} &= -p + 2\eta\left(\dot{\xi}_{xx} - \tfrac{1}{3}\dot{\xi}_{kk}\right) \\
p_{yy} &= -p + 2\eta\left(\dot{\xi}_{yy} - \tfrac{1}{3}\dot{\xi}_{kk}\right) \\
p_{xy} &= p_{yx} = 2\eta\dot{\xi}_{xy} \\
\dot{\xi}_{kk} &= \dot{\xi}_{xx} + \dot{\xi}_{yy}
\end{aligned}\right\} \tag{11.2.3}$$

With these relations and by using (10.3.3)–(10.3.7) we can obtain the equations of motion and the equation of state for 9-fold symmetry quasicrystals in the soft matter as follows

$$\left.\begin{aligned}
&\frac{\partial\rho}{\partial t} + \nabla \cdot (\rho\mathbf{V}) = 0 \\
&\frac{\partial(\rho V_x)}{\partial t} + \frac{\partial(V_x\rho V_x)}{\partial x} + \frac{\partial(V_y\rho V_x)}{\partial y} = -\frac{\partial p}{\partial x} + \eta\nabla^2 V_x + \tfrac{1}{3}\eta\frac{\partial}{\partial x}\nabla \cdot \mathbf{V} + M\nabla^2 u_x \\
&\qquad\qquad + (L + M - B)\frac{\partial}{\partial x}\nabla \cdot \mathbf{u} - (A - B)\frac{1}{\rho_0}\frac{\partial\delta\rho}{\partial x} \\
&\frac{\partial(\rho V_y)}{\partial t} + \frac{\partial(V_x\rho V_y)}{\partial x} + \frac{\partial(V_y\rho V_y)}{\partial y} = -\frac{\partial p}{\partial y} + \eta\nabla^2 V_y + \tfrac{1}{3}\eta\frac{\partial}{\partial y}\nabla \cdot \mathbf{V} + M\nabla^2 u_y \\
&\qquad\qquad + (L + M - B)\frac{\partial}{\partial y}\nabla \cdot \mathbf{u} - (A - B)\frac{1}{\rho_0}\frac{\partial\delta\rho}{\partial y} \\
&\frac{\partial u_x}{\partial t} + V_x\frac{\partial u_x}{\partial x} + V_y\frac{\partial u_x}{\partial y} = V_x + \Gamma_{\mathbf{u}}\left[M\nabla^2 u_x + (L + M)\frac{\partial}{\partial x}\nabla \cdot \mathbf{u}\right] \\
&\frac{\partial u_y}{\partial t} + V_x\frac{\partial u_y}{\partial x} + V_y\frac{\partial u_y}{\partial y} = V_y + \Gamma_{\mathbf{u}}\left[M\nabla^2 u_y + (L + M)\frac{\partial}{\partial y}\nabla \cdot \mathbf{u}\right] \\
&\frac{\partial v_x}{\partial t} + V_x\frac{\partial v_x}{\partial x} + V_y\frac{\partial v_x}{\partial y} = \Gamma_v\left[T_1\nabla^2 v_x + G\left(\frac{\partial^2 w_x}{\partial x^2} - \frac{\partial^2 w_x}{\partial y^2}\right) + 2G\frac{\partial^2 w_y}{\partial x\partial y}\right] \\
&\frac{\partial v_y}{\partial t} + V_x\frac{\partial v_y}{\partial x} + V_y\frac{\partial v_y}{\partial y} = \Gamma_v\left[T_1\nabla^2 v_y + 2G\frac{\partial^2 w_x}{\partial x\partial y} + G\left(\frac{\partial^2 w_y}{\partial x^2} - \frac{\partial^2 w_y}{\partial y^2}\right)\right] \\
&\frac{\partial w_x}{\partial t} + V_x\frac{\partial w_x}{\partial x} + V_y\frac{\partial w_x}{\partial y} = \Gamma_{\mathbf{w}}\left[K_1\nabla^2 w_x + G\left(\frac{\partial^2 v_x}{\partial x^2} - \frac{\partial^2 v_x}{\partial y^2}\right) - 2G\frac{\partial^2 v_y}{\partial x\partial y}\right] \\
&\frac{\partial w_y}{\partial t} + V_x\frac{\partial w_y}{\partial x} + V_y\frac{\partial w_y}{\partial y} = \Gamma_{\mathbf{w}}\left[K_1\nabla^2 w_y - 2G\frac{\partial^2 v_x}{\partial x\partial y} + G\left(\frac{\partial^2 v_y}{\partial x^2} - \frac{\partial^2 v_y}{\partial y^2}\right)\right] \\
&p = f(\rho)
\end{aligned}\right\} \tag{11.2.4}$$

which are similar to those of (10.3.13), in which $\mathbf{V} = \mathbf{i}V_x + \mathbf{j}V_y$, $\mathbf{u} = \mathbf{i}u_x + \mathbf{j}u_y$,

$\nabla = \mathbf{i}\frac{\partial}{\partial x} + \mathbf{j}\frac{\partial}{\partial y}$, $\nabla^2 = \frac{\partial^2}{\partial x^2} + \frac{\partial^2}{\partial y^2}$, and $L = C_{12}$, $M = (C_{11} - C_{12})/2$ are phonon elastic constants, T_1, K_1 the elastic constants of the first and second phasons, $R = 0$, G the coupling elastic constant between first and second phasons, η the fluid viscosity, Γ_u, Γ_v, and Γ_w the phonon, first phason and second phason dissipation coefficients, A and B the material constants (named LRT constants) due to the variation of mass density, respectively.

Equations (11.2.4) consist of 10 field variables, i.e., phonon field $\boldsymbol{u} = (u_x, u_y)$, first phason field $\boldsymbol{v} = (v_x, v_y)$ and second phason field $\boldsymbol{w} = (w_x, w_y)$, fluid velocity field $V = (V_x, V_y)$, mass density ρ, and fluid pressure p, respectively. The amount of the equations is also 10: the first one is the mass conservation equation, the second and third ones the momentum conservation equations, i.e., the generalized Navier–Stokes equations, the fourth and fifth the equations of motion of phonons due to symmetry breaking, the sixth and seventh the first phason dissipation equations, the eighth and ninth ones the second phason dissipation equations, the tenth the equation of state, respectively. If there is no equation of state, the equation system is not closed, the importance of the equation is evident, which is just introduced by Chap. 4, further verification by experiments must be done.

In the present case, Eqs. (11.2.4) are consistent mathematically and solvable.

11.3 Dislocation Solutions of the Possible 9-fold Symmetrical Quasicrystals of Soft Matter

Here we first consider a special mathematical solution of Eqs. (11.2.4) in the static case and discuss dislocation in a 9-fold symmetry quasicrystal of soft matter. For a dislocation in the second kind of two-dimensional quasicrystals, the Burgers vector can be expressed by $\mathbf{b} = \mathbf{b}^{\parallel} \oplus \mathbf{b}_1^{\perp} \oplus \mathbf{b}_2^{\perp} = \left(b_1^{\parallel}, b_2^{\parallel}, b_{11}^{\perp}, b_{12}^{\perp}, b_{21}^{\perp}, b_{22}^{\perp}\right)$ where

$$\int_{\Gamma} du_j = b_j, \quad \int_{\Gamma} dv_j = b_{1j}^{\perp}, \quad \int_{\Gamma} dw_j = b_{2j}^{\perp}, \quad j = 1, 2. \tag{11.3.1}$$

In the above, the first two components are in the physical space or parallel space, and the last four components are in the perpendicular space, in which the first and second phason fields are existed, respectively.

The equation system (11.2.4) offers a basis and possibility to solve various solutions for the matter distribution, deformation, motion, and reconstruction of the new phase. As a simplified model, we first consider a static dislocation problem, i.e., the fluid effects can be omitted for the time being. In this case, the above equations reduce to

$$M\nabla^2 u_x + (L + M)\frac{\partial}{\partial x}\nabla \cdot \mathbf{u} = 0, \tag{11.3.2}$$

$$M\nabla^2 u_y + (L + M)\frac{\partial}{\partial y}\nabla \cdot \mathbf{u} = 0, \tag{11.3.3}$$

and

$$T_1\nabla^2 v_x + G\left(\frac{\partial^2 w_x}{\partial x^2} - \frac{\partial^2 w_x}{\partial y^2}\right) + 2G\frac{\partial^2 w_y}{\partial x \partial y} = 0, \tag{11.3.4}$$

$$T_1\nabla^2 v_y - 2G\frac{\partial^2 w_x}{\partial x \partial y} + G\left(\frac{\partial^2 w_y}{\partial x^2} - \frac{\partial^2 w_y}{\partial y^2}\right) = 0, \tag{11.3.5}$$

$$K_1\nabla^2 w_x + G\left(\frac{\partial^2 v_x}{\partial x^2} - \frac{\partial^2 v_x}{\partial y^2}\right) - 2G\frac{\partial^2 v_y}{\partial x \partial y} = 0, \tag{11.3.6}$$

$$K_1\nabla^2 w_y + 2G\frac{\partial^2 v_x}{\partial x \partial y} + G\left(\frac{\partial^2 v_y}{\partial x^2} - \frac{\partial^2 v_y}{\partial y^2}\right) = 0, \tag{11.3.7}$$

From the above, for the 9-fold symmetry quasicrystals, the displacement fields in the parallel space and perpendicular space are decoupled. For those in the parallel space, we can directly write the dislocation solution according to the classical one (the phonon one, i.e., the solution (7.3.1), which holds for the present case) and so, in what follows we neglect the part given by (7.3.1) and only solve the phason part.

To obtain the dislocation-induced phason field, we introduce two biharmonic functions f_1 and f_2 choose

$$v_x = -G\left(\frac{\partial^2 f_1}{\partial x^2} - \frac{\partial^2 f_1}{\partial y^2}\right) - 2G\frac{\partial^2 f_2}{\partial x \partial y}, \tag{11.3.8}$$

$$v_y = 2G\frac{\partial^2 f_1}{\partial x \partial y} - G\left(\frac{\partial^2 f_2}{\partial x^2} - \frac{\partial^2 f_2}{\partial y^2}\right), \tag{11.3.9}$$

$$w_x = T_1\nabla^2 f_1, \quad w_y = T_1\nabla^2 f_2, \tag{11.3.10}$$

where

$$\nabla^2\nabla^2 f_j = 0, \quad j = 1, 2. \tag{11.3.11}$$

It is easily found that the coupled partial differential equations are all automatically satisfied. Of course, it is mentioned that other general representations are given and omitted. Based on the above general representation, to fulfill the dislocation circuit condition, omitting concrete detail we readily obtain the dislocation phason displacement as follows [4]

$$v_x = \frac{b_{11}^{\perp}}{2\pi}arctan\frac{y}{x} + \frac{b_{21}^{\perp}G}{2\pi T_1}\frac{xy}{x^2 + y^2} - \frac{b_{22}^{\perp}G}{2\pi T_1}\frac{x^2 - y^2}{2(x^2 + y^2)}, \tag{11.3.12}$$

$$v_y = \frac{b_{12}^{\perp}}{2\pi}arctan\frac{y}{x} + \frac{b_{21}^{\perp}G}{2\pi T_1}\frac{x^2 - y^2}{2(x^2 + y^2)} + \frac{b_{22}^{\perp}G}{2\pi T_1}\frac{xy}{x^2 + y^2}, \tag{11.3.13}$$

$$w_x = \frac{b_{11}^{\perp}G}{2\pi K_1}\frac{xy}{x^2 + y^2} + \frac{b_{12}^{\perp}G}{2\pi K_1}\frac{x^2 - y^2}{2(x^2 + y^2)} + \frac{b_{21}^{\perp}}{2\pi}arctan\frac{y}{x}, \tag{11.3.14}$$

$$w_y = -\frac{b_{11}^{\perp}G}{2\pi K_1}\frac{x^2 - y^2}{2(x^2 + y^2)} + \frac{b_{12}^{\perp}G}{2\pi K_1}\frac{xy}{x^2 + y^2} + \frac{b_{22}^{\perp}}{2\pi}arctan\frac{y}{x}. \tag{11.3.15}$$

The above displacement fields have no logarithmic singularity near the dislocation core. Moreover, using the following constitutive equations

$$\tau_{xx} = T_1\frac{\partial v_x}{\partial x} + T_2\frac{\partial v_y}{\partial y} + G\left(\frac{\partial w_x}{\partial x} + \frac{\partial w_y}{\partial y}\right), \tag{11.3.16}$$

$$\tau_{yy} = T_2\frac{\partial v_x}{\partial x} + T_1\frac{\partial v_y}{\partial y} - G\left(\frac{\partial w_x}{\partial x} + \frac{\partial w_y}{\partial y}\right), \tag{11.3.17}$$

$$\tau_{xy} = T_1\frac{\partial v_y}{\partial x} - T_2\frac{\partial v_x}{\partial y} + G\left(\frac{\partial w_y}{\partial x} - \frac{\partial w_x}{\partial y}\right), \tag{11.3.18}$$

$$\tau_{yx} = -T_2\frac{\partial v_x}{\partial y} + T_1\frac{\partial v_y}{\partial x} + G\left(\frac{\partial w_y}{\partial x} - \frac{\partial w_x}{\partial y}\right), \tag{11.3.19}$$

$$H_{xx} = K_1\frac{\partial w_x}{\partial x} + K_2\frac{\partial w_y}{\partial y} + G\left(\frac{\partial v_x}{\partial x} - \frac{\partial v_y}{\partial y}\right), \tag{11.3.20}$$

$$H_{yy} = K_2\frac{\partial w_x}{\partial x} + K_1\frac{\partial w_y}{\partial y} + G\left(\frac{\partial v_x}{\partial x} - \frac{\partial v_y}{\partial y}\right), \tag{11.3.21}$$

$$H_{xy} = K_1\frac{\partial w_x}{\partial y} - K_2\frac{\partial w_y}{\partial x} - G\left(\frac{\partial v_x}{\partial y} + \frac{\partial v_y}{\partial x}\right), \tag{11.3.22}$$

$$H_{yx} = -K_2\frac{\partial w_x}{\partial y} + K_1\frac{\partial w_y}{\partial x} + G\left(\frac{\partial v_x}{\partial y} + \frac{\partial v_y}{\partial x}\right). \tag{11.3.23}$$

We get the corresponding stress field as follows

$$\tau_{xx} = -\frac{b_{11}^{\perp}(K_1T_1 + G^2)}{2\pi K_1}\frac{y}{x^2 + y^2} + \frac{b_{12}^{\perp}(K_1T_2 + G^2)}{2\pi K_1}\frac{x}{x^2 + y^2}$$
$$- \frac{b_{21}^{\perp}(T_1 + T_2)G}{\pi T_1}\frac{x^2 y}{(x^2 + y^2)^2} + \frac{b_{22}^{\perp}(T_1 + T_2)G}{2\pi T_1}\frac{x(x^2 - y^2)}{(x^2 + y^2)^2}, \tag{11.3.24}$$

$$\tau_{yy} = -\frac{b_{11}^{\perp}(K_1T_2 + G^2)}{2\pi K_1}\frac{y}{x^2 + y^2} + \frac{b_{12}^{\perp}(K_1T_1 - G^2)}{2\pi K_1}\frac{x}{x^2 + y^2}$$

$$- \frac{b_{21}^{\perp}(T_1 + T_2)G}{2\pi T_1}\frac{y(x^2 - y^2)}{(x^2 + y^2)^2} - \frac{b_{22}^{\perp}(T_1 + T_2)G}{\pi T_1}\frac{xy^2}{(x^2 + y^2)^2}, \qquad (11.3.25)$$

$$\tau_{xy} = \frac{b_{11}^{\perp}(K_1 T_1 - G^2)}{2\pi K_1}\frac{x}{x^2 + y^2} + \frac{b_{12}^{\perp}(K_1 T_2 + G^2)}{2\pi K_1}\frac{y}{x^2 + y^2}$$

$$- \frac{b_{21}^{\perp}(T_1 + T_2)G}{\pi T_1}\frac{xy^2}{(x^2 + y^2)^2} + \frac{b_{22}^{\perp}(T_1 + T_2)G}{2\pi T_1}\frac{y(x^2 - y^2)}{(x^2 + y^2)^2}, \qquad (11.3.26)$$

$$\tau_{yx} = -\frac{b_{11}^{\perp}(K_1 T_2 + G^2)}{2\pi K_1}\frac{x}{x^2 + y^2} - \frac{b_{12}^{\perp}(K_1 T_1 - G^2)}{2\pi K_1}\frac{y}{x^2 + y^2}$$

$$- \frac{b_{21}^{\perp}(T_1 + T_2)G}{2\pi T_1}\frac{x(x^2 - y^2)}{(x^2 + y^2)^2} - \frac{b_{22}^{\perp}(T_1 + T_2)G}{\pi T_1}\frac{x^2 y}{(x^2 + y^2)^2}, \qquad (11.3.27)$$

$$H_{yy} = -\frac{b_{11}^{\perp}(K_1 - K_2)G}{\pi K_1}\frac{x^2 y}{(x^2 + y^2)^2} - \frac{b_{12}^{\perp}(K_1 - K_2)G}{2\pi K_1}\frac{x(x^2 - y^2)}{(x^2 + y^2)^2}$$

$$- \frac{b_{21}^{\perp}(K_1 T_1 - G^2)}{2\pi T_1}\frac{y}{x^2 + y^2} + \frac{b_{22}^{\perp}(K_2 T_1 - G^2)}{2\pi T_1}\frac{x}{x^2 + y^2}, \qquad (11.3.28)$$

$$H_{yy} = \frac{b_{11}^{\perp}(K_1 - K_2)G}{2\pi K_1}\frac{y(x^2 - y^2)}{(x^2 + y^2)^2} - \frac{b_{12}^{\perp}(K_1 - K_2)G}{\pi K_1}\frac{xy^2}{(x^2 + y^2)^2}$$

$$- \frac{b_{21}^{\perp}(K_2 T_1 - G^2)}{2\pi T_1}\frac{y}{x^2 + y^2} + \frac{b_{22}^{\perp}(K_1 T_1 - G^2)}{2\pi T_1}\frac{x}{x^2 + y^2} \qquad (11.3.29)$$

$$H_{yx} = -\frac{b_{11}^{\perp}(K_1 - K_2)G}{\pi K_1}\frac{xy^2}{(x^2 + y^2)^2} - \frac{b_{12}^{\perp}(K_1 - K_2)G}{2\pi K_1}\frac{y(x^2 - y^2)}{(x^2 + y^2)^2}$$

$$+ \frac{b_{21}^{\perp}(K_1 T_1 - G^2)}{2\pi T_1}\frac{x}{x^2 + y^2} + \frac{b_{22}^{\perp}(K_2 T_1 - G^2)}{2\pi T_1}\frac{y}{x^2 + y^2} \qquad (11.3.30)$$

$$H_{yx} = \frac{b_{11}^{\perp}(K_1 - K_2)G}{2\pi K_1 r}\frac{x(x^2 - y^2)}{(x^2 + y^2)^2} - \frac{b_{12}^{\perp}(K_1 - K_2)G}{\pi K_1}\frac{x^2 y}{(x^2 + y^2)^2}$$

$$- \frac{b_{21}^{\perp}(K_2 T_1 - G^2)}{2\pi T_1}\frac{x}{x^2 + y^2} - \frac{b_{22}^{\perp}(K_1 T_1 - G^2)}{2\pi T_1}\frac{y}{x^2 + y^2}. \qquad (11.3.31)$$

It is easily checked that the above these stress fields obey equilibrium equations. Figure 11.1 shows the contour lines of the stress distribution induced by a component of the Burgers vector.

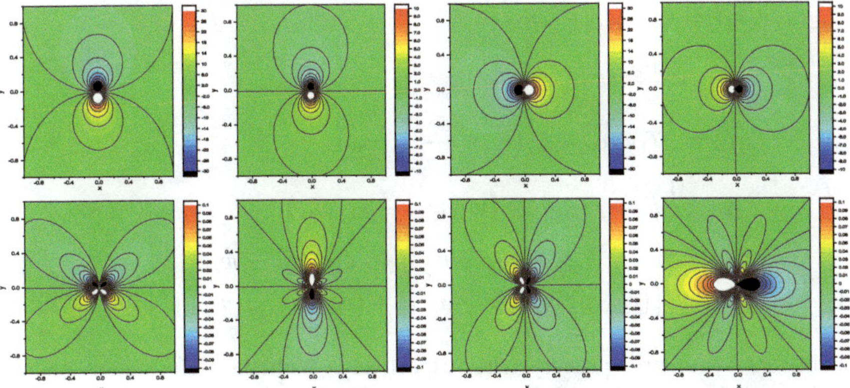

Fig. 11.1 Contour lines of the stress distribution $2\pi Y 10^{-6}/b_{11}^{\perp}$, Y : $\tau_{xx}, \tau_{yy}, \tau_{xy}, \tau_{yx}$, $H_{xx}, H_{yy}, H_{xy}, H_{yx}$, induced by a component $b_{11}^{\perp}$ of the Burgers vector, where the material properties [5] $K_1 = 5$ MPa, $K_2 = -1$ MPa, $T_1 = 4$ MPa, $T_2 = 1$ MPa, $G = 0.04$ MPa

11.4   The Possible 14-fold Symmetrical Quasicrystals with Point Group 14*mm* of Soft Matter and Their Dynamics

The quasicrystals of 14-fold symmetry have not been observed that far yet, so there is no diffraction pattern of the matter like that was shown by Fig. 2.7. Hu et al. [3] predicted the existence of a 14-fold symmetry structure. They gave a point group 14*mm* to describe the possible solid quasicrystals, but the solid quasicrystals have not been observed too. Although there is a lack of experimental results, the prediction of Hu et al. is meaningful. We should point out that apart from the point group 14*mm* there might be other point groups of 14-fold symmetry quasicrystals according to the group theory, refer to Chap. 2 or Ref. [6] but the discussion here is only for point group 14*mm* quasicrystals.

Based on the point group 14*mm* we can discuss the phonon and phason constitutive laws.

According to the theory of group representation, the possible 14-fold symmetrical quasicrystals of soft matter have similar constitutive equations to those of 7-fold symmetrical quasicrystals, but there are some distinctions. The elastic deformation energy density is

$$f_{\text{def}}(u, v, w) = f_{\text{def}}(\varepsilon_{ij}, v_{ij}, w_{ij}) = f_u + f_v + f_w + f_{uw} + f_{vw}$$

$$f_u = \frac{1}{2} L \varepsilon_{ii} \varepsilon_{ii} + M \varepsilon_{ij} \varepsilon_{ij}$$

$$f_v = \frac{1}{2} T_1 v_{ij} v_{ij} + T_2 \left(v_{xx} v_{yy} - v_{xy} v_{yx} \right)$$

$$f_w = \frac{1}{2} K_1 w_{ij} w_{ij} + K_2 \left(w_{xx} w_{yy} - w_{xy} w_{yx} \right)$$

$$f_{uw} = R\left[(\varepsilon_{xx} - \varepsilon_{yy})(w_{xx} + w_{yy}) + 2\varepsilon_{xy}(w_{yx} - w_{xy}) \right]$$

$$f_{vw} = G\left[(v_{xx} + v_{yy})(w_{xx} - w_{yy}) + (v_{yx} - v_{xy})(w_{yx} + w_{xy}) \right] \quad (11.4.1)$$

for 14-fold symmetry quasicrystals, respectively. With the deformation energy density and the definition of the generalized Hooke's law the concrete relation between stresses and strains can be obtained, and add the relations between fluid stresses and deformation rate components, we have

$$
\left.
\begin{aligned}
\sigma_{xx} &= (L + 2M)\varepsilon_{xx} + L\varepsilon_{yy} + R\left(w_{xx} + w_{yy}\right) \\
\sigma_{yy} &= L\varepsilon_{xx} + (L + 2M)\varepsilon_{yy} - R\left(w_{xx} + w_{yy}\right) \\
\sigma_{xy} &= \sigma_{yx} = 2M\varepsilon_{xy} + R\left(w_{yx} - w_{xy}\right) \\
\tau_{xx} &= T_1 v_{xx} + T_2 v_{yy} + G\left(w_{xx} - w_{yy}\right) \\
\tau_{yy} &= T_2 v_{xx} + T_1 v_{yy} + G\left(w_{xx} - w_{yy}\right) \\
\tau_{xy} &= T_1 v_{xy} - T_2 v_{yx} - G\left(w_{yx} + w_{xy}\right) \\
\tau_{yx} &= -T_2 v_{xy} + T_1 v_{yx} + G\left(w_{yx} + w_{xy}\right) \\
H_{xx} &= K_1 w_{xx} + K_2 w_{yy} + R\left(\varepsilon_{xx} - \varepsilon_{yy}\right) + G\left(v_{xx} + v_{yy}\right) \\
H_{yy} &= K_2 w_{xx} + K_1 w_{yy} + R\left(\varepsilon_{xx} - \varepsilon_{yy}\right) - G\left(v_{xx} + v_{yy}\right) \\
H_{xy} &= K_1 w_{xy} - K_2 w_{yx} - 2R\varepsilon_{xy} + G\left(v_{yx} - v_{xy}\right) \\
H_{yx} &= K_1 w_{yx} - K_2 w_{xy} + 2R\varepsilon_{xy} + G\left(v_{yx} - v_{xy}\right) \\
p_{xx} &= -p + 2\eta\left(\dot{\xi}_{xx} - \tfrac{1}{3}\dot{\xi}_{kk}\right) \\
p_{yy} &= -p + 2\eta\left(\dot{\xi}_{yy} - \tfrac{1}{3}\dot{\xi}_{kk}\right) \\
p_{xy} &= p_{yx} = 2\eta\dot{\xi}_{xy} \\
\dot{\xi}_{kk} &= \dot{\xi}_{xx} + \dot{\xi}_{yy}
\end{aligned}
\right\} \quad (11.4.2)
$$

Completely similar, utilizing the procedure of Sect. 11.1 the equations of motion 14-fold symmetry quasicrystals, in addition, plus the equation of state, we have the final governing equations as follows

$$
\left.
\begin{aligned}
&\frac{\partial \rho}{\partial t} + \nabla \cdot (\rho \mathbf{V}) = 0 \\
&\frac{\partial(\rho V_x)}{\partial t} + \frac{\partial(V_x \rho V_x)}{\partial x} + \frac{\partial(V_y \rho V_x)}{\partial y} = -\frac{\partial p}{\partial x} + \eta \nabla^2 V_x + \tfrac{1}{3}\eta\frac{\partial}{\partial x}\nabla \cdot \mathbf{V} + M\nabla^2 u_x \\
&\qquad\qquad +(L + M - B)\frac{\partial}{\partial x}\nabla \cdot \mathbf{u} - (A - B)\frac{1}{\rho_0}\frac{\partial \delta\rho}{\partial x} \\
&\frac{\partial(\rho V_y)}{\partial t} + \frac{\partial(V_x \rho V_y)}{\partial x} + \frac{\partial(V_y \rho V_y)}{\partial y} = -\frac{\partial p}{\partial y} + \eta \nabla^2 V_y + \tfrac{1}{3}\eta\frac{\partial}{\partial y}\nabla \cdot \mathbf{V} + M\nabla^2 u_y \\
&\qquad\qquad +(L + M - B)\frac{\partial}{\partial y}\nabla \cdot \mathbf{u} - (A - B)\frac{1}{\rho_0}\frac{\partial \delta\rho}{\partial y} \\
&\frac{\partial u_x}{\partial t} + V_x\frac{\partial u_x}{\partial x} + V_y\frac{\partial u_x}{\partial y} = V_x \\
&\qquad\qquad +\Gamma_{\mathbf{u}}\left[M\nabla^2 u_x + (L + M)\frac{\partial}{\partial x}\nabla \cdot \mathbf{u} \right. \\
&\qquad\qquad\qquad \left. +R\left(\frac{\partial^2 w_x}{\partial x^2} + 2\frac{\partial^2 w_y}{\partial x \partial y} - \frac{\partial w_x}{\partial y^2}\right) \right] \\
&\frac{\partial u_y}{\partial t} + V_x\frac{\partial u_y}{\partial x} + V_y\frac{\partial u_y}{\partial y} = V_y \\
&\qquad\qquad +\Gamma_{\mathbf{u}}\left[M\nabla^2 u_y + (L + M)\frac{\partial}{\partial y}\nabla \cdot \mathbf{u} \right. \\
&\qquad\qquad\qquad \left. +R\left(\frac{\partial^2 w_y}{\partial x^2} - 2\frac{\partial^2 w_x}{\partial x \partial y} - \frac{\partial^2 w_y}{\partial y^2}\right) \right] \\
&\frac{\partial v_x}{\partial t} + V_x\frac{\partial v_x}{\partial x} + V_y\frac{\partial v_x}{\partial y} = \Gamma_v\left[T_1\nabla^2 v_x + G\left(\frac{\partial^2 w_x}{\partial x^2} - \frac{\partial^2 w_x}{\partial y^2}\right) - 2G\frac{\partial^2 w_y}{\partial x \partial y}\right] \\
&\frac{\partial v_y}{\partial t} + V_x\frac{\partial v_y}{\partial x} + V_y\frac{\partial v_y}{\partial y} = \Gamma_v\left[T_1\nabla^2 v_y + 2G\frac{\partial^2 w_x}{\partial x \partial y} + G\left(\frac{\partial^2 w_y}{\partial x^2} - \frac{\partial^2 w_y}{\partial y^2}\right)\right] \\
&\frac{\partial w_x}{\partial t} + V_x\frac{\partial w_x}{\partial x} + V_y\frac{\partial w_x}{\partial y} = \Gamma_{\mathbf{w}}\left[K_1\nabla^2 w_x + R\left(\frac{\partial^2 u_x}{\partial x^2} - 2\frac{\partial^2 u_y}{\partial x \partial y} - \frac{\partial^2 u_x}{\partial y^2}\right) \right. \\
&\qquad\qquad\qquad \left. +G\left(\frac{\partial^2 v_x}{\partial x^2} - \frac{\partial^2 v_x}{\partial y^2}\right) + 2G\frac{\partial^2 v_y}{\partial x \partial y}\right] \\
&\frac{\partial w_y}{\partial t} + V_x\frac{\partial w_y}{\partial x} + V_y\frac{\partial w_y}{\partial y} = \Gamma_{\mathbf{w}}\left[K_1\nabla^2 w_y + R\left(\frac{\partial^2 u_y}{\partial x^2} + 2\frac{\partial^2 u_x}{\partial x \partial y} - \frac{\partial^2 u_y}{\partial y^2}\right) \right. \\
&\qquad\qquad\qquad \left. -2G\frac{\partial^2 v_x}{\partial x \partial y} - G\left(\frac{\partial^2 v_y}{\partial x^2} - \frac{\partial^2 v_y}{\partial y^2}\right)\right] \\
&p = f(\rho)
\end{aligned}
\right\}
$$

$$(11.4.3)$$

in which $\mathbf{V} = \mathbf{i}V_x + \mathbf{j}V_y$, $\mathbf{u} = \mathbf{i}u_x + \mathbf{j}u_y$, $\nabla = \mathbf{i}\frac{\partial}{\partial x} + \mathbf{j}\frac{\partial}{\partial y}$, $\nabla^2 = \frac{\partial^2}{\partial x^2} + \frac{\partial^2}{\partial y^2}$, and $L = C_{12}$, $M = (C_{11} - C_{12})/2$ are phonon elastic constants, T_1, K_1 the elastic constants of the first and second phasons, R, G the coupling elastic constants between the phonons and second phasons and between first and second phasons, η the fluid viscosity, Γ_u, Γ_v, and Γ_w the phonon, first phason and second phason dissipation coefficients, A and B the material constants called LRT constants due to the variation of mass density, respectively.

Equations (11.4.3) consist of 10 field variables, i.e., phonon field $\mathbf{u} = (u_x, u_y)$, first phason field $\mathbf{v} = (v_x, v_y)$ and second phason field $\mathbf{w} = (w_x, w_y)$, fluid velocity field $V = (V_x, V_y)$, mass density ρ, and fluid pressure p, respectively. The amount of the equations is also 10: the first one is the mass conservation equation, the second and third ones the momentum conservation equations, or named the generalized Navier–Stokes equations, the fourth and fifth the equations of motion of phonons due to symmetry breaking, the sixth and seventh the first phason dissipation equations, the eighth and ninth ones the second phason dissipation equations, the tenth the equation of state, respectively. If there is no equation of state, the equation system is not closed, the importance of the equation is evident, which is introduced in Chap. 4, the further verification by experiments must be done.

In the present case, Eqs. (12.4.3) are consistent mathematically and solvable.

11.5   The Numerical Solution of Dynamics of 14-fold Symmetrical Quasicrystals of Soft Matter

Equations (11.4.3) are a basis for the dynamics of soft-matter quasicrystals with 14-fold symmetry. If we want to obtain further information on the distribution, deformation, and motion of the material, these equations must be solved under appropriate initial and boundary conditions. For this purpose, a specimen, as shown by Fig. 11.2, is optioned which is subjected to the following initial and boundary conditions:

$$t = 0 : V_x = V_y = 0, u_x = u_y = 0, w_x = w_y = 0, \rho = \rho_0$$

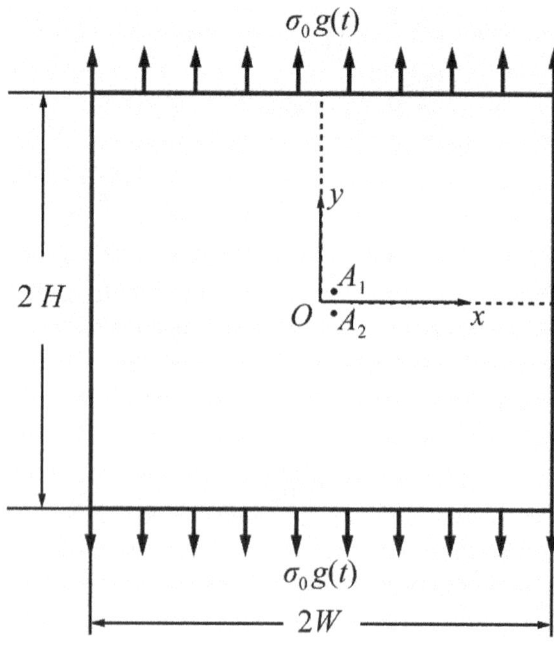

Fig. 11.2 Specimen of soft-matter quasicrystals of 14-fold symmetries under dynamic loading

$$y = \pm H, |x| \leq W, \sigma_{yy} = \sigma_0 g(t), \sigma_{yx} = 0, H_{yy} = H_{yx} = 0,$$
$$\tau_{yy} = \tau_{yx} = 0, V_x = V_y = 0;$$
$$x = \pm W, |y| \leq H, \sigma_{xx} = \sigma_{xy} = 0, H_{xx} = H_{xy} = 0,$$
$$\tau_{xx} = \tau_{xy} = 0, V_x = V_y = 0; \tag{11.5.1}$$

where $\sigma_0 g(t)$ represents a dynamic loading, as shown in Fig. 14.1, here $g(t)$ is taken as the Heaviside function of time and σ_0 the applied stress amplitude, respectively.

In the present computation, we take $2H = 0.01$ m, $2W = 0.01$ m, $\sigma_0 = 0.01$ MPa,

$\rho_0 = 1.5 \times 10^3 \text{kg/m}^3$, $\eta = 1$ Poise, $L = 10$ MPa, $M = 4$ MPa, $K_1 = 0.5L$, $K_2 = -0.1L$, $K_3 = 0.05L$, $R = 0.04 M$, $\Gamma_v = \Gamma_w = 4.8 \times 10^{-19} \text{m}^3 \cdot \text{s/kg}$, $\Gamma_u = 4.8 \times 10^{-17} \text{m}^3 \cdot \text{s/kg}$, $T_1 = T_2 = 0.5 L$, $G = 0.004 M$.

The problem of (11.4.3) and (14.5.1) is the initial-boundary value problem of nonlinear partial differential equations which is consistent mathematically, but the existence and uniqueness of the solution have not been proved yet due to the complexity of the problem. We can solve it by numerical method and the stability and correctness of the solution can be verified by the numerical results only.

Here we chose the finite difference method to solve the problem, and some results are given through the following illustrations, the computational point is A_1 $(10^{-4}\text{m}, 10^{-4}\text{m})$ (or A_2 $(10^{-4}\text{m}, -10^{-4}\text{m})$):

The results for the second kind of soft-matter quasicrystals are obtained for the first time, among them the most interesting ones are the first phason filed, which are never been observed before. Here we obtain the first phason displacements, in which Fig. 11.4 shows the time variation of displacement component v_y, which is compared with the value of the second phason displacement component w_y shown in Fig. 11.5. It is evident that the values of v_y are very smaller than those of w_y. The reason for this lies in the coupling between phonons $u = (u_x, u_y)$ and second phasons $w = (w_x, w_y)$, while there is decoupling between phonons $u = (u_x, u_y)$ and first phasons $v = (v_x, v_y)$, although there is coupling between $v = (v_x, v_y)$ and $w = (w_x, w_y)$. From Fig. 11.3 we can find that the phonon field is very strong than the second phason field, so that the coupling between $v = (v_x, v_y)$ and $w = (w_x, w_y)$, the influence of $w = (w_x, w_y)$ to $v = (v_x, v_y)$ is very weak.

The above results are given by Ref. [7].

11.6   Incompressible Complex Fluid Model

Similar to that in Chap. 10, we can suggest the incompressible model for soft-matter quasicrystals with 7-,9-, and 14-fold symmetries, in which the governing equations of the generalized dynamics can be obtained by simplifying the Eqs. (11.1.4), (11.2.4), and (11.4.3) respectively if taking $\rho = $ const and omitting the equation of state. These derivations are similar to those given in Chap. 10 and Chaps. 7–9 respectively. So the details are omitted here.

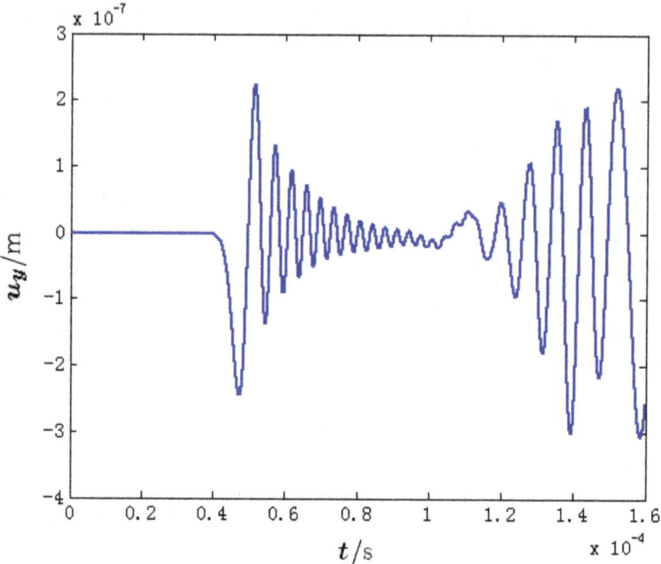

Fig. 11.3 Phonon displacement u_y at the point A_1 (or A_2) versus time

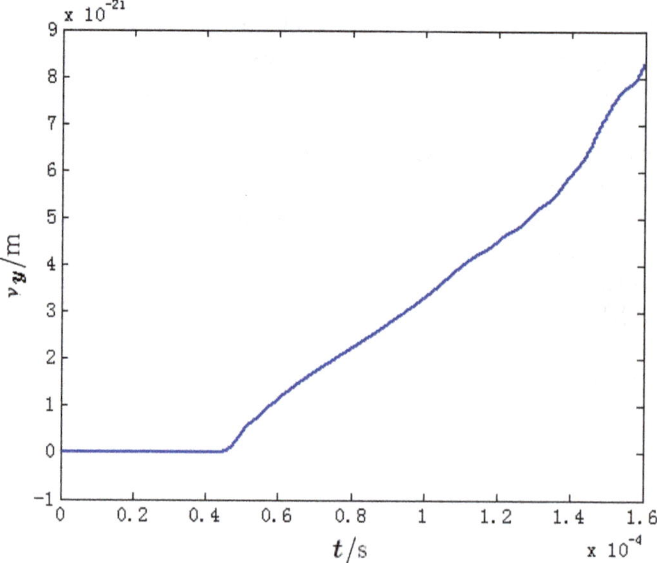

Fig. 11.4 First phason displacement v_y at the point A_1 (or A_2) versus time

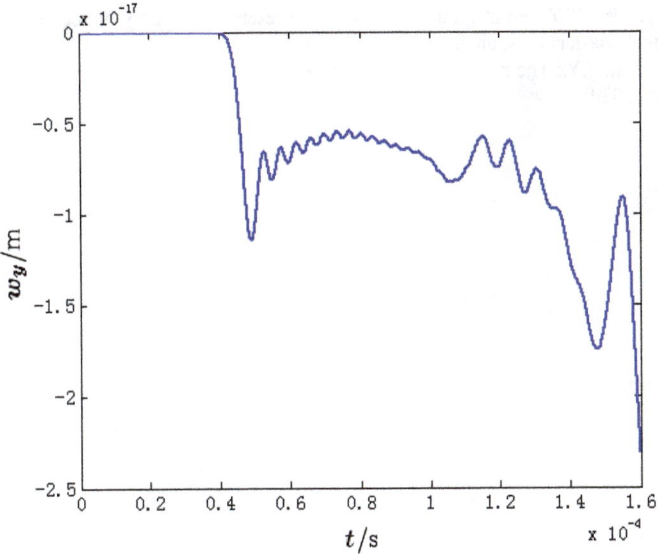

Fig. 11.5 Second phason displacement w_y at the point A_1 (or A_2) versus time

11.7 Conclusion and Discussion

This chapter gives a simplest introduction to the quasicrystals with 7-, 9-, and 14-fold symmetries in soft matter. The main part of the contents is concerned with the governing equations of these types of quasicrystals. The solutions have been reported only for the dislocation of the 9-fold symmetry and numerical analysis of 14-fold symmetry quasicrystals. These results may help us understand the structure and a part of properties of the second kind of soft-matter quasicrystals.

References

1. Fan, T.Y.: Equation system of generalized hydrodynamics of soft-matter quasicrystals. Appl. Math. Mech. **37**, 331–347 (2016), in Chinese (2016). arXiv:1908.06425 [cond-mat.soft], Oct 15, 2019
2. Fan, T.Y.: Generalized hydrodynamics of the second kind of two-dimensional quasicrystals in soft matter. Appl. Math. Mech. **38**, 189–199 (2017), in Chinese; arXiv:1908.06430 [cond-mat.soft], Oct 15, 2019
3. Hu, C.Z., Ding, D.H., Yang, W.G., Wang, R.H.: Possible two-dimensional quasicystal structure. Phys. Rev. B **49**, 9423–9427 (1994)
4. Li, X.F., Fan, T.Y.: Dislocations in the second kind two-dimensional quasicrystals of soft matter. Physica B **52**, 175–180 (2016)
5. Fan, T.Y.: Mathematical Theory and Relevant Topics of Solid and Soft-Matter Quasicrystals and Its Applications. Beijing Institute of Technology Press, Beijing, Chinese (2014)

6. Tang, Z.Y., Fan, T.Y.: Point groups and group representation theory of second kind of two-dimensional quasicrystals, unpublished work (2017)
7. Wang, F., Fan, T.Y.: The new results of second kind soft-matter quasicrystals of point group , submitted (2020)

Chapter 12
Re-Discussion on Symmetry Breaking and Elementary Excitations

In the first 11 chapters, to establish the generalized dynamics theory of soft-matter quasicrystals, we used the general concepts from the conservation laws and symmetry breaking principle. Based on that some applications have been successfully demonstrated in Chaps. 7–11 via solving the initial- or boundary- or initial and boundary-condition problems of the governing equations of the dynamics. These results validated our theory and explored the effects of elementary excitations phonons, phasons, fluid phonon, and their interactions. So far, we have not systematically introduced symmetry breaking and elementary excitations. Here we like to take a chance to introduce the Landau theory and these elementary excitations, which will be beneficial to the theoretical framework and applications to solid quasicrystals and soft-matter quasicrystals.

The symmetry breaking or broken symmetry concept has been widely used to describe the symmetry change, such as reducing the high symmetry to a lower one spontaneously triggered by certain thermodynamic state parameters. The phenomena can occur in nature, especially in solid-state physics.

In 1900, Planck [1] published his work on electromagnetic radiation of the black body, where he suggested that the energy can be quantized, and be used successfully to explain the experimental observations, which led to the establishment of the classical-quantum theory. Five years later, Einstein [2] developed this theory and proposed the photon concept as the quantized energy of the electromagnetic wave, from which he explained the photoelectric effect. The result can also be derived from the Maxwell equations [3]. Photon now becomes an essential elementary excitation in modern physics.

In 1912, Debye applied the Planck quantum theory to study the specific heat of crystals [4] (the detailed derivation refers to Sommerfeld [5]). Later in 1913, Born and von Kármán [6] further treated elastic waves in periodic structures and set up the phenomenological theory of lattice dynamics with great success in condensed matter physics. Their theoretical prediction is in excellent agreement with the experimental results and improved Einstein's theory of crystal-specific heat. In the Debye and Born theory framework, the propagation of the lattice vibration is called a lattice

© The Author(s), under exclusive license to SPRINGER Nature Singapore Pte Ltd. 2022
T.-Y. Fan et al., *Generalized Dynamics of Soft-Matter Quasicrystals*, Springer Series in Materials Science 260, https://doi.org/10.1007/978-981-16-6628-5_12

wave. As the long-wavelength, the lattice vibration can be seen approximately as continuum elastic vibration. Although the motion is mechanical, Debye and Born assumed the energy can be quantized based on Planck's hypothesis. The quantum of the elastic vibration or the smallest unit of energy of the elastic wave is named as phonon because the elastic wave is one of the acoustic waves. Similar to photon, phonon is another elementary excitation in modern physics, which belongs to the theory of the classical-quantum theory.

After the establishment of quantum mechanics, the quantization theory of fields was developed.

In the 1930s Landau [7] paid attention to study symmetry breaking on continuous phase transition and proposed the elementary excitation or quasiparticle concepts. Anderson [8] further developed the Landau theory. According to Lifshitz and Pitaevskii [9], the quasiparticles are quantum mechanics descriptions of the collective motion of massive atoms, and the quasiparticles could not be seen as conventional atoms or molecules. Due to the field quantization, the derivation of elementary excitations is more complicated than those for photon and phonon by the classical-quantum theory as Planck and Debye mentioned above. For the detail on the elementary excitation discussion, please refer to Chap. 4 of Kittel [10], where a simple derivation on the quantization of lattice wave field is given.

Regarding crystals, one assumes the solid has an infinite periodic ideal structure without boundary, while the Bloch theorem is key considering the periodic symmetry. For finite-sized periodic structures, the boundary effect needs to be considered. Recently, Ren [11] studied the electronic state of finite-sized periodic crystals. The existence of finite-sized boundaries leads to another symmetry breaking to the solid.

In quasicrystals, there is no periodic symmetry in the physical space, so the Bloch theorem for crystalline structure does not hold. Since the quasicrystals can be described as a high dimensional periodic structure, in that sense, we can inherit the phonon concept, but more elementary excitations are needed.

Bak [12, 13] and Lubensky et al. [14] for the first time proposed a new type of elementary excitation phason, in addition to the phonon to describe the lattice structure. In the 1970s, Bak studied the incommensurate phase, where the phason has been introduced already in the phase. In the Landau continuous phase transition theory, there is an order parameter. Anderson [8] suggested the order parameter as $\rho(\mathbf{r})$—the mass density, which can be expressed as the periodicity in lattice or reciprocal lattice in a higher dimensional periodic space

$$\rho(\mathbf{r}) = \sum_{\mathbf{G} \in L_R} \rho_{\mathbf{G}} \exp\{i\mathbf{G} \cdot \mathbf{r}\} = \sum_{\mathbf{G} \in L_R} |\rho_{\mathbf{G}}| \exp\{-i\Phi_{\mathbf{G}} + i\mathbf{G} \cdot \mathbf{r}\} \tag{12.1}$$

where $\mathbf{G}$ is a reciprocal vector, and L_R the reciprocal lattice (the concepts on the reciprocal vector and reciprocal lattice refer to [10]), $\rho_{\mathbf{G}}$ is a complex number

$$\rho_{\mathbf{G}} = |\rho_{\mathbf{G}}| e^{-i\Phi_{\mathbf{G}}} \tag{12.2}$$

with an amplitude $|\rho_\mathbf{G}|$ and phase angle $\Phi_\mathbf{G}$, due to $\rho(r)$ being real, $|\rho_\mathbf{G}|=|\rho_{-\mathbf{G}}|$ and $\Phi_\mathbf{G} = -\Phi_{-\mathbf{G}}$.

For quasicrystal, a convenient parametrization of the phase angle is given by

$$\Phi_n = \mathbf{G}_n^{\|} \cdot \mathbf{u} + \mathbf{G}_n^{\perp} \cdot \mathbf{w} \tag{12.3}$$

in which $\mathbf{u}$ represents wave propagation, similar to the phonon like that in conventional crystals; while $\mathbf{w}$ represents diffusion, a phason degree of freedom in quasicrystals, which describes the local rearrangement of unit cell description based on the Penrose tiling. Both are functions of the position vector in the physical space only, where $\mathbf{G}_n^{\|}$ is the reciprocal vector in the physical space $E_{\|}^3$ just mentioned and $\mathbf{G}_n^{\perp}$ is the conjugate vector in the perpendicular space $E_{\perp}^3$. People can realize that the above-mentioned Bak's hypothesis [12, 13] is a natural development of Anderson's theory.

The effect of phason excitation can be realized through the rearrangements of the local atoms in quasicrystals. The two-dimensional quasicrystals can be classified into two different kinds: the first and second kinds, in which the first one consists of 5-, 8-, 10-, and 12-fold symmetry quasicrystals, and the second one consists of 7-, 9-, 14-, and 18-fold symmetry quasicrystals, respectively. For the simplest first kind of two-dimensional quasicrystals, i.e., 12-fold symmetry quasicrystals, the phonons and phasons are decoupled to each other; for the 5-, 8-, and 10-fold symmetry two-dimensional quasicrystals, the phonons and phasons are coupled to each other (Check Ding et al. [15] and Hu et al. [16] for detail). For the second kind of two-dimensional quasicrystals, Hu et al. [17] pointed out there are two different types of phasons, they introduced the so-called six-dimensional embedding space, which consists of two-dimensional parallel space $E_{\|}^2$ and two two-dimensional perpendicular spaces $E_{\perp 1}^2$ and $E_{\perp 2}^2$, i.e.,

$$E^6 = E_{\|}^2 \oplus E_{\perp 1}^2 \oplus E_{\perp 2}^2 \tag{12.4}$$

where $E_{\perp 1}^2$ as the first perpendicular space, $E_{\perp 2}^2$ the second one. Based on the concept, the Landau-Anderson expansion (12.1) can be used still but the extended phase angular (12.3) can be extended as

$$\Phi_n = \mathbf{G}_n^{\|} \cdot \mathbf{u} + \mathbf{G}_n^{\perp 1} \cdot \mathbf{v} + \mathbf{G}_n^{\perp 2} \cdot \mathbf{w} \tag{12.5}$$

where $\mathbf{G}_n^{\|}$ represents reciprocal lattice vector in parallel space $E_{\|}^2$, and $\mathbf{G}_n^{\perp 1}$, and $\mathbf{G}_n^{\perp 2}$ the reciprocal lattice vectors in the first and second perpendicular spaces $E_{\perp 1}^2$ and $E_{\perp 2}^2$, $\mathbf{u}$ the phonon displacement field in parallel space, and $\mathbf{v}$, and $\mathbf{w}$ the first and second phason displacement fields in the two perpendicular spaces, respectively.

For the 9- and 18-fold symmetry quasicrystals, the phonons, and phasons are decoupled to each other, but the first and second phasons are coupled. For the other second kind of two-dimensional quasicrystals, i.e., 7- and 14-fold symmetry quasicrystals cases, the phonons and the second phasons are coupled and the first and second phasons are coupled to each other too, refer to Hu et al. [17].

In Chaps. 7–11 we find that the coupling or decoupling between phonons and phasons influence strongly the motion of the matter.

For soft-matter quasicrystals, apart from phonon and phason elementary excitations, there is another elementary excitation, the fluid phonon, firstly introduced in liquid [9]. Fan for the first time introduced it to quasicrystal study [18]. In the applications described in Chaps. 7–11, the fluid phonon term played a great role in soft-matter quasicrystals. The solutions of the fluid phonon field in soft-matter quasicrystal dynamics study have arrived in an excellent agreement with those of the classical fluid dynamics, which demonstrates the importance of the fluid phonon elementary excitation in soft-matter quasicrystals.

The successful introduction of symmetry breaking and elementary excitations into quasicrystals were largely initiated by the pioneers Bak, Lubensky, etc., following the Landau and Anderson theory [7, 8, 19]. We expect more progress with the breaking symmetry in the quasicrystal field.

References

1. Planck, M.: Ueber irreversible Strahlungsvorgaenge. Ann. Phys. **1**, 89–96 (1900)
2. Einstein, A.: Ueber einen die Erzeugung und Verwandlung des Lichits betreffenden heuristschen Gesichtspunkt. Ann. Phys. **17**, 132–148 (1905)
3. Wang, Z.X.: Statistical Physics. Peking University Press, Beijing, Chinese (2010)
4. Debye, P.: Die Eigentuemlichkeit der spezifischen Waermen bei tiefen Temperaturen. Arch de Genéve **33**(4), 256–258 (1912)
5. Sommerfeld, A.: Vorlesungen ueber theoretische Physik, Band II, Mechanik der deformierbaren Medium. Verlag Harri Deutsch, Thun. Frankfurt/M (1992)
6. Born, M., von Kármán, Th.: Zur Theorie der spezifischen Waermen. Physikalische Zeitschrift **14**(1), 15–19 (1913); Born, M., Huang, K.: Dynamic Theory of Crystal Lattices. Clarendon Press, Oxford (1954)
7. ter Harr, D. (ed.): Collected Papers of L D Landau. Pergamon Press, New York (1965)
8. Anderson, P.W.: Basic Notations of Condensed Matter Physics. Benjamin-Cummings, Menlo-Park (1984)
9. Lifshitz, M.E., Pitaevskii, L.: Statistical Physics, Part 2. Pergamon, Oxford (1980)
10. Kittel, C.: Introduction to Solid State Physics, 5th edn. John, New York (1976)
11. Ren, S.Y.: Electronic State in Crystals of Finite Size. Springer, New York (2017)
12. Bak, P.: Phenomenological theory of icosahedral incommensurate ("quaisiperiodic") order in Mn-Al alloys. Phys. Rev. Lett. **54**, 1517–1519 (1985)
13. Bak, P.: Symmetry, stability and elastic properties of icosahedral incommensurate crystals. Phys. Rev. B **32**(9), 5764–5772 (1985)
14. Lubensky, T.C., Ramaswamy, S., Toner, J.: Hydrodynamics of icosahedral quasicrystals. Phys. Rev. B **32**(11), 7444–7452 (1985)
15. Ding, D.H., Yang, W.G., Hu, C.Z., et al.: Generalized elasticity theory of quasicrystals. Phys. Rev. B **48**(10), 7003–7010 (1993)
16. Hu, C.Z., Wang, R.H., Ding, D.H.: Symmetry groups, physical property tensors, elasticity and dislocations in quasicrystals. Rep. Prog. Phys. **63**(1), 1–39 (2000)
17. Hu, C.Z., Ding, D.H., Yang, W.G., Wang, R.H.: Possible two-dimensional quasicystal structure. Phys. Rev. B **49**, 9423–9427 (1994)

18. Fan, T.Y.: Equation system of generalized hydrodynamics of soft-matter quasicrystals. Appl. Math. Mech. **37**, 331–347. in Chinese (2016); arXiv:1908.06425 [cond-mat.soft], Oct 15 (2019)
19. Anderson, P.W.: More is different—broken symmetry and the nature of the hierarchical structure of science. Science **177**, 393–396 (1972)

Chapter 13
An Application to the Thermodynamic Stability of Soft-Matter Quasicrystals

In Chaps. 7–11 we discussed several quasicrystal systems in soft matter, in which these quasicrystals must be stable thermodynamically, but the validity of this stability is held under certain conditions.

The present chapter presents a study on the thermodynamic stability of soft-matter quasicrystals from the perspective of thermodynamics and the matter structure. The results are quantitative, depending only upon the material constants of the novel phase, which can be measured easily by experiments. This style of research is quite different from Ref. [1] as well as the subsequent works, conducted by the group that may be the leading school in the field of stability of soft-matter quasicrystals.

13.1 Introduction

Soft-matter quasicrystals were first observed over 16 years ago in liquid crystals, polymers, colloids, nanoparticles, and surfactants, etc. Among them, the most frequently observed are the 12-fold symmetry ones, which have become the most important soft-matter quasicrystals up to now. It is noted that the 18-fold symmetry soft-matter quasicrystals were found in colloids only. More recently, tenfold symmetry quasicrystals were also observed, but have not been reported yet.

We mentioned previously that soft-matter quasicrystals are formed through the self-assembly of spherical building blocks by supramolecules, compounds, block copolymers, etc. The self-assembly is associated with a chemical process and is quite different from that of solid quasicrystals (i.e., the binary and ternary metallic alloy quasicrystals), which are formed under rapid cooling. These two thermodynamic environments are completely different from each other. The stability of the soft-matter quasicrystals was discussed by Lifshitz and Diamant [1] in 2007, soon after the discovery of the new phase. They pointed out that because of the quite different formation mechanisms of soft-matter quasicrystals from the solid ones, the source of stability of soft-matter quasicrystals remains a question of great debate to this

T.-Y. Fan et al., *Generalized Dynamics of Soft-Matter Quasicrystals*, Springer Series in Materials Science 260, https://doi.org/10.1007/978-981-16-6628-5_13

day. Up to now, the topic of stability of soft-matter quasicrystals has attracted a lot of interest from researchers [2–6]. They gave various treatments, showing the problem is complicated. Lifshitz and Diamant suggested studying the problem using the effective free energy proposed by Lifshitz and Petrich [7] which is connected with the mass density ρ such as

$$E_{\text{eff}}[\rho] = \int F(\rho)\mathrm{d}x\mathrm{d}y \qquad (13.1.1)$$

in which $F(\rho)$ denotes a function of mass density ρ, the detail was given by Lifshitz and Diamant; they and their co-workers did many analyses on the stability of 12-fold symmetry soft-matter. The other discussions on the stability in Refs. [1–6] concern the two natural length scales and two wavenumbers, which are the continuation and development of Lifshitz's pioneering work.

Different from the effective free energy approach in studying the stability suggested by Lifshitz and other physicists and mathematicians, we would like to give a probe by an alternative version. The first author and his group have worked in the generalized dynamics of the soft-matter quasicrystals over the years and found that some results of the dynamics are collected with thermodynamics. This simplifies the discussion on stability and is easy to find some quantitative results. For this purpose, we suggest extended free energy of the quasicrystal system in the soft matter which is the key of the following discussion.

13.2   Extended Free Energy of the Quasicrystal System in Soft Matter

In Sects. 13.2, 13.3, 13.4, 13.5 and 13.6, the discussion is limited to the first kind of two-dimensional quasicrystals.

The soft-matter quasicrystal is a complex viscous and compressible fluid with quasiperiodic symmetry, consisting of elementary excitations phonon displacement field $\mathbf{u}(u_x, u_y, u_z)$, phason displacement field $\mathbf{w}(w_x, w_y, w_z)$, and fluid phonon velocity field $\mathbf{V}(V_x, V_y, V_z)$. Accordingly

$$\varepsilon_{ij} = \frac{1}{2}\left(\frac{\partial u_i}{\partial x_j} + \frac{\partial u_j}{\partial x_i}\right), w_{ij} = \frac{\partial w_i}{\partial x_j}, \dot{\xi}_{ij} = \frac{1}{2}\left(\frac{\partial V_i}{\partial x_j} + \frac{\partial V_j}{\partial x_i}\right) \qquad (13.2.1)$$

are tensors of phonon strain, phason strain, and fluid deformation rate, respectively. Then, we define the extended internal energy density

$$U_{ex} = \frac{1}{2}A\left(\frac{\delta\rho}{\rho_0}\right)^2 + B\left(\frac{\delta\rho}{\rho_0}\right)\nabla \cdot \mathbf{u} + C\left(\frac{\delta\rho}{\rho_0}\right)\nabla \cdot \mathbf{w} + U_{el} \qquad (*)$$

where the first term denotes energy density due to the mass density variation, with the quantity $\frac{\delta\rho}{\rho_0}$ describing the variation of the mass density, in which $\delta\rho = \rho - \rho_0$ and ρ_0 the initial mass density. According to our computation in the motion of transient response and so on of soft-matter quasicrystals, $\frac{\delta\rho}{\rho_0} = 10^{-7} \sim 10^{-5}$ for the matter, which describes the fluid effect of the matter. It is 6–8 orders of magnitude greater than that of solid quasicrystals, for the latter its effect is very weak and hence not so important. The second term in formula (*) is the one caused by the coupling between the mass density variation and the phonons; the third represents the coupling between the mass density variation and the phasons. $A, B,$ and C are the Lubensky-Ramaswamy-Toner (or LRT) constants, respectively. Due to the extremely smaller value of $\nabla \cdot \mathbf{w}$ than that of $\nabla \cdot \mathbf{u}$ based on our computation, the term $C\left(\frac{\delta\rho}{\rho_0}\right)\nabla \cdot \mathbf{w}$ can be omitted hereafter. Thus, equation (*) is reduced to

$$U_{ex} = \frac{1}{2}A\left(\frac{\delta\rho}{\rho_0}\right)^2 + B\left(\frac{\delta\rho}{\rho_0}\right)\nabla \cdot \mathbf{u} + U_{el} \tag{13.2.2}$$

in which the third term in (13.2.2) represents the elastic internal energy of the matter.

We should point out energy density of mass density variation, mass density variation coupling phonons, and phasons were suggested by Lubensky et al. [8] for the first time from the hydrodynamics of solid quasicrystals. The present generalized dynamics of soft-matter quasicrystals is developed based on the theory of Lubensky et al. In particular, the Hamiltonian

$$H = H[\Psi(r, t)] = \int \frac{\mathbf{g}^2}{2\rho}d^d r + \int \left[\frac{1}{2}A\left(\frac{\delta\rho}{\rho_0}\right)^2 + B\left(\frac{\delta\rho}{\rho_0}\right)\nabla \cdot \mathbf{u}\right]d^d r + F_{el}$$

$$= H_{kin} + H_\rho + F_{el}$$

$$F_{el} = F_{\mathbf{u}} + F_{\mathbf{w}} + F_{\mathbf{uw}}, \mathbf{g} = \rho\mathbf{V}$$

$$\tag{13.2.3}$$

is drawn from Lubensky et al. [8], in which H_{kin} denotes the kinetic energy, H_ρ the energy due to the variation of mass density, F_{el} the elastic deformation energy consisting of contributions from phonons, phasons, and phonon-phason coupling, respectively. The detailed definitions will be given by (13.2.4)–(13.2.6) in the following. However, the dynamics equations of soft-matter quasicrystals are different from those of hydrodynamics of solid quasicrystals, and some results of Lubensky equations originated from [8] are not the same as those in soft-matter quasicrystals. For example, the equation (A2) in [8]

$$\frac{\delta\rho}{\rho_0} = -\frac{B}{A}\nabla \cdot \mathbf{u}$$

may not be held for the dynamics of soft-matter quasicrystals, etc. In addition, in
Eq. (13.2.2)

$$U_{el} = U_{\mathbf{u}} + U_{\mathbf{w}} + U_{\mathbf{uw}} \tag{13.2.4}$$

represents the elastic free energy density coming from phonons, phasons, and
phonon-phason coupling as

$$
\begin{aligned}
U_{\mathbf{u}} &= \frac{1}{2} C_{ijkl}\varepsilon_{ij}\varepsilon_{kl} \\
U_{\mathbf{w}} &= \frac{1}{2} K_{ijkl}w_{ij}w_{kl} \\
U_{\mathbf{uw}} &= R_{ijkl}\varepsilon_{ij}w_{kl} + R_{klij}\varepsilon_{ij}w_{ij}\varepsilon_{kl}
\end{aligned}
\tag{13.2.5}
$$

according to the constitutive law of soft-matter quasicrystals (please refer to (5.6.8))
in which

$$
\left.
\begin{aligned}
\sigma_{ij} &= C_{ijkl}\varepsilon_{ik} + R_{ijkl}w_{kl}, \\
H_{ij} &= K_{ijkl}w_{ij} + R_{klij}\varepsilon_{kl}, \\
p_{ij} &= -p\delta_{ij} + \sigma'_{ij} = -p\delta_{ij} + \eta_{ijkl}\dot{\xi}_{kl},
\end{aligned}
\right\}
\tag{13.2.6}
$$

where C_{ijkl} denotes the phonon elastic constants, K_{ijkl} the phason elastic constants,
and R_{ijkl}, R_{klij} the phonon-phason coupling elastic constants, and η_{ijkl} the fluid
viscous constants, respectively.

According to thermodynamics, the extended free energy density is defined by

$$F_{ex} = U_{ex} - TS \tag{13.2.7}$$

where U_{ex} the extended internal energy density defined by (13.2.2), and T the
absolute temperature and S the entropy, respectively.

By defining the vector of an extended strain

$$\varepsilon^{\text{ext}} = \left(\frac{\delta\rho}{\rho_0}, \varepsilon_{11}, \varepsilon_{22}, \varepsilon_{33}, \varepsilon_{31}, \varepsilon_{23}, \varepsilon_{12}, w_{11}, w_{22}, w_{23}, w_{12}, w_{13}, w_{21} \right) \tag{13.2.8}$$

then the extended internal energy density is

$$U_{ex} = \varepsilon^{\text{ext}} M \left(\varepsilon^{\text{ext}} \right)^T \tag{13.2.9}$$

where $\left(\varepsilon^{\text{ext}} \right)^T$ is the transpose of the vector of the extended strain, and M is the
so-called rigidity matrix that will be defined as follows.

Lemma 1 *From (13.2.7) and (13.2.2), (13.2.4)–(13.2.6), we have*

$$S = -\frac{\partial F_{ex}}{\partial T}, \quad \sigma_{ij} = \frac{\partial F_{ex}}{\partial \varepsilon_{ij}}, \quad H_{ij} = \frac{\partial F_{ex}}{\partial w_{ij}}, \quad \delta^2 F_{ex} \geq 0 \qquad (13.2.10)$$

in which the second and third terms are equivalent to the elastic constitutive law of the material for degrees of freedom of phonons and phasons, and the last one is the stability condition of the matter, respectively. Because the formula (13.2.2) introduces an extended internal energy density, the variational principle defined by the last term in Eq. (13.2.10) is an extended or generalized variational principle.

13.3   The Positive Definite Nature of the Rigidity Matrix and the Stability of the Soft-Matter Quasicrystals with 12-Fold Symmetry

For point group $12mm$ of the 12-fold symmetry quasicrystals in soft matter, the concrete constitutive law is as follows

$$\left.\begin{aligned}
\sigma_{xx} &= C_{11}\varepsilon_{xx} + C_{12}\varepsilon_{yy} + C_{13}\varepsilon_{zz} \\
\sigma_{yy} &= C_{12}\varepsilon_{xx} + C_{11}\varepsilon_{yy} + C_{13}\varepsilon_{zz} \\
\sigma_{zz} &= C_{13}\varepsilon_{xx} + C_{13}\varepsilon_{yy} + C_{33}\varepsilon_{zz} \\
\sigma_{yz} &= \sigma_{zy} = 2C_{44}\varepsilon_{yz} \\
\sigma_{zx} &= \sigma_{xz} = 2C_{44}\varepsilon_{zx} \\
\sigma_{xy} &= \sigma_{yx} = 2C_{66}\varepsilon_{xy}
\end{aligned}\right\} \qquad (13.3.1a)$$

$$\left.\begin{aligned}
H_{xx} &= K_1 w_{xx} + K_2 w_{yy} \\
H_{yy} &= K_2 w_{xx} + K_1 w_{yy} \\
H_{yz} &= K_4 w_{yz} \\
H_{xy} &= (K_1 + K_2 + K_3)w_{xy} + K_3 w_{yx} \\
H_{xz} &= K_4 w_{xz} \\
H_{yx} &= K_3 w_{xy} + (K_1 + K_2 + K_3)w_{yx}
\end{aligned}\right\} \qquad (13.3.1b)$$

$$
\left.\begin{aligned}
p_{xx} &= -p + 2\eta\dot{\xi}_{xx} - \frac{2}{3}\eta\dot{\xi}_{kk} \\
p_{yy} &= -p + 2\eta\dot{\xi}_{yy} - \frac{2}{3}\eta\dot{\xi}_{kk} \\
p_{zz} &= -p + 2\eta\dot{\xi}_{zz} - \frac{2}{3}\eta\dot{\xi}_{kk} \\
p_{yz} &= 2\eta\dot{\xi}_{yz},\ p_{zx} = 2\eta\dot{\xi}_{zx},\ p_{xy} = 2\eta\dot{\xi}_{xy} \\
\dot{\xi}_{kk} &= \dot{\xi}_{xx} + \dot{\xi}_{yy} + \dot{\xi}_{zz}
\end{aligned}\right\}
\tag{13.3.1c}
$$

in which

$$
\left.\begin{aligned}
&C_{1111} = C_{11},\ C_{1122} = C_{12},\ C_{3333} = C_{33}, \\
&C_{1133} = C_{13},\ C_{2323} = C_{44},\ C_{1212} = C_{66}, \\
&(C_{11} - C_{12})/2 = C_{66},
\end{aligned}\right\}
\tag{13.3.2a}
$$

$$
\left.\begin{aligned}
&K_{1111} = K_{2222} = K_{2121} = K_{1212} = K_1, \\
&K_{1122} = K_{2211} = -K_{2112} = -K_{1221} = K_2, \\
&K_{1122} = K_{1221} = K_{2112} = K_3, \\
&K_{2323} = K_{1313} = K_4
\end{aligned}\right\}
\tag{13.3.2b}
$$

the phonon-phason coupling constants $R_{ijkl} = R_{klij} = 0$ due to decoupling between phonons and phasons for 12-fold symmetry quasicrystals, i.e.,

$$
U_{uw} = R_{ijkl}\varepsilon_{ij}w_{kl} + R_{klij}w_{ij}\varepsilon_{kl} = 0
$$

for this type of quasicrystals.

According to the above definition on internal energy density (13.2.9) and constitutive law (13.3.1), there is a rigidity matrix for point group $12mm$ soft-matter quasicrystals

$$
M_{12} =
\begin{pmatrix}
A & B & B & B & 0 & 0 & 0 & 0 & 0 & 0 & 0 & 0 & 0 \\
B & C_{11} & C_{12} & C_{13} & 0 & 0 & 0 & 0 & 0 & 0 & 0 & 0 & 0 \\
B & C_{12} & C_{11} & C_{13} & 0 & 0 & 0 & 0 & 0 & 0 & 0 & 0 & 0 \\
B & C_{13} & C_{13} & C_{33} & 0 & 0 & 0 & 0 & 0 & 0 & 0 & 0 & 0 \\
0 & 0 & 0 & 0 & C_{44} & 0 & 0 & 0 & 0 & 0 & 0 & 0 & 0 \\
0 & 0 & 0 & 0 & 0 & C_{44} & 0 & 0 & 0 & 0 & 0 & 0 & 0 \\
0 & 0 & 0 & 0 & 0 & 0 & \tfrac{1}{2}(C_{11}-C_{12}) & 0 & 0 & 0 & 0 & 0 & 0 \\
0 & 0 & 0 & 0 & 0 & 0 & 0 & K_1 & K_2 & 0 & 0 & 0 & 0 \\
0 & 0 & 0 & 0 & 0 & 0 & 0 & K_2 & K_1 & 0 & 0 & 0 & 0 \\
0 & 0 & 0 & 0 & 0 & 0 & 0 & 0 & 0 & K_4 & 0 & 0 & 0 \\
0 & 0 & 0 & 0 & 0 & 0 & 0 & 0 & 0 & 0 & K_1+K_2+K_3 & 0 & K_3 \\
0 & 0 & 0 & 0 & 0 & 0 & 0 & 0 & 0 & 0 & 0 & K_4 & 0 \\
0 & 0 & 0 & 0 & 0 & 0 & 0 & 0 & 0 & 0 & K_3 & 0 & K_1+K_2+K_3
\end{pmatrix}
\tag{13.3.3}
$$

The positive definite condition of the rigidity matrix (13.3.3) is equivalent to the variation formula given by the last term in (13.2.10), i.e.,

$$\delta^2 F_{ex} \geq 0 \qquad\qquad (13.3.4)$$

this non-negative condition of the second-order variation of the extended internal energy density functional requires the extended rigidity matrix must be positive definite. We have the theorem for describing the stability of the soft-matter quasicrystals with 12-fold symmetry as below:

Theorem 1 *Under the condition (13.2.2), the validity of variation (13.3.4) is equivalent to the positive definite nature of the rigidity matrix (13.3.1) and leads to*

$$\left.\begin{array}{l} A > 0,\, C_{11} - C_{12} > 0,\, A(C_{11}C_{33} + C_{12}C_{33} - 2C_{13}^2) \\[4pt] \quad - B^2(C_{11} + C_{12} - 4C_{13} + 2C_{33}) > 0, \\[4pt] C_{44} > 0,\, K_4 > 0,\, K_1 - K_2 > 0,\, K_1 + K_2 > 0,\, K_1 + K_2 + 2K_3 > 0 \end{array}\right\} \qquad (13.3.5)$$

The proof of the theorem is straightforward so is omitted.

When the conditions (13.3.5) are satisfied, the soft-matter quasicrystals are stable thermodynamically. This stability takes into account the effects of phonons, phasons, and fluid phonon and their interaction of soft-matter quasicrystals of 12-fold symmetry. More exactly speaking the stability depends upon the material constants concerning only fluid, fluid coupling phonons, and phasons of the soft-matter quasicrystals. These constants can be measured by experiments that are similar to that in crystallography [9] and solid quasicrystallography [10] and presents evidence and intuitive character of the complexity of stability of soft-matter quasicrystals. This shows the substantive nature of the stability of soft-matter quasicrystals. Substantively it explores the structure of the matter because it comes from the constitutive law (13.3.1), which is the result of the symmetry of the structure—i.e., the result is obtained by the theory of group and group representation of the quasicrystals, which can be referred to the relevant chapters of this book.

13.4 Comparison and Examination of Results of Soft-Matter Quasicrystals with 12-Fold Symmetry

It is well-known that the 12-fold symmetry quasicrystals in soft matter belong to a type of two-dimensional quasicrystals, and in which the phonon field structure presents the character of the hexagonal crystals.

In the special case when the phason field is absent, i.e.,

$$K_1 = K_2 = K_3 = K_4 = 0 \qquad\qquad (13.4.1)$$

and at the same time one takes $B = 0$ and A is any nonzero positive value, then (13.3.3) reduces to

$$C_{11} - C_{12} > 0, \ C_{44} > 0, \ C_{11}C_{33} + C_{12}C_{33} - 2C_{13}^2 > 0 \qquad (13.4.2)$$

Here constant A can be any nonzero positive value, and the value can be arbitrarily small, due to $\left(\frac{\delta\rho}{\rho_0}\right)^2 \sim 10^{-10}$ (because $\frac{\delta\rho}{\rho_0} \sim 10^{-5}$, according to our computation). Thus $A\left(\frac{\delta\rho}{\rho_0}\right)^2$ is very small, and this may be understood that if the fluid effect is very weak and no phason field, the stability condition (13.3.5) is reduced to that of hexagonal crystal system (13.4.2), which was derived by Cowley [9]. This, in one angle, is explored to examine if the dynamics and thermodynamics of soft-matter quasicrystals are correct; but in another angle, it shows the constant A could not be zero, except a structural phase transition appears, indicating that the soft-matter phase cannot be reduced to a solid phase as required by the stability of soft-matter.

While in another case, (13.3.3) reduces to

$$\left. \begin{aligned} &C_{11} - C_{12} > 0, C_{44} > 0, C_{11}C_{33} + C_{12}C_{33} - 2C_{13}^2 > 0 \\ &K_1 - K_2 > 0, K_1 + K_2 > 0, K_1 + K_2 + K_3 > 0, K_4 > 0 \end{aligned} \right\} \qquad (13.4.3)$$

under the conditions that $B=0$ and A is any nonzero positive value, the inequalities (13.4.3) is the stability condition of solid quasicrystals of 12-fold symmetry, which explores the correctness of the dynamics and thermodynamics of soft-matter quasicrystals once again. Again, this result shows the soft-matter phase cannot be reduced to a solid phase, except a structural phase transition appears, due to the requirement of soft-matter stability.

The stability is connected with the positive definite nature of the mathematical structure of the matter dynamics. This is useful for the numerical solution (e.g., the finite element method), which will be discussed in other places.

The stability is connected to the phase transition, which is a more important problem. For crystals, Cowley [9] gave an analysis.

13.5   The Stability of 8-Fold Symmetry Soft-Matter Quasicrystals

The constitutive law for point group $8mm$ soft-matter quasicrystals can be referred to Chap. 9 and is given by

$$\left.\begin{aligned}
\sigma_{xx} &= C_{11}\varepsilon_{xx} + C_{12}\varepsilon_{yy} + C_{13}\varepsilon_{zz} + R\left(w_{xx} + w_{yy}\right) \\
\sigma_{yy} &= C_{12}\varepsilon_{xx} + C_{11}\varepsilon_{yy} + C_{13}\varepsilon_{zz} - R\left(w_{xx} + w_{yy}\right) \\
\sigma_{zz} &= C_{13}\varepsilon_{xx} + C_{13}\varepsilon_{yy} + C_{33}\varepsilon_{zz} \\
\sigma_{yz} &= \sigma_{zy} = 2C_{44}\varepsilon_{yz} \\
\sigma_{zx} &= \sigma_{xz} = 2C_{44}\varepsilon_{zx} \\
\sigma_{xy} &= \sigma_{yx} = 2C_{66}\varepsilon_{xy} - Rw_{xy} + Rw_{yx}
\end{aligned}\right\} \tag{13.5.1a}$$

$$\left.\begin{aligned}
H_{xx} &= K_1 w_{xx} + K_2 w_{yy} + R\left(\varepsilon_{xx} - \varepsilon_{yy}\right) \\
H_{yy} &= K_2 w_{xx} + K_1 w_{yy} + R\left(\varepsilon_{xx} - \varepsilon_{yy}\right) \\
H_{yz} &= K_4 w_{yz} \\
H_{xy} &= (K_1 + K_2 + K_3)w_{xy} + K_2 w_{yz} - 2R\varepsilon_{xy} \\
H_{xz} &= K_4 w_{xz} \\
H_{yx} &= K_3 w_{xy} + (K_1 + K_2 + K_3)w_{yx} + 2R\varepsilon_{xy}
\end{aligned}\right\} \tag{13.5.1b}$$

$$\left.\begin{aligned}
p_{xx} &= -p + 2\eta\dot{\xi}_{xx} - \frac{2}{3}\eta\dot{\xi}_{kk} \\
p_{yy} &= -p + 2\eta\dot{\xi}_{yy} - \frac{2}{3}\eta\dot{\xi}_{kk} \\
p_{zz} &= -p + 2\eta\dot{\xi}_{zz} - \frac{2}{3}\eta\dot{\xi}_{kk} \\
p_{yz} &= 2\eta\dot{\xi}_{yz}, \; p_{zx} = 2\eta\dot{\xi}_{zx}, \; p_{xy} = 2\eta\dot{\xi}_{xy} \\
\dot{\xi}_{kk} &= \dot{\xi}_{xx} + \dot{\xi}_{yy} + \dot{\xi}_{zz}
\end{aligned}\right\} \tag{13.5.1c}$$

in which the evident difference of this quasicrystal system from the 12-fold symmetry quasicrystals lies in the coupling between phonons and phasons. We have the rigidity matrix for the point group $8mm$ soft-matter quasicrystals as follows

$$M_8 = \begin{pmatrix}
A & B & B & B & 0 & 0 & 0 & 0 & 0 & 0 & 0 & 0 & 0 \\
B & C_{11} & C_{12} & C_{13} & 0 & 0 & 0 & R & R & 0 & 0 & 0 & 0 \\
B & C_{12} & C_{11} & C_{13} & 0 & 0 & 0 & -R & -R & 0 & 0 & 0 & 0 \\
B & C_{13} & C_{13} & C_{33} & 0 & 0 & 0 & 0 & 0 & 0 & 0 & 0 & 0 \\
0 & 0 & 0 & 0 & C_{44} & 0 & 0 & 0 & 0 & 0 & 0 & 0 & 0 \\
0 & 0 & 0 & 0 & 0 & C_{44} & 0 & 0 & 0 & 0 & 0 & 0 & 0 \\
0 & 0 & 0 & 0 & 0 & 0 & \frac{1}{2}(C_{11}-C_{12}) & 0 & 0 & 0 & -R & 0 & R \\
0 & R & -R & 0 & 0 & 0 & 0 & K_1 & K_2 & 0 & 0 & 0 & 0 \\
0 & R & -R & 0 & 0 & 0 & 0 & K_2 & K_1 & 0 & 0 & 0 & 0 \\
0 & 0 & 0 & 0 & 0 & 0 & 0 & 0 & 0 & K_4 & 0 & 0 & 0 \\
0 & 0 & 0 & 0 & 0 & 0 & -R & 0 & 0 & 0 & K_1+K_2+K_3 & 0 & K_3 \\
0 & 0 & 0 & 0 & 0 & 0 & 0 & 0 & 0 & 0 & 0 & K_4 & 0 \\
0 & 0 & 0 & 0 & 0 & 0 & R & 0 & 0 & 0 & K_3 & 0 & K_1+K_2+K_3
\end{pmatrix} \tag{13.5.2}$$

and we have the theorem

Theorem 2 *The positive definite condition of the rigidity matrix (13.5.2) is equivalent to the variation inequality (13.3.4) (i.e., the fourth formula of (13.2.10)), which leads to the stability criterion of soft-matter quasicrystals of point group $8mm$ as* *follows*

$$\left.\begin{array}{l} A > 0, \ C_{11} - C_{12} > 0, \ A(C_{11}C_{33} + C_{12}C_{33} - 2C_{13}^2) \\ -B^2(C_{11} + C_{12} - 4C_{13} + 2C_{33}) > 0, \ C_{44} > 0, \\ K_4 > 0, \ K_1 - K_2 > 0, \ K_1 + K_2 > 0, \ K_1 \\ +K_2 + 2K_3 > 0, \ (C_{11} - C_{12})(K_1 + K_2) - 4R^2 > 0 \end{array}\right\} \tag{13.5.3}$$

The coupling constant R has been taken into account in the stability criterion (13.5.3), which is the most evident difference from that of the 12-fold symmetry quasicrystals.

The proof detail of the theorem is omitted because it is very straightforward.

13.6   The Stability of 10-Fold Symmetry Soft-Matter Quasicrystals

We consider the point group $10mm$ soft-matter quasicrystals, whose constitutive law is [1]

$$\left.\begin{array}{l} \sigma_{xx} = C_{11}\varepsilon_{xx} + C_{12}\varepsilon_{yy} + C_{13}\varepsilon_{zz} + R\left(w_{xx} + w_{yy}\right) \\ \sigma_{yy} = C_{12}\varepsilon_{xx} + C_{11}\varepsilon_{yy} + C_{13}\varepsilon_{zz} - R\left(w_{xx} + w_{yy}\right) \\ \sigma_{zz} = C_{13}\varepsilon_{xx} + C_{13}\varepsilon_{yy} + C_{33}\varepsilon_{zz} \\ \sigma_{yz} = \sigma_{zy} = 2C_{44}\varepsilon_{yz} \\ \sigma_{zx} = \sigma_{xz} = 2C_{44}\varepsilon_{zx} \\ \sigma_{xy} = \sigma_{yx} = 2C_{66}\varepsilon_{xy} - R\left(w_{xy} - w_{yx}\right) \end{array}\right\} \tag{13.6.1a}$$

$$\left.\begin{array}{l} H_{xx} = K_1 w_{xx} + K_2 w_{yy} + R\left(\varepsilon_{xx} - \varepsilon_{yy}\right) \\ H_{yy} = K_2 w_{xx} + K_1 w_{yy} + R\left(\varepsilon_{xx} - \varepsilon_{yy}\right) \\ H_{yz} = K_4 w_{yz} \\ H_{xy} = K_1 w_{xy} - K_2 w_{yx} \\ H_{xz} = K_4 w_{xz} \\ H_{yx} = -K_2 w_{xy} + K_1 w_{yx} + 2R\varepsilon_{xy} \end{array}\right\} \tag{13.12.6.1b}$$

$$
\left.\begin{aligned}
p_{xx} &= -p + 2\eta\dot{\xi}_{xx} - \frac{2}{3}\eta\dot{\xi}_{kk} \\
p_{yy} &= -p + 2\eta\dot{\xi}_{yy} - \frac{2}{3}\eta\dot{\xi}_{kk} \\
p_{zz} &= -p + 2\eta\dot{\xi}_{zz} - \frac{2}{3}\eta\dot{\xi}_{kk} \\
p_{yz} &= 2\eta\dot{\xi}_{yz}, \ p_{zx} = 2\eta\dot{\xi}_{zx}, \ p_{xy} = 2\eta\dot{\xi}_{xy} \\
\dot{\xi}_{kk} &= \dot{\xi}_{xx} + \dot{\xi}_{yy} + \dot{\xi}_{zz}
\end{aligned}\right\}
\tag{13.6.1c}
$$

and we obtain the rigidity matrix for the $10mm$ point group soft-matter quasicrystals

$$
M_{10} =
\begin{pmatrix}
A & B & B & B & 0 & 0 & 0 & 0 & 0 & 0 & 0 & 0 & 0 \\
B & C_{11} & C_{12} & C_{13} & 0 & 0 & 0 & R & R & 0 & 0 & 0 & 0 \\
B & C_{12} & C_{11} & C_{13} & 0 & 0 & 0 & -R & -R & 0 & 0 & 0 & 0 \\
B & C_{13} & C_{13} & C_{33} & 0 & 0 & 0 & 0 & 0 & 0 & 0 & 0 & 0 \\
0 & 0 & 0 & 0 & C_{44} & 0 & 0 & 0 & 0 & 0 & 0 & 0 & 0 \\
0 & 0 & 0 & 0 & 0 & C_{44} & 0 & 0 & 0 & 0 & 0 & 0 & 0 \\
0 & 0 & 0 & 0 & 0 & 0 & \frac{1}{2}(C_{11}-C_{12}) & 0 & 0 & 0 & -R & 0 & R \\
0 & R & -R & 0 & 0 & 0 & 0 & K_1 & K_2 & 0 & 0 & 0 & 0 \\
0 & R & -R & 0 & 0 & 0 & 0 & K_2 & K_1 & 0 & 0 & 0 & 0 \\
0 & 0 & 0 & 0 & 0 & 0 & 0 & 0 & 0 & K_4 & 0 & 0 & 0 \\
0 & 0 & 0 & 0 & 0 & 0 & -R & 0 & 0 & 0 & K_1 & 0 & -K_2 \\
0 & 0 & 0 & 0 & 0 & 0 & 0 & 0 & 0 & 0 & 0 & K_4 & 0 \\
0 & 0 & 0 & 0 & 0 & 0 & R & 0 & 0 & 0 & -K_2 & 0 & K_1
\end{pmatrix}
\tag{13.6.2}
$$

The corresponding theorem for stability is

Theorem 3 *Due to the equivalence of the positive definite condition of the matrix (13.6.2) to the variation given by the fourth formula of (13.2.10), or inequality (13.3.4), the stability criterion of soft-matter quasicrystals of point group 10mm can be derived as follows*

$$
\left.\begin{aligned}
&A > 0, \ C_{11} - C_{12} > 0, \ A(C_{11}C_{33} + C_{12}C_{33} - 2C_{13}^2) \\
&\quad - B^2(C_{11} + C_{12} - 4C_{13} + 2C_{33}) > 0, \\
&C_{44} > 0, \ K_4 > 0, \ K_1 - K_2 > 0, \\
&\quad K_1 + K_2 > 0, \ (C_{11} - C_{12})(K_1 + K_2) - 4R^2 > 0
\end{aligned}\right\}
\tag{13.6.3}
$$

The proof of the theorem is again omitted.

13.7 The Stability of the 18-Fold Symmetry Soft-Matter Quasicrystals

The 12-, 8-, and ten-fold symmetry quasicrystals belong to the first kind of two-dimensional quasicrystals, and their stabilities have been discussed in the above sections. It is well-known the 18-fold symmetry soft-matter quasicrystals were observed in colloids. The 18-fold and possibly discovered 7-, 9-, and 14-fold symmetry quasicrystals belong to the second kind of soft-matter quasicrystals. So the study on the stability of 18-fold symmetry soft-matter quasicrystals is significant and different from that of the first kind of soft-matter quasicrystals. To the authors' knowledge, there were no reported results of the stability for 18-fold symmetry soft-matter quasicrystals yet; perhaps the present work is the first probe on the topic.

The lack of research on the stability of 18-fold symmetry soft-matter quasicrystals may be due to the incomplete data of three-dimensional dynamics of the matter. However, this difficulty has been overcome by the work introduced in Chap. 10 of this book.

13.7.1 A Brief Review on Some Fundamental Relations from the Dynamics of the Second Kind of Soft-Matter Quasicrystals

We have mentioned in the previous section that the 18-fold and possibly discovered 7-, 9-, and 14-fold symmetry quasicrystals belong to the second kind of soft-matter quasicrystals, see Chaps. 10 and 11, which are different from those of the first kind ones. They exhibit four elementary excitations as phonons, first phasons, second phasons, and fluid phonon. The relevant fields are the phonon displacement field $\mathbf{u}(u_x, u_y, u_z)$, the first phason displacement field $\mathbf{v}(v_x, v_y)$, the second phason displacement field $\mathbf{w}(w_x, w_y)$, and fluid phonon velocity field $\mathbf{V}(V_x, V_y, V_z)$, respectively, in which the $z-$ axis represents the 18- or 7-, 9-,14-fold symmetry axis.

We define the tensors of phonon strain, phason strain, and fluid phonon deformation rate as follows

$$\varepsilon_{ij} = \frac{1}{2}\left(\frac{\partial u_i}{\partial x_j} + \frac{\partial u_j}{\partial x_i}\right), v_{ij} = \frac{\partial v_i}{\partial x_j}, w_{ij} = \frac{\partial w_i}{\partial x_j}, \dot{\xi}_{ij} = \frac{1}{2}\left(\frac{\partial V_i}{\partial x_j} + \frac{\partial V_j}{\partial x_i}\right) \quad (13.7.1)$$

respectively, in which

$$x = x_1, y = x_2, z = x_3, \quad i, j = 1, 2, 3$$

and the corresponding constitutive law [2]

$$\left.\begin{aligned}
\sigma_{ij} &= C_{ijkl}\varepsilon_{kl} + r_{ijkl}v_{kl} + R_{ijkl}w_{kl} \\
\tau_{ij} &= T_{ijkl}v_{kl} + r_{klij}\varepsilon_{kl} + G_{ijkl}w_{kl} \\
H_{ij} &= K_{ijkl}w_{kl} + R_{klij}\varepsilon_{kl} + G_{klij}v_{kl} \\
p_{ij} &= -p\delta_{ij} + \sigma'_{ij} = -p\delta_{ij} + \eta_{ijkl}\dot{\xi}_{kl}
\end{aligned}\right\} \tag{13.7.2}$$

and the internal energy densities

$$\left.\begin{aligned}
U_{\mathbf{u}} &= \frac{1}{2}C_{ijkl}\varepsilon_{ij}\varepsilon_{kl}, \; U_{\mathbf{v}} = \frac{1}{2}T_{ijkl}v_{ij}v_{kl}, \; U_{\mathbf{w}} = \frac{1}{2}K_{ijkl}w_{ij}w_{kl} \\
U_{\mathbf{uv}} &= r_{ijkl}\varepsilon_{ij}v_{kl} + r_{klij}v_{ij}\varepsilon_{kl} \\
U_{\mathbf{uw}} &= R_{ijkl}\varepsilon_{ij}w_{kl} + R_{klij}w_{ij}\varepsilon_{kl} \quad i,j,k,l = 1,2,3 \\
U_{\mathbf{vw}} &= G_{ijkl}v_{ij}w_{kl} + G_{klij}w_{ij}v_{kl}
\end{aligned}\right\} \tag{13.7.3}$$

in which C_{ijkl} denote the phonon elastic constants, T_{ijkl} the first phason elastic
constants, K_{ijkl} the second phason elastic constants, r_{ijkl}, r_{klij} and R_{ijkl}, R_{klij}
the phonon-first phason and phonon-second phason coupling elastic constants,
G_{ijkl}, G_{klij} the first-second phason coupling elastic constants respectively. They are
different from those for the 12-, 8-, and ten-fold symmetry soft-matter quasicrystals
given in Sects. 13.2, 13.3, 13.4, 13.5 and 13.6.

13.7.2   Extended Free Energy of the Quasicrystals System of Second Kind

We define the extended internal energy density

$$U_{ex} = \frac{1}{2}A\left(\frac{\delta\rho}{\rho_0}\right)^2 + B\left(\frac{\delta\rho}{\rho_0}\right)\nabla\cdot\mathbf{u} + C\left(\frac{\delta\rho}{\rho_0}\right)\nabla\cdot\mathbf{v} + D\left(\frac{\delta\rho}{\rho_0}\right)\nabla\cdot\mathbf{w} + U_{el} \quad (**)$$

where the first term denotes the energy density due to mass density variation, the
quantity $\delta\rho/\rho_0$ describes the variation of the mass density, in which $\delta\rho = \rho - \rho_0$ and
ρ_0 the initial mass density; the second term is the one due to the coupling between the
mass density variation and the phonons; the third and fourth represent the couplings
between the mass density variation and the first and second phasons, respectively, and
A, B, C, D are the LRT constants, respectively. For quasicrystals C and D should
be zero. So that the equation (**) is reduced to

$$U_{ex} = \frac{1}{2}A\left(\frac{\delta\rho}{\rho_0}\right)^2 + B\left(\frac{\delta\rho}{\rho_0}\right)\nabla\cdot\mathbf{u} + U_{el} \tag{13.7.4}$$

We now extend the Hamiltonian discussed in the previous sections to the present case

$$
\left.
\begin{aligned}
H &= H[\Psi(\mathbf{r}, t)] \\
&= \int \frac{\mathbf{g}^2}{2\rho} d^d\mathbf{r} + \int \left[\frac{1}{2} A \left(\frac{\delta\rho}{\rho_0} \right)^2 + B \left(\frac{\delta\rho}{\rho_0} \right) \nabla \cdot \mathbf{u} \right] d^d\mathbf{r} + F_{el} \\
&= H_{kin} + H_\rho + F_{el} \\
\mathbf{g} &= \rho \mathbf{V}
\end{aligned}
\right\}
\tag{13.7.5}
$$

in which H_{kin} denotes the kinetic energy, H_ρ the energy due to the variation of mass density, F_{el} the elastic deformation energy consisting of contributions from phonons, phasons, and phason-phason coupling, respectively. The detailed definition will be given by (13.7.6)–(13.7.8) in the following.

From the Eq. (13.7.2) one can obtain the total elastic free energy density

$$
U_{el} = U_{\mathbf{u}} + U_{\mathbf{v}} + U_{\mathbf{w}} + U_{\mathbf{uv}} + U_{\mathbf{uw}} + U_{\mathbf{vw}}
\tag{13.7.6}
$$

which come from phonons, phasons, phonon-phason coupling, and phason-phason coupling. And the elastic energy will be

$$
F_{el} = F_{\mathbf{u}} + F_{\mathbf{v}} + F_{\mathbf{w}} + F_{\mathbf{uv}} + F_{\mathbf{uw}} + F_{\mathbf{vw}}
\tag{13.7.7}
$$

where $F_{\mathbf{u}}, F_{\mathbf{v}}, F_{\mathbf{w}}, F_{\mathbf{uv}}, F_{\mathbf{uw}}, F_{\mathbf{vw}}$ represent the strain energies of phonons, first phasons, second phasons, phonon-first phason coupling, phonon-second phason coupling, and first phason-second phason coupling, respectively:

$$
\left.
\begin{aligned}
F_{\mathbf{u}} &= \int \frac{1}{2} C_{ijkl} \varepsilon_{ij} \varepsilon_{kl} d^d\mathbf{r} \\
F_{\mathbf{v}} &= \int \frac{1}{2} T_{ijkl} v_{ij} v_{kl} d^d\mathbf{r} \\
F_{\mathbf{w}} &= \int \frac{1}{2} K_{ijkl} w_{ij} w_{kl} d^d\mathbf{r} \\
F_{\mathbf{uv}} &= \int \left(r_{ijkl} \varepsilon_{ij} v_{kl} + r_{klij} v_{ij} \varepsilon_{kl} \right) d^d\mathbf{r} \\
F_{\mathbf{uw}} &= \int \left(R_{ijkl} \varepsilon_{ij} w_{kl} + R_{klij} w_{ij} \varepsilon_{kl} \right) d^d\mathbf{r} \\
F_{\mathbf{vw}} &= \int \left(G_{ijkl} v_{ij} w_{kl} + G_{klij} w_{ij} v_{kl} \right) d^d\mathbf{r}
\end{aligned}
\right\}
\tag{13.7.8}
$$

The integrations are taken into account on the domain.

Therefore, the problem of the second kind of quasicrystals is more complex than that of the first kind.

According to thermodynamics, the extended free energy density is defined by

$$F_{ex} = U_{ex} - TS \qquad (13.7.9)$$

where U_{ex} the extended internal energy density defined by (13.7.4), T absolute temperature, and S the entropy, respectively.

Lemma 2 *From (13.7.9) and (13.7.4)-(13.7.6), we have*

$$S = -\frac{\partial F_{ex}}{\partial T}, \sigma_{ij} = \frac{\partial F_{ex}}{\partial \varepsilon_{ij}}, \tau_{ij} = \frac{\partial F_{ex}}{\partial v_{ij}}, H_{ij} = \frac{\partial F_{ex}}{\partial w_{ij}}, \delta^2 F_{ex} \geq 0 \qquad (13.7.10)$$

in which the second to fourth equations are equivalent to the elastic constitutive law of the material for degrees of freedom of phonons, first and second phasons and the last one is the stability condition of the matter, respectively. Because the formula (13.7.4) introduces an extended internal energy density, the variational principle in Eq. (13.7.10) is an extended or generalized variation. Note that $C = D = 0$ in (13.7.4).

13.7.3   The Positive Definite Nature of the Rigidity Matrix and the Stability of the Soft-Matter Quasicrystals with 18-Fold Symmetry

For point group $18mm$ symmetry soft-matter quasicrystals the concrete constitutive law is as follows

$$\left.\begin{array}{l} \sigma_{xx} = C_{11}\varepsilon_{xx} + C_{12}\varepsilon_{yy} + C_{13}\varepsilon_{zz} \\[4pt] \sigma_{yy} = C_{12}\varepsilon_{xx} + C_{11}\varepsilon_{yy} + C_{13}\varepsilon_{zz} \\[4pt] \sigma_{zz} = C_{13}\varepsilon_{xx} + C_{13}\varepsilon_{yy} + C_{33}\varepsilon_{zz} \\[4pt] \sigma_{yz} = \sigma_{zy} = 2C_{44}\varepsilon_{yz} \\[4pt] \sigma_{zx} = \sigma_{xz} = 2C_{44}\varepsilon_{zx} \\[4pt] \sigma_{xy} = \sigma_{yx} = 2C_{66}\varepsilon_{xy} \end{array}\right\} \qquad (13.7.11a)$$

$$\left.\begin{array}{l} \tau_{xx} = T_1 v_{xx} + T_2 v_{yy} + G\left(w_{xx} - w_{yy}\right) \\[4pt] \tau_{yy} = T_2 v_{xx} + T_1 v_{yy} + G\left(w_{xx} - w_{yy}\right) \\[4pt] \tau_{xy} = T_1 v_{xy} - T_2 v_{yx} - G\left(w_{xy} + w_{yx}\right) \\[4pt] \tau_{yx} = T_1 v_{yx} - T_2 v_{xy} + G\left(w_{xy} + w_{yx}\right) \\[4pt] \tau_{xz} = T_3 v_{xz}, \tau_{yz} = T_3 v_{yz} \end{array}\right\} \qquad (13.7.11b)$$

$$
\left.\begin{aligned}
H_{xx} &= K_1 w_{xx} + K_2 w_{yy} + G\left(v_{xx} + v_{yy}\right) \\
H_{yy} &= K_2 w_{xx} + K_1 w_{yy} - G\left(v_{xx} + v_{yy}\right) \\
H_{xy} &= K_1 w_{xy} - K_2 w_{yx} - G\left(v_{xy} - v_{yx}\right) \\
H_{yx} &= K_1 w_{yx} - K_2 w_{xy} - G\left(v_{xy} - v_{yx}\right) \\
H_{xz} &= K_3 w_{xz}, \; H_{yz} = K_3 w_{yz}
\end{aligned}\right\} \tag{13.7.11c}
$$

$$
\left.\begin{aligned}
p_{xx} &= -p + 2\eta\dot{\xi}_{xx} - \frac{2}{3}\eta\dot{\xi}_{kk} \\
p_{yy} &= -p + 2\eta\dot{\xi}_{yy} - \frac{2}{3}\eta\dot{\xi}_{kk} \\
p_{zz} &= -p + 2\eta\dot{\xi}_{zz} - \frac{2}{3}\eta\dot{\xi}_{kk} \\
p_{yz} &= p_{zy} = 2\eta\dot{\xi}_{yz} \\
p_{zx} &= p_{xz} = 2\eta\dot{\xi}_{zx} \\
p_{xy} &= p_{yx} = 2\eta\dot{\xi}_{xy} \\
\dot{\xi}_{kk} &= \dot{\xi}_{xx} + \dot{\xi}_{yy} + \dot{\xi}_{zz}
\end{aligned}\right\} \tag{13.7.11d}
$$

the phonon-phason coupling constants $r_{ijkl}=r_{klij} = R_{ijkl} = R_{klij} = 0$ due to the decoupling between phonons and phasons, i.e.,

$$
U_{\mathbf{uv}} = r_{ijkl}\varepsilon_{ij}v_{kl} + r_{klij}v_{ij}\varepsilon_{kl} = 0
$$
$$
U_{\mathbf{uw}} = R_{ijkl}\varepsilon_{ij}w_{kl} + R_{klij}w_{ij}\varepsilon_{kl} = 0
$$

for this type of quasicrystals.

Just as the conventional internal energy density, where between the stress and strain tensors there is a rigidity matrix, for the extended internal energy density (12.7.4) the extended rigidity matrix such as

$$
M_{18} = \left(\begin{array}{ccccccccccccccccccc}
A & B & B & B & 0 & 0 & 0 & 0 & 0 & 0 & 0 & 0 & 0 & 0 & 0 & 0 & 0 & 0 & 0 \\
B & C_{11} & C_{12} & C_{13} & 0 & 0 & 0 & 0 & 0 & 0 & 0 & 0 & 0 & 0 & 0 & 0 & 0 & 0 & 0 \\
B & C_{12} & C_{11} & C_{13} & 0 & 0 & 0 & 0 & 0 & 0 & 0 & 0 & 0 & 0 & 0 & 0 & 0 & 0 & 0 \\
B & C_{13} & C_{13} & C_{33} & 0 & 0 & 0 & 0 & 0 & 0 & 0 & 0 & 0 & 0 & 0 & 0 & 0 & 0 & 0 \\
0 & 0 & 0 & 0 & 2C_{44} & 0 & 0 & 0 & 0 & 0 & 0 & 0 & 0 & 0 & 0 & 0 & 0 & 0 & 0 \\
0 & 0 & 0 & 0 & 0 & 2C_{44} & 0 & 0 & 0 & 0 & 0 & 0 & 0 & 0 & 0 & 0 & 0 & 0 & 0 \\
0 & 0 & 0 & 0 & 0 & 0 & 2C_{66} & 0 & 0 & 0 & 0 & 0 & 0 & 0 & 0 & 0 & 0 & 0 & 0 \\
0 & 0 & 0 & 0 & 0 & 0 & 0 & T_1 & T_2 & 0 & 0 & 0 & 0 & G & -G & 0 & 0 & 0 & 0 \\
0 & 0 & 0 & 0 & 0 & 0 & 0 & T_2 & T_1 & 0 & 0 & 0 & 0 & G & -G & 0 & 0 & 0 & 0 \\
0 & 0 & 0 & 0 & 0 & 0 & 0 & 0 & 0 & T_3 & 0 & 0 & 0 & 0 & 0 & 0 & 0 & 0 & 0 \\
0 & 0 & 0 & 0 & 0 & 0 & 0 & 0 & 0 & 0 & T_1 & 0 & -T_2 & 0 & 0 & 0 & -G & 0 & -G \\
0 & 0 & 0 & 0 & 0 & 0 & 0 & 0 & 0 & 0 & 0 & T_3 & 0 & 0 & 0 & 0 & 0 & 0 & 0 \\
0 & 0 & 0 & 0 & 0 & 0 & 0 & 0 & 0 & 0 & -T_2 & 0 & T_1 & 0 & 0 & 0 & G & 0 & G \\
0 & 0 & 0 & 0 & 0 & 0 & 0 & G & G & 0 & 0 & 0 & 0 & K_1 & K_2 & 0 & 0 & 0 & 0 \\
0 & 0 & 0 & 0 & 0 & 0 & 0 & -G & -G & 0 & 0 & 0 & 0 & K_2 & K_2 & 0 & 0 & 0 & 0 \\
0 & 0 & 0 & 0 & 0 & 0 & 0 & 0 & 0 & 0 & 0 & 0 & 0 & 0 & 0 & K_3 & 0 & 0 & 0 \\
0 & 0 & 0 & 0 & 0 & 0 & 0 & 0 & 0 & 0 & -G & 0 & G & 0 & 0 & 0 & K_1 & 0 & -K_2 \\
0 & 0 & 0 & 0 & 0 & 0 & 0 & 0 & 0 & 0 & 0 & 0 & 0 & 0 & 0 & 0 & 0 & K_3 & 0 \\
0 & 0 & 0 & 0 & 0 & 0 & 0 & 0 & 0 & 0 & -G & 0 & G & 0 & 0 & 0 & -K_2 & 0 & K_1
\end{array}\right) \tag{13.7.12}
$$

Due to the condition in (13.7.10)

$$\delta^2 F_{ex} \geq 0 \tag{13.7.13}$$

this non-negative condition of the second-order variation of the extended free energy density functional requires the extended rigidity matrix must be positive definite. We have the theorem for describing the stability of the soft-matter quasicrystals with 18-fold symmetry as follows:

Theorem 4 *Under the condition (13.7.4), the validity of variation (13.7.13) is equivalent to the positive definite nature of matrix (13.7.12). It then gives*

$$\left.\begin{aligned}
&A > 0,\, C_{11} - C_{12} > 0,\, C_{44} > 0,\, A(C_{11}C_{33} + C_{12}C_{33} - 2C_{13}^2) \\
&-B^2(C_{11} + C_{12} - 4C_{13} + 2C_{33}) > 0, \\
&K_1 + K_2 > 0,\, T_1 - T_2 > 0,\, (K_1 - K_2)(T_1 + T_2) - 4G^2 > 0, \\
&K_1 - K_2 > 0,\, T_3 > 0,\, K_3 > 0
\end{aligned}\right\} \tag{13.7.14}$$

The proof of the theorem is straightforward and is omitted.

When the conditions (13.7.14) are satisfied, the 18-fold symmetry soft-matter quasicrystals are stable. Naturally, it explores the structure of the matter, because it comes from the constitutive law (13.7.11), which is the result of the symmetry of the structure—i.e., the result obtained by the theory of group and group representation of the quasicrystals, refer to the previous chapters.

13.7.4   Comparison and Examination

It is well-known that the 18-fold symmetry quasicrystals in soft matter belong to a type of two-dimensional quasicrystals, and in which the phonon field structure presents the character of the hexagonal crystals [9].

In the special case where the phason field is absent, i.e.,

$$K_1 = K_2 = T_1 = T_2 = 0,\, G = 0 \tag{13.7.15}$$

and at the same time one takes $B = 0$ and A is any positive value, then (13.7.14) reduces

$$C_{11} - C_{12} > 0,\, C_{44} > 0,\, C_{11}C_{33} + C_{12}C_{33} - 2C_{13}^2 > 0 \tag{13.7.16}$$

While for another case, i.e., in the solid quasicrystals of 18-fold symmetry, although this type of solid quasicrystals has not been observed by experiments so far,

the possible structure can be estimated from the group theory, then (13.7.14) reduces
to

$$\left.\begin{array}{l} C_{11} - C_{12} > 0, \ C_{44} > 0, \ C_{11}C_{33} + C_{12}C_{33} - 2C_{13}^2 > 0, \\[4pt] K_1 + K_2 > 0, \ T_1 - T_2 > 0, \ (K_1 - K_2)(T_1 + T_2) - 4G^2 > 0, \\[4pt] K_1 - K_2 > 0, \ T_3 > 0, \ K_3 > 0 \end{array}\right\} \qquad (13.7.17)$$

13.7.5   Some Discussions

Theorem 4 can also be given in accordance with the first law of thermodynamics.
Omitting the details, we have

$$\sigma_{ij} = \frac{\partial U_{ex}}{\partial \varepsilon_{ij}}, \ \tau_{ij} = \frac{\partial U_{ex}}{\partial v_{ij}}, \ H_{ij} = \frac{\partial U_{ex}}{\partial w_{ij}}, \ \delta^2 U_{ex} \geq 0 \qquad (13.7.18)$$

where U_{ex} is defined by (13.7.4), which is a quadratic form. The condition $\delta^2 U_{ex} \geq 0$
requires the matrix (13.7.12) must be positive definite, giving rise to the results
(13.7.14), i.e., the theorem holds.

By a completely different approach compared with Refs. [11, 12] this section
discussed the stability of soft-matter quasicrystals, which is directly based on the
thermodynamics with the help of generalized dynamics of the matter, and offered
some quantitative and very concise results. The stability depends only upon the
material constants, which can be measured by experiments. The correctness and
precision of the theoretical prediction are examined by results of crystals and solid
quasicrystals qualitatively as well as quantitatively (refer to (13.7.16) and (13.7.17)).
In the examination through the crystals and solid quasicrystals, we find that the
constant A can be an arbitrary positive number but does not equal zero. This shows
the soft-matter state cannot be reduced to any solid-state phase, so the constant
presents an important meaning for soft-matter. The introduction of the constant A
in studying the generalized dynamics of soft-matter quasicrystals can refer to Refs.
[11, 12] for more details. The thermodynamics of the matter shows that the constant
A becomes more important.

For possible soft-matter quasicrystals with 7-, 9-, and 14-fold symmetry the
stabilities are similar to that given by (13.7.14), which we will report in other cases.

The work on the stability of 12-(including 8- and 10-) and 18-fold symmetry
quasicrystals found that

(1)   The stability of the 12-fold symmetry quasicrystals depends upon the fluid
      effect, described by constant A, fluid coupling to phonons described by constant
      B, phonons described by C_{ijkl} and phasons described by K_{ijkl};

(2)   The stability of the 18-fold symmetry quasicrystals depends upon the fluid effect, fluid coupling to phonons and phasons (which are similar to those in [11]), and coupling between the first and second phasons describing by G_{ijkl};

(3)   These constants can be measured by experiments;

(4)   The differences between those of the first and second kinds of soft-matter quasicrystals lie in their different structures. This is explored through the constitutive laws (i.e., the Eq. (13.7.9) in Ref. [11] and (13.7.11) in this paper) determined by the theory of group representation.

(5)   Due to the combination between thermodynamics and generalized dynamics of the matter, the approach determining the stability developed in Ref. [11] and this paper presents systematic, direct, and simple features. It does not need to use any simulation treatment. The results obtained present bright physical meaning. Of course, it needs to be verified by experiments further.

13.8   Conclusion

The discussion on thermodynamic stability of soft-matter quasicrystals based on the extended free energy belongs to a phenomenological description. One of the physical bases is the importance of quantity $\frac{\delta\rho}{\rho_0}$ in the soft matter which is much greater than that in solid quasicrystals. The effect of $\frac{\delta\rho}{\rho_0}$ can be described through the LRT constants A and B. This phenomenological treatment greatly reduces the difficulty compared with the method based on the effective free energy, which is a complex function of mass density ρ. The evaluation of the function is very hard. The present method is supported by the generalized dynamics of soft-matter quasicrystals. Due to the advances in the theory, the constitutive laws of the matter are quite complete, providing all the material constants needed in constructing the rigidity matrixes for all quasicrystal systems observed and possibly observed. Of course, the variational principle provides a useful and powerful mathematical tool. All these beneficial conditions help us to find quantitatively the stability criteria for individual soft-matter quasicrystal systems.

References

1. Lifshitz, R., Diamant, H.: Soft quasicrystals—why are they stable? Phil. Mag. **87**, 3021–3030 (2007)
2. Barkan, K., Diamant, H., Lifshitz, R.: Stability of quasicrystals composed of soft isotropic particles. Phys. Rev. B **83** 172201 (2011)
3. Jiang, K., Tong, J., Zhang, P., Shi, A.-C.: Stability of two-dimensional soft quasicrystals in systems with two length scales. Phys. Rev. E **92**, 042159 (2015)
4. Savitz, S., Babadi, M., Lifshitz, R.: Multiple-scalestructures: from Faraday waves to soft-matter quasicrystals. IUCrJ **5**, 247–268 (2018)
5. Jiang, K., Si, W.: Stability of three-dimensional icosahedral quasicrystals in multi-component systems. arXiv preprint arXiv:1903.07859, (2019)

6. Ratli, D.J., Archer, A.J., Subramanian, P., Rucklidge, A.M.: Which wavenumbers determine the thermodynamic stability of soft matter quasicrystals? arXiv:1907.058501,12 Jul (2019)
7. Lifshitz, R., Petrich, D.M.: Theoretical model for faraday waves with multiple-frequency forcing. arXiv:cond-mat/9704060v2[cond-mat.soft] Jan 29 (1998)
8. Lubensky, T.C., Ramaswamy, S., Toner, J.: Hydrodynamics of icosahedral quasicrystals. Phys. Rev. B **32**, 7444–7452 (1985)
9. Cowley, R.A.: Acoustic phonon instabilities and structural phase transitions. Phys. Rev. B **13**, 4877–4885 (1976)
10. Fan, T.Y., Tang, Z.Y.: Three-dimensional generalized dynamics of soft-matter quasicrystals, Appl. Math. Mech. **38**, 1195–1207 (2017), in Chinese; Advances in materials science and engineering, **2020**, Article 1D 4875854 (2020)
11. Fan, T.Y., Tang, Z.Y.: The stability of soft-matter quasicrystals of 12-fold symmetry. arXiv:1909.00312[cond-mat.soft] 15 Oct 2019 (2019)
12. Tang, Z.Y., Fan, T.Y.: The stability of the second kind of soft matter quasicrystals, unpublished work (2019)

Chapter 14
Applications to Device Physics—Photon Band Gap of Holographic Photonic Quasicrystals

The most attractive aspect of the application of soft-matter quasicrystals may be in photon band gap. The soft-matter quasicrystals observed so far are two-dimensional structures with quasiperiodic symmetry, and higher fold of orientational symmetry being greater than that of solid one appeared, there is superiority than solid quasicrystals in this respect.

In this chapter, we construct two-dimensional photonic quasicrystals (PQCs) with different rotational symmetries by using multi-beam holographic interference. Their photonic band-gap (PBG) properties are investigated by the finite element method (FEM). The results show that 10-fold PQC formed by five-beam interference and 12-fold PQC are easier to form PBG. On the other hand, 10-fold PQCs were prepared in the SU8 photoresist film by using single-prism holographic interfering lithography. This will give guidance for selecting appropriate PQC structures to form photonic integrated devices.

14.1 Introduction

Compared with periodic photonic crystals (PCs), PQCs has many advantages, such as the easy appearance of a complete PBG [1–3], higher PBG isotropy [4, 5], and richer defect modes [6]. Because of abundant PBG and localization properties, PQCs have been used for designing waveguides [4, 7], sensors [8], laser microcavities [9], filters [2, 10], etc. This shows that PQC has great potential in the design of small size and multi-functional optical integrated circuits.

Research of PQC band gap is the basis of applications. Searching for wider complete PBG characteristics [11] and developing more application fields of PQC have been the goal pursued by scientists. Therefore, various PQC structures are designed and implemented by different methods. Among all these methods, the multiple-beam holographic interfering method is considered to be the simplest method for fabricating PQC structures with large areas and without defects [12,

© The Author(s), under exclusive license to Springer Nature Singapore Pte Ltd. 2022

T.-Y. Fan et al., *Generalized Dynamics of Soft-Matter Quasicrystals*, Springer Series in Materials Science 260, https://doi.org/10.1007/978-981-16-6628-5_14

13]. In this chapter, the multi-beam interfering method is used to construct different PQCs structures and the FEM method is utilized to investigate their PBG characteristics and compare the ability to form band-gaps. At the same time, 10-fold PQCs were prepared by using single-prism holographic lithography.

14.2   Design and Formation of Holographic PQCs

The multi-beam interference system is shown in Fig. 14.1. By using this method, two-dimensional PQC structures can be obtained [12]. The interfering patterns in the medium are expressed as follows,

$$I(r) = \sum_{j,l} E_j E_l * \exp\left\{-i\left[\left(\vec{k_j} - \vec{k_l}\right) \cdot \vec{r}\right]\right\}, \qquad (14.2.1)$$

where E_j and $\vec{k_j}$ are the amplitude and wave vector of beam j, respectively. $\vec{r} = (x, y, z)$ is the position vector. The wave vectors of the mth beam are:

$$\vec{k_m} = \frac{2\pi n_w}{\lambda}\left(\cos\frac{2(m-1)\pi}{N}\sin\varphi, \sin\frac{2(m-1)\pi}{N}\sin\varphi, \cos\varphi\right) \qquad (14.2.2)$$

where $m = 1 \rightarrow N$ is the number of beams, n_w represents the refractive index of the medium, φ is the angle between each beam and the central axis. We can get the interferograms by substituting $\varphi = 44.3°$, $n_w = 1.5$, $\lambda = 355$ nm into Eq. (14.2.1) and Eq. (14.2.2). The two-dimension different PQCs are shown in Fig. 14.2. Among these figures, Fig. 14.2b, e–h are formed by interfering 5, 7, 9, 11, and 13 beams, while Fig. 14.2a, c, and d are obtained by 8, 10, 12 beams interference. All of these

Fig. 14.1 Formation of multi-beam interfering system

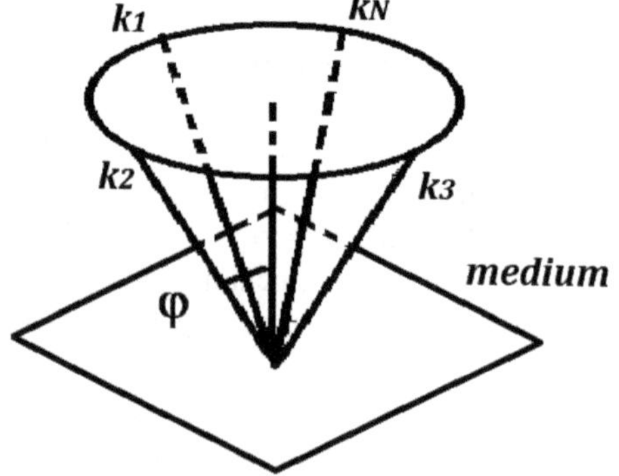

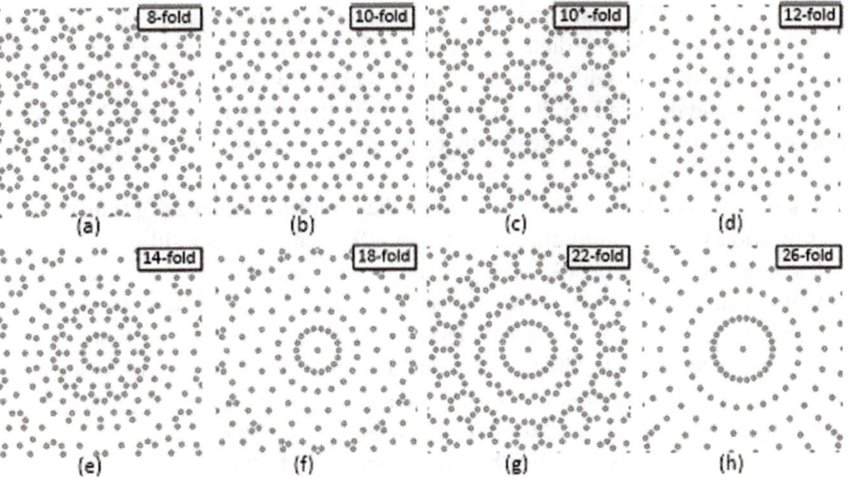

Fig. 14.2 Different two-dimensional PQC structures by interference of **a** 8 beams **b** 5 beams **c** 10 beams **d** 12 beams **e** 7 beams **f** 9 beams **g** 11 beams **h** 13 beams

structures possess even-fold rotational symmetry. From Fig. 14.2b and Fig. 14.2c, we can see that 10-fold PQC structures are different by the interference of five beams and ten beams, with the increase of beams, a more sparse structure has appeared in Fig. 14.2c.

14.3 Band Gap of 8-fold PQCs

In Fig. 14.3a, the 8×8 μm central regions in Fig. 14.2a is selected for calculating

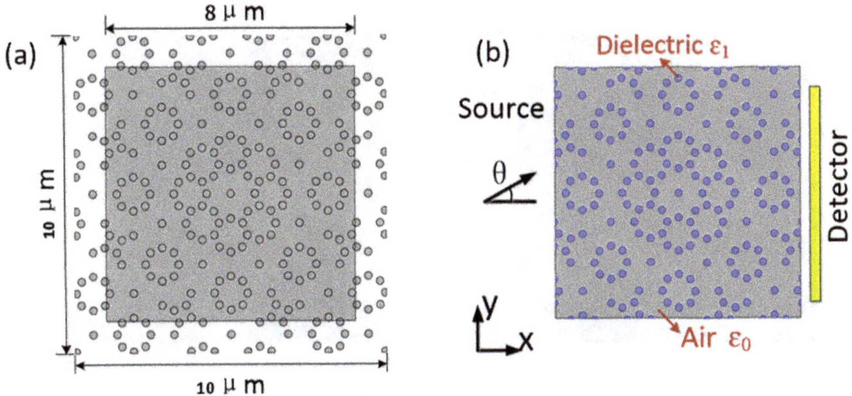

Fig. 14.3 **a** The selected 8-fold PQC, **b** The modeling setup

the PBG of 8-fold PQC. The more detailed setting is shown in Fig. 14.3b. Blue spots stand for medium rods with dielectric constant $\varepsilon_1 = 14$ and radius $r = 0.12\,\mu\text{m}$, and gray background is air with $\varepsilon_0 = 1$. Light is incident on the sample at an angle θ to the x-axis and detected on the right side. θ is in the range from 0 to 22.5°. PQCs band-gap properties are investigated by calculating the transmittance and reflection spectra of the sample [14].

When a TE polarized wave is incident on the sample, the change of transmission and reflection with the normalized frequency a/λ is calculated at different incident angles θ, as shown in Fig. 14.4. The lattice constant is chosen as $a = 1\,\mu\text{m}$ and λ is the wavelength. From the figure, we can see that the biggest PBGs happen in the frequency range from 0.476 to 0.523. The electric field distribution at the frequency $a/\lambda = 0.507$ and $a/\lambda = 0.730$ are also calculated and shown in Fig. 14.5. They are corresponding to the frequency in the band gap and outside the band gap. We can see that the light in the band gap is forbidden to propagate in the PQC, as in Fig. 14.5 a.

Then the maximum band gap of 8-fold quasicrystal was further investigated when dielectric contrast $\varepsilon_1/\varepsilon_0$ and fill factor r/a changed. The result is shown in Fig. 14.6.

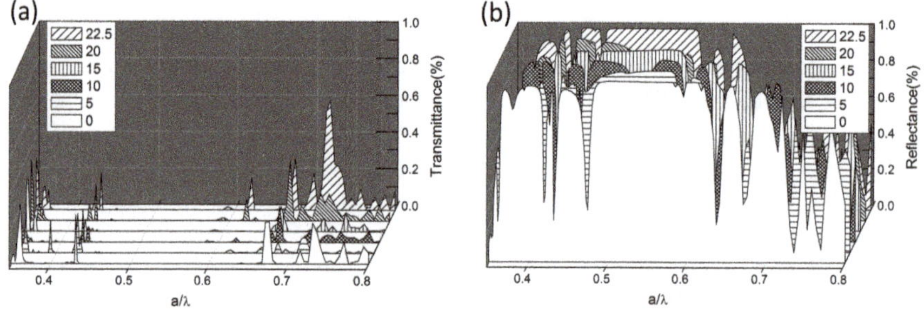

Fig. 14.4 **a** Transmission spectrum and **b** reflection spectrum of 8-fold PQC for different incidence angles

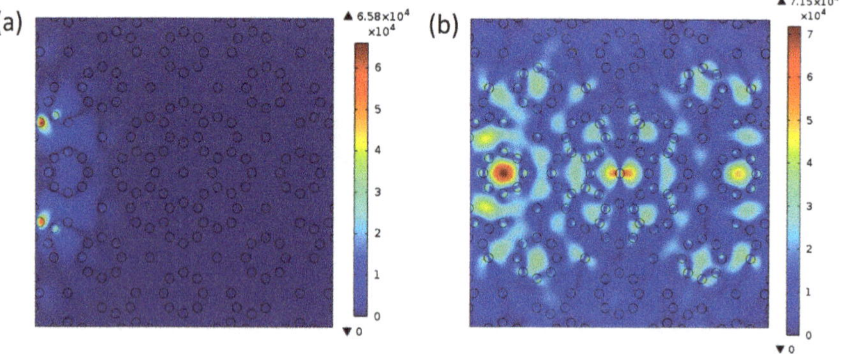

Fig. 14.5 The electric field distribution **a** $a/\lambda = 0.507$ in the band gap, **b** $a/\lambda = 0.730$ outside the band gap

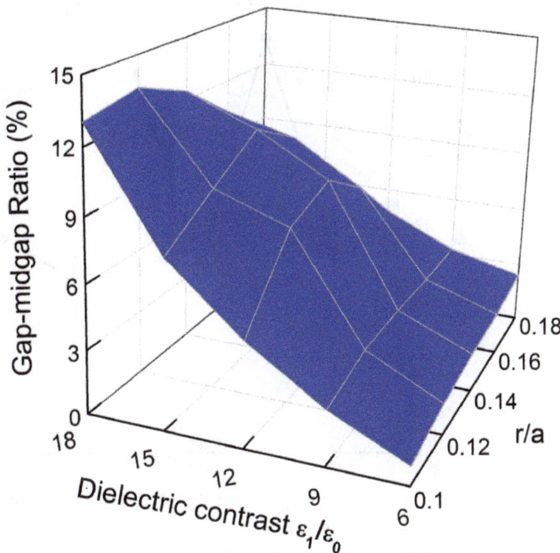

Fig. 14.6 Relative PBG versus parameters $\varepsilon_1/\varepsilon_0$ and r/a of 8-fold PQC

The relative PBG is expressed by the Gap-midgap Ratio, which is the ratio of band-gap width to the central band. From the figure, we can see that the maximum PBG happens at $r/a = 0.12$ and $\varepsilon_1/\varepsilon_0 = 18$.

14.4 Band Gap of Multi-fold Complex PQCs

In the same way, the relative band gap $\Delta\omega/\omega_c$ is optimized for all the PQCs formed by interference, where $\Delta\omega$ and ω_c are the width and central frequency of the PBG, respectively. The results are shown in Fig. 14.7. The figure reveals that five-beam-interference 10-fold PQC and 12-fold PQC are easier to generate PBG than others, PQC above 12-fold symmetry has smaller PBG. This is consistent with the viewpoint in the reference [15].

14.5 Fabrication of 10-Fold Holographic PQCs

14.5.1 Material and Writing System

The sample preparation includes four steps, clean glass substrates, spin coating of the SU8 photoresist film, holographic lithography of the film, and post-processing of the exposed film. The main material used in the sample preparation is SU-8 photoresist, a high contrast, epoxy-based photoresist resin from MICRO CHEM Cop., which

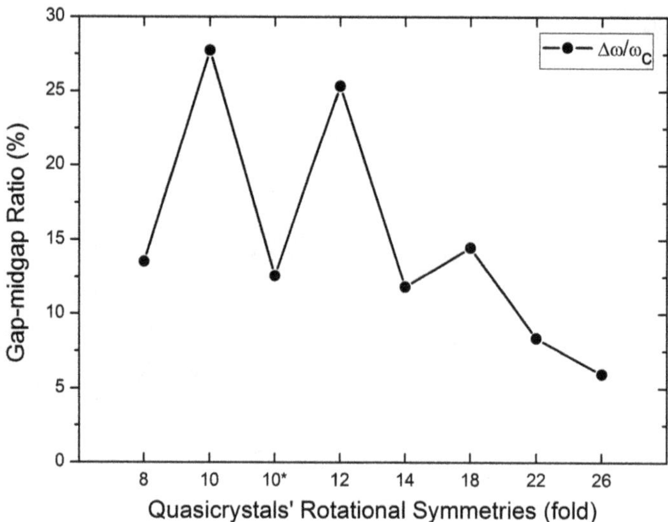

Fig. 14.7 Comparison of relative band gap for different PQCs

can produce photosensitive reactions under ultraviolet light and record interference light field. After the SU8 film is irradiated by the interfering laser, Propylene Glycol Monomethyl Ether Acetate (PGMEA, $C_6H_{12}O_3$) is used to develop the exposed SU8 film to obtain PQC structures [16].

The writing system is shown in Fig. 14.8. A 355 nm laser beam from a triple-frequency Nd:YAG laser (YG980, Quantel Co. France) was expanded and collimated, then split into six umbrella-like beams by using a top-cut prism invented by our group. The cutting angle of the prism is $\alpha = 54.7°$. The central beam was covered and the other beams irradiate the photoresist medium. The 2D interference patterns were recorded in the films. $n_\omega = 1.6$ is the refractive index of the SU-8 film at the writing wavelength.

14.5.2   Experimental Results

Scanning Electronic Microscopy (SEM S-4800, Hitachi Limited, Japan) and Scanning Probe Microscopy (Multimode 8 SPM, Bruker) are utilized to observe the two-dimensional PQC micro-structures. Before SEM measurement, the 1×1 cm sample needs to be coated with a thin gold film. For SPM, the film topography was obtained by using the tapping mode.

SEM micrographs of two samples with different exposure conditions are measured and shown in Fig. 14.9 and Fig. 14.10, respectively. Figure 14.9 indicates the patterns with different magnifications when the film is irradiated by five beams with the same intensity of 40 mJ/cm^2. From the figure, we can see a 10-fold rotational symmetrical

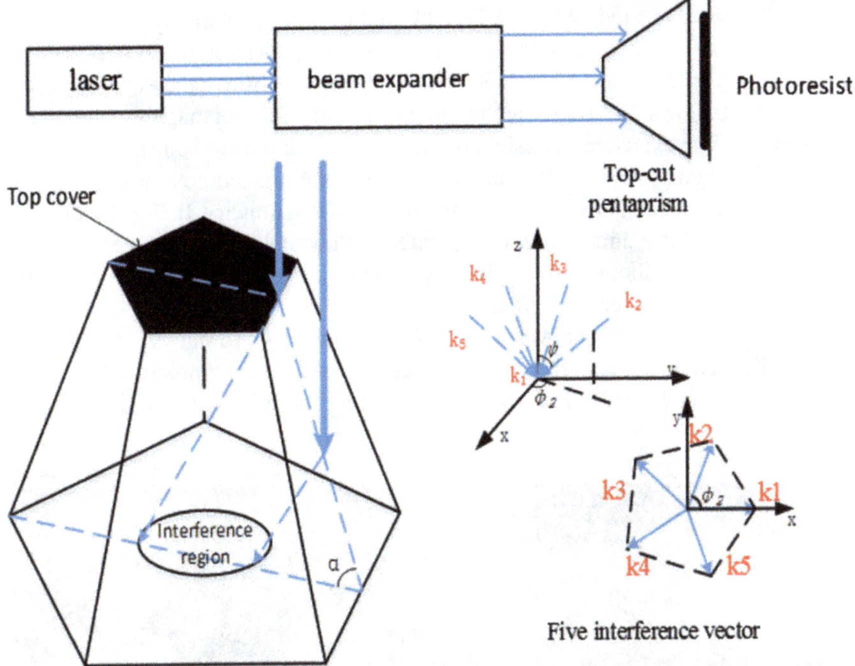

Fig. 14.8 The diagram of the interfering system

Fig. 14.9 SEM micrographs of PQC prepared with the same beam intensity

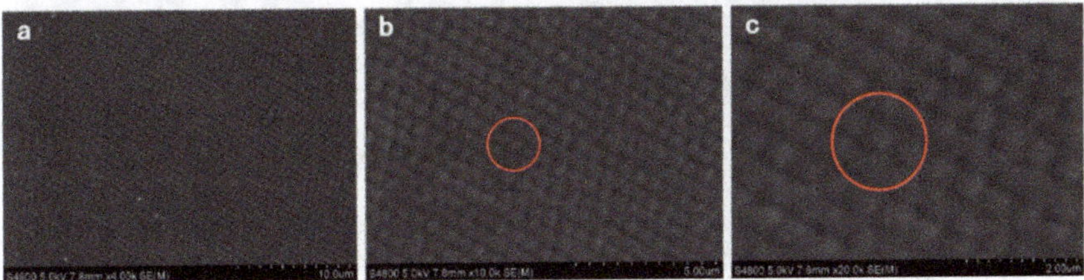

Fig. 14.10 SEM micrograph of PQC structure with two beams weaker

structure, as circled in red. While in Fig. 14.10, because the intensity of two beams is changed as 10 mJ/cm^2, the symmetry is broken and the structure is not perfect.

At the same time, we also provide SPM photographs to observe the PQC structures more clearly. Figure 14.11a and b are photographs with different magnification of the same sample. The exposure intensity and time of the sample are 20 mJ/cm^2 and 4 s, respectively. Figure 14.11b is the magnification of the marked area in Fig. 14.11a. From this figure, we can observe the 10-fold rotational symmetry. In Fig. 14.11b, the red circle stands for a unit cell, whose diameter is about 1.6 μm. The gap between two adjacent dots is about 490 nm. To reveal the rotational symmetry more clearly, a 532 nm Nd:YVO$_4$ laser is incident on the sample perpendicularly to observe the diffraction pattern, which is shown in Fig. 14.11c. Figure 14.11d shows the calculated intensity distributions. From this figure, we can see the same symmetrical structure as the experimental result. The size of a unit cell and adjacent dots is 1.62 μm

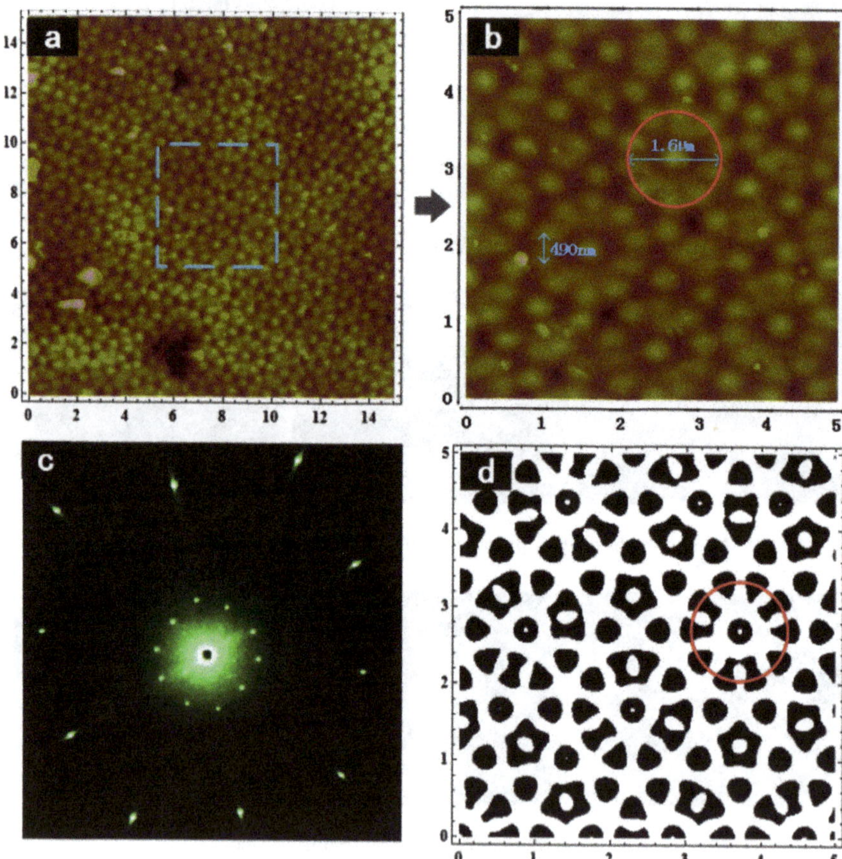

Fig. 14.11 SPM micrographs **a** 14 × 14 μm, **b** 5 × 5 μm, **c** diffraction patterns, **d** the calculated intensity distribution

and 500 nm, respectively. These two values are a little bigger than the experimental results. This is caused by the shrinkage of the photoresist film in the developing and drying process.

14.6 Band Gap of Cholesteric Liquid Crystal

Cholesteric liquid crystal (CLC) is fascinating material due to its unique optical properties. CLC is also called chiral nematic liquid crystal. It can be considered as a one-dimensional photonic crystal (PC) with a regularly modulated refractive index along the helix axis because of the particular arrangement of the molecules, and it has a one-dimensional photonic pseudo-band gap like PC. In planarly aligned cells, the macroscopic orientation of its helical superstructures will be perpendicular to the substrates and lead to film, which will selectively reflect the same handedness radiation as itself [17–21].

When light is incident along the helix axis or on the substrate perpendicularly, the central wavelength is expressed as $\lambda = np$, where p is the cholesteric pitch and $n = \frac{n_o + n_e}{2}$ is the average of the ordinary (n_o) and extraordinary (n_e) refractive indices. The spectral width $\Delta\lambda$ of the band gap is defined as $\Delta\lambda = p\Delta n$, where $\Delta n = n_e - n_o$ is the birefringence. Within $\Delta\lambda$, an incident unpolarized or linearly polarized light parallel to the helix axis is split into two opposite circularly-polarized components, one of which is transmitted whereas the other is reflected. The polarization rotation direction of reflected light is consistent with the spiral direction of CLC. A wavelength out of $\Delta\lambda$ is simply transmitted. Compared to conventional pigmented color filters, a CLC filter has the advantage to integrate several optical properties in one layer: it is not only a color filter but also a reflector and a polarizer. Since Δn values are limited for colorless organic compounds, $\Delta\lambda$ is often less than 100 nm in the visible spectrum.

When the pitch length p is on the same order as visible light, vibrant, highly colored films are formed which have been examined for numerous displays, sensing, and photonic applications. A large amount of work has been performed on making this coloration dynamic, through the use of heat, light, or electrical fields. Both tunable and switchable constructs are continuing to be explored and methodologies to increase both the response and relaxation speeds are at the forefront of this research.

Experimentally, a polymer-stabilized cholesteric liquid crystal (PSCLC) is explored to tune and broaden the band gap. PSCLC is obtained by in situ photopoly-merizations of reactive liquid–crystal molecules in the presence of non-reactive liquid–crystal molecules in an oriented Bragg planar texture by combining the UV-curing with a thermally induced pitch variation.

14.7 Conclusions

The multi-beam holographic interfering method has been used to form various PQCs structures with different rotational symmetries. By using the FEM method to calculating the transmission and reflection characteristics, we compare and optimize their PBGs. On the other hand, 10-fold PQCs were prepared by using single-prism holographic interference lithography. This will be helpful for the design and fabrication of PQC based photonic integrated circuits.

At the same time, to help readers better understand the relationship between quasicrystals and liquid crystals, the characteristics of CLC and the relevance between CLC and PC are also introduced.

References

1. Rechtsman, M.C., Jeong, H.C., Chaikin, P.M., Torquato, S., Steinhardt, P.J.: Optimized structures for photonic quasicrystals. Phys. Rev. Lett. **101**, 073902 (2008)
2. Romero-Vivas, J., Chigrin, D.N., Lavrinenko, A.V., Torres, C.S.: Resonant add-drop filter based on a photonic quasicrystal. Opt. Express **13**, 826–835 (2005)
3. Hase, M., Miyazaki, H., Egashira, M., Shinya, N., Kojima, K.M., Uchida, S.I.: Isotropic photonic band-gap and anisotropic structures in transmission spectra of two-dimensional fivefold and eightfold symmetric quasiperiodic photonic crystals. Phys. Rev. B **66**, 214205 (2002)
4. Jin, C., Cheng, B., Man, B., Li, Z., Zhang, Z., Ban, S., Sun, B.: Band-gap and wave guiding effect in a quasiperiodic photonic crystal. Appl. Phys. Lett. **75**, 1848–1850 (1999)
5. Yin, J., Huang, X., Liu, S., Hu, S.: Photonic band-gap properties of 8-fold symmetric photonic quasicrystals. Opt. Commun. **269**, 385–388 (2007)
6. Mnaymneh, K., Gauthier, R.C.: Mode localization and band-gap formation in defect-free photonic quasicrystals. Opt. Express **14**, 5089–5099 (2007)
7. Rose, P., Zito, G., Di Gennaro, E., Abbate, G., Andreone, A.: Control of the light transmission through a quasiperiodic waveguide. Opt. Express **20**, 26056–26061 (2012)
8. Wang, S., Sun, X., Wang, C., Peng, G., Qi, Y., Wang, X.: Liquid refractive index sensor based on a 2D 10-fold photonic quasicrystal. J. Phys. D Appl. Phys. **50**, 365102 (2017)
9. Ren, J., Sun, X., Wang, S.: A low threshold nanocavity in a two-dimensional 12-fold photonic quasicrystal. Opt. Laser Technol. **101**, 42–48 (2018)
10. Ren, J., Sun, X., Wang, S.: A narrowband filter based on 2D 8-fold photonic quasicrystal. Superlattices Microstruct. **116**, 221–226 (2018)
11. Florescu, M., Torquato, S., Steinhardt, P.J.: Complete band-gaps in two-dimensional photonic quasicrystals. Phys. Rev. B **80**, 145112 (2009)
12. Sun, X., Wang, S., Liu, W., Jiang, L.: A simple configuration for fabrication of 2D and 3D photonic quasicrystals with complex structures. Opt. Commun. **369**, 138–144 (2016)
13. Sun, X., Liu, W., Wang, G., Tao, X.: Optics design of a top-cut prism interferometer for holographic photonic quasicrystals. Opt. Commun. **285**, 4593–4598 (2012)
14. Xi, X.Y., Sun, X.H.: Photonic band-gap properties of two-dimensional photonic quasicrystals with multiple complex structures. Superlattices Microstruct. **129**, 247–251 (2019)
15. Sellers, S., Man, W., Sahba, S., Florescu, M.: Local self-uniformity in photonic networks. Nat. Commun. **8**, 14439 (2017)
16. Wang, S., Sun, X.H., Li, W.Y., Liu, W., Jiang, L., Han, J.: Fabrication of photonic quasicrystalline structures in the sub-micrometer scale. Superlattices Microstruct. **93**, 122–127 (2016)

17. Dunmur, D., Toriyama, K.: In: Demus, D., et al. (ed.) Physical Properties of Liquid Crystals. Wiley-VCH, Weinheim, pp. 124–128 (1999)
18. Mitova, M., Nouvet, E., Dessaud, N.: Polymer-stabilized cholesteric liquid crystals as switchable photonic broad band-gaps. Eur. Phys. J. E **14**, 413–419 (2004)
19. Hrozhyk, U.A., Serak, S.V., Tabiryan, N.V., White, T.J., Bunning, T.J.: Nonlinear optical properties of fast, photo-switchable cholesteric liquid crystal band-gaps. Opt. Mater. Express **1**, 943–952 (2011)
20. Hwang, J., Ha, N.Y., Chang, H.J., Park, B., Wu, J.W.: Enhanced optical nonlinearity near the photonic band-gap edges of a cholesteric liquid crystal. Opt. Lett. **29**, 2644–2646 (2004)
21. Costello, M.J., Meiboom, S., Sammon, M.: Electron microscopy of a cholesteric liquid crystal and its blue phase. Phys. Rev. A **29**, 2957–2959 (1984)

Chapter 15
Possible Applications to General Soft Matter

In Chaps. 7–11, we have introduced the dynamics of soft-matter quasicrystals from a unified point of view, where the quasiperiodic symmetry has been specially considered for quasicrystal applications in soft matter, such as liquid crystals, polymers, colloids, nanoparticles, surfactants, and macromolecules, etc. Soft-matter quasicrystal is a subclass of soft matter, whose dynamics are inherited from the complex fluid dynamics of general soft matter. If we do not consider the quasiperiodic symmetry, the soft-matter quasicrystals would be reduced to general soft matter. In this case, the phason fields are excluded, and the interactions among phasons and phonons do not exist. We only need to consider the effects of elasticity and fluidity and their interactions under macroscopic continuum mechanics, or a phenomenological point of view. In fact, in Chap. 16 we will present smectic A liquid crystals in this way.

This chapter will mainly focus on the discussion of structures and dynamic properties of the general soft matter, which includes the fluid and elastic fields and their interactions. The governing equations of the dynamics have not been developed before, and we will simplify them based on the discussion in this book for soft-matter quasicrystal cases by excluding the phason fields.

15.1 A Basis of Dynamics of Two-Dimensional Soft Matter

Soft matter is an intermediate phase between simple liquid and conventional solid, which can be characterized by fluid field (with fluid velocity $\mathbf{V}$) and elastic field (with elastic displacement $\mathbf{u}$). From the previous discussion in Chaps. 7–11, one can simply omit the phason degree, and draw the dynamic equations for general soft matter.

© The Author(s), under exclusive license to Springer Nature Singapore Pte Ltd. 2022

T.-Y. Fan et al., *Generalized Dynamics of Soft-Matter Quasicrystals*, Springer Series in Materials Science 260, https://doi.org/10.1007/978-981-16-6628-5_15

Based on the Poisson bracket method of condensed matter physics [1–5], we have the mass conservation equation or continuum equation

$$\frac{\partial \rho}{\partial t} + \nabla_k(\rho V_k) = 0 \tag{15.1.1}$$

the momentum conservation equations or generalized Navier–Stokes equations

$$\frac{\partial g_i(r,t)}{\partial t} = -\nabla_k(r)(V_k g_i) + \nabla_j(r)\left(-\rho\delta_{ij} + \eta_{ijkl}\nabla_k(r)g_1\right)$$

$$- \left(\delta u_i - \nabla_i u_j\right)\frac{\delta H}{\delta u_i(r,t)}$$

$$\left(\nabla_i w_j\right)\frac{\delta H}{\delta w_i(r,t)} = -\rho\nabla_i(r)\frac{\delta H}{\delta w_i(r,t)}, \, g_j = \rho V_j \tag{15.1.2}$$

the equations of motion of phonons due to the symmetry breaking

$$\frac{\partial u_i(r,t)}{\partial t} = -V_j\nabla_j(r)u_i - \Gamma_u\frac{\delta H}{\delta u_i(r,t)} + V_i, \tag{15.1.3}$$

Here Γ_u is the phonon dissipation coefficient.

We must supplement more equations to keep the system close. The equation of state, the relation between fluid pressure and mass density $p = f(\rho)$ can be selected as we discussed starting from Chap. 4 which comes from the classical thermodynamics [6, 7].

$$p = f(\rho) = \frac{1}{\kappa_T}\left(1 - \frac{\rho_0}{\rho}\right) \tag{15.1.4}$$

where κ_T is a key compressibility parameter, which is in the order of

$$\kappa_T = 10^{-5}/\text{Pa} \tag{15.1.5}$$

based on our systematical and large scale computations.

The above Eqs. (15.1.1) and (15.1.2) correspond to conservation laws for mass and momentum, while (15.1.3) to the symmetry breaking rules, respectively, in which H denotes the Hamiltonian defined by

$$H = H[\Psi(r, t)] = \int \frac{g^2}{2\rho} d^d r + \int \left[\frac{1}{2} A \left(\frac{\delta\rho}{\rho_0} \right)^2 + B \left(\frac{\delta\rho}{\rho_0} \right) \nabla \cdot u \right] d^d r + F_{el}$$

$$= H_{kin} + H_{density} + F_{el}$$

$$F_{el} = F_u, g = \rho V \tag{15.1.6}$$

and $\mathbf{V}$ represents the fluid velocity field, A, B the constants describing density variation named as Lubensky- Ramaswamy-Toner (LRT) constants. The last term of (15.1.6) represents elastic energies, which consists of only phonon contribution,

$$F_u = \int \frac{1}{2} C_{ijkl} \varepsilon_{ij} \varepsilon_{kl} \mathrm{d}r \tag{15.1.7}$$

where C_{ijkl} is the phonon elastic constant tensor, and ε_{ij} the Cauchy strain tensor defined by

$$\varepsilon_{ij} = \frac{1}{2} \left(\frac{\partial u_i}{\partial x_j} + \frac{\partial u_j}{\partial x_i} \right) \tag{15.1.8}$$

and constitutive law including fluid is given by

$$\sigma_{ij} = C_{ijkl} \varepsilon_{ik}, \, p_{ij} = -p\delta_{ij} + \sigma'_{ij} = -p\delta_{ij} + \eta_{ijkl} \dot{\xi}_{kl} \tag{15.1.9}$$

which were discussed in detail in the previous chapters.

15.2 The Outline on Governing Equations of Dynamics of Soft Matter

From the general Eqs. (15.1.1)–(15.1.9) the dynamic governing equations for soft matter can be derived. As an example, the equations for two-dimensional basic equations can be simplified as follows:

$$
\left.
\begin{aligned}
&\frac{\partial \rho}{\partial t} + \nabla \cdot (\rho V) = 0 \\
&\frac{\partial(\rho V_x)}{\partial t} + \frac{\partial(V_x \rho V_x)}{\partial x} + \frac{\partial(V_y \rho V_x)}{\partial y} = -\frac{\partial p}{\partial x} + \eta \nabla^2 V_x + \frac{1}{3}\eta \frac{\partial}{\partial x} \nabla \cdot V \\
&\quad + M\nabla^2 u_x + (L + M - B)\frac{\partial}{\partial x}\nabla \cdot u - (A - B)\frac{1}{\rho_0}\frac{\partial \delta\rho}{\partial x} \\
&\frac{\partial(\rho V_y)}{\partial t} + \frac{\partial(V_x \rho V_y)}{\partial x} + \frac{\partial(V_y \rho V_y)}{\partial y} = -\frac{\partial p}{\partial y} + \eta \nabla^2 V_y \\
&\quad + \frac{1}{3}\eta \frac{\partial}{\partial y}\nabla \cdot V + M\nabla^2 u_y + (L + M - B)\frac{\partial}{\partial y}\nabla \cdot u \\
&\quad - (A - B)\frac{1}{\rho_0}\frac{\partial \delta\rho}{\partial y} \\
&\frac{\partial u_x}{\partial t} + V_x \frac{\partial u_x}{\partial x} + V_y \frac{\partial u_x}{\partial y} = V_x + \Gamma_u\left[M\nabla^2 u_x + (L + M)\frac{\partial}{\partial x}\nabla \cdot u\right] \\
&\frac{\partial u_y}{\partial t} + V_x \frac{\partial u_y}{\partial x} + V_y \frac{\partial u_y}{\partial y} = V_y + \Gamma_u\left[M\nabla^2 u_y + (L + M)\frac{\partial}{\partial y}\nabla \cdot u\right] \\
&p = f(\rho)
\end{aligned}
\right\} \quad (15.2.1)
$$

Equations (15.2.1) are an outline of possible complex fluid dynamics for general two-dimensional soft matter derived from (15.1.1)–(15.1.9). In fact, for different types of soft matter, they have different structural characters and should be treated differently. For example, liquid crystals are different from surfactants in soft matter, the differences between them have not been explored yet in (15.2.1). In the smectic A liquid crystals, there is Frank strain tensor apart from the Cauchy strain tensor [8]. The former is induced from the local curvature, and we will discuss it further in Chap. 16. With this connection, there are two sets of elasticity constitutive equations, which leads to more complicated dynamic equations than those given by (15.2.1). General liquid crystals and polymers have more complicated structures than those of smectic A liquid crystals [8, 9]. These discussions are beyond the scope of the book.

15.3 The Modification and Supplement to Eq. (15.2.1)

To describe concretely the individual soft matter, we need to provide some physical parameters regarding the specific soft matter into Eq. (15.2.1), i.e., the equations should be specified. One example of smectic A liquid crystals will be given in Chap. 16.

15.4 Solution of the Dynamics of Soft Matter

After the modification and supplement to Eq. (15.2.1), one can obtain the governing equations for complex fluid dynamics of individual soft matter. With appropriate initial- or boundary- or initial, and boundary-value conditions and solving the equations, one can obtain approximate results about the motion of soft matter. These equations are similar to those of soft-matter quasicrystals without all terms related to phason field. Since the initial- or boundary- or initial, and boundary-value conditions are similar to those for problems concerning soft-matter quasicrystals, the related solution process is also similar. With the experiences for soft-matter quasicrystals, we can easily treat the initial- or boundary- or initial, and boundary-value problems in general soft matter. The discussion in Chap. 16 can be referenced as an example, the discussions there would set up a model system for other general soft matter.

15.5 Conclusion and Discussion

This chapter provides a probe for studying complex fluid dynamics of general soft matter. The in-depth study should consider the specific structure of individual soft matter, and one can modify and supplement Eq. (15.2.1) accordingly. Chapter 16 will give an example for the purpose.

According to our practice, the equation of state in the dynamics of soft matter plays a very important role. For different soft matters, the equation of state should be treated differently. At the moment, the precise form of the equation of state in different soft matters is still an open question. If one uses Eq. (15.1.4), then an approximate estimation can be provided. Another practical problem is the material constants for different soft matters, due to very limited data available.

We have focused out discussion on two-dimensional compressible soft matter. The three-dimensional and incompressible cases can be considered similarly. Referring to the experience in soft-matter quasicrystals given by Chaps. 7–11 and 13, the equations for these regimes can be obtained, and the solving procedure can also be learned from those given in Chaps. 7–11, 13, and Chap. 16. We need to make specific justification for concrete soft matter system. The description in Chap. 16 gives us a successful example. Even though without phason field, the analysis is still quite difficult.

References

1. Dzyaloshinskii, I.E., Volovick, G.E.: Poisson brackets in condensed matter physics. Ann. Phys. (NY) **125**, 67–97 (1980)
2. Dzyaloshinskii, I.E., Volovick, G.E.: On the concept of local invariance in spin glass theory. J. de Phys. **39**, 693–700 (1978)

3. Volovick, G.E.: Additional localized degrees of freedom in spin glasses. ZhEkspTeorFiz **75**, 1102–1109 (1978)
4. Stark, H., Lubensky, T.C.: Poisson bracket approach to the dynamics of nematic liquid crystals. Phys. Rev. E **67**, 061709 (2003)
5. Fan, T.Y.: Poisson bracket method and its applications to quasicrystals, liquid crystals and a class of soft matter. Chin. J. Theor. Appl. Mech. **45**, 548–559 (2013), in Chinese
6. Cheng, H., Fan, T.Y., Yao, Y.G., Xing, Y.X.: Equation of state and complex fluid dynamics of soft-matter quasicrystals, Phys. Rev. B. (2021) to be submitted
7. Wang, Z.C.: Thermodynamics and Statistics. Higher Education Press, Beijing (2014), in Chinese
8. de Gennes, P.D., Prost, J.: The Physics of Liquid Crystals. Clarendon, London (1993)
9. Landau, L.D., Lifshitz, E.M.: Theory of Elasticity. Pergamon Press, Oxford (1980)

Chapter 16
An Application to Smectic A Liquid Crystals, Dislocation, and Crack

In the previous chapter, we discuss general soft matter using the theory and method developed for solving soft-matter quasicrystals, where we emphasized one must consider the structure of concrete soft matter. In this chapter, we study a concrete soft matter, i.e., the smectic A liquid crystal and its dislocation and crack problem. These are interesting topics in soft matter. Apart from this, we hope to explore a longstanding puzzle, perhaps a paradox. The solution to the paradox may yield some beneficial results and lessons.

16.1 Basic Equations

The structure of smectic A liquid crystals is well-known and typical in soft matter, which has been studied for quite a long time. The free energy due to deformation is obtained, i.e., the well-known Landau-Ginzburg-de Gennes free energy [1, 2].

$$
\begin{cases}
F_d = F - F_0(T) = \dfrac{1}{2}(A/\rho_0)(\rho - \rho_0)^2 + C(\rho - \rho_0)\dfrac{\partial u}{\partial z} + \dfrac{1}{2}B\rho_0\left(\dfrac{\partial u}{\partial z}\right)^2 \\[2ex]
\qquad + \dfrac{1}{2}K_1(\nabla^2 u)^2 = \dfrac{1}{2}\rho_0 B'\left(\dfrac{\partial u}{\partial z}\right)^2 + \dfrac{1}{2}K_1(\nabla^2 u)^2 \\[2ex]
\nabla^2 = \dfrac{\partial^2}{\partial x^2} + \dfrac{\partial^2}{\partial y^2},\ \rho - \rho_0 = -\rho_0 m\dfrac{\partial u}{\partial z},\ m = \rho_0\dfrac{C}{A},\ B' = B - \dfrac{C^2}{A}
\end{cases}
$$

$$(16.1.1)$$

in which the contribution of kinetic energy is not included, the deformation caused by bulk deformation with Cauchy energy density or elastic energy

$$
f_e = \frac{1}{2}C_{ijkl}\varepsilon_{ij}\varepsilon_{kl}
$$

© The Author(s), under exclusive license to Springer Nature Singapore Pte Ltd. 2022
T.-Y. Fan et al., *Generalized Dynamics of Soft-Matter Quasicrystals*, Springer Series in Materials Science 260, https://doi.org/10.1007/978-981-16-6628-5_16

and one of curvature with Frank energy density

$$f_c = \frac{1}{2}K_1(\mathrm{div}\,\mathbf{n})^2 + \frac{1}{2}K_2(\mathbf{n}\cdot\mathrm{rot}\,\mathbf{n})^2 + \frac{1}{2}K_3(\mathbf{n}\times\mathrm{rot}\,\mathbf{n})^2$$

where $\mathbf{n} = (n_x, n_y, n_z)$ is the so-called director vector, and for smectic A liquid crystals.

$$n_x \approx \frac{\partial u_z}{\partial x}, n_y \approx \frac{\partial u_z}{\partial y}, n_z \approx 1,$$

and $\mathbf{u} = (0, 0, u_z)$, $u \equiv u_z$ and $u_x = u_y = 0$, the displacement normally to layer of the smectic A liquid crystals, $\rho_0 B'$ denotes the shear Young's modulus of bulk deformation and K_1 the splay modulus corresponding to deformation due to curvature, K_2 and K_3 corresponding twisting and bending have no contribution, respectively. The mass and momentum conservation equations and the equation of motion of displacement discussed in previous chapters are still valid for the present case, only the phason dissipation equation is not needed, i.e., we have the governing equations with the simplified version [1].

$$\frac{\partial \rho}{\partial t} = -\nabla_i(\rho V_i) \tag{16.1.2}$$

$$\rho\frac{\partial V_i}{\partial t} = -\rho V_k(\nabla_k V_i) + \nabla_j\left(\sigma_{ij} + p_{ij}\right) \tag{16.1.3}$$

$$\frac{\partial u_i(r, t)}{\partial t} = -V_j\nabla_j u_i - \Gamma_u\frac{\partial \sigma_{ij}}{\partial x_j} + V_i \tag{16.1.4}$$

$$p = f(\rho) \tag{16.1.5}$$

and $(\sigma_{ij})^{\mathrm{total}} = p_{ij} + \sigma_{ij}$ in which p_{ij} the fluid stresses

$$p_{ij} = -p\delta_{ij} + \sigma'_{ij} = -p\delta_{ij} + \eta_{ijkl}\dot{\xi}_{ij}, \dot{\xi}_{ij} = \frac{1}{2}\left(\frac{\partial V_i}{\partial x_j} + \frac{\partial V_j}{\partial x_i}\right) \tag{16.1.6}$$

and σ_{ij} the elastic ones:

$$
\begin{cases}
\sigma_{xx} = \sigma_{yy} = K_1 \nabla^2 \dfrac{\partial u}{\partial z} \\[2ex]
\sigma_{zz} = \rho_0 B' \dfrac{\partial u}{\partial z} \\[2ex]
\sigma_{zx} = \sigma_{xz} = -K_1 \nabla^2 \dfrac{\partial u}{\partial x} \\[2ex]
\sigma_{zy} = \sigma_{yz} = -K_1 \nabla^2 \dfrac{\partial u}{\partial y} \\[2ex]
\sigma_{xy} = \sigma_{yx} = 0
\end{cases}
\tag{16.1.7}
$$

In the equation of displacement, a term connecting thermal conductivity is omitted. The equation set is reduced by omitting nonlinear terms

$$
\frac{\partial \rho}{\partial t} = -\rho \frac{\partial V}{\partial z} - V \frac{\partial \rho}{\partial z}
$$

$$
\rho \frac{\partial V}{\partial t} = -\frac{\partial p}{\partial z} + \rho_0 B' \frac{\partial^2 u}{\partial z^2} - K_1 \nabla^2 \nabla^2 u - \eta \nabla^2 V
$$

$$
\frac{\partial u}{\partial t} = V + \Gamma_u \nabla^2 \nabla^2 u
$$

$$
p = f(\rho)
$$

$$
\frac{\partial(\rho V_z)}{\partial t} + \frac{\partial(V_x \rho V_z)}{\partial x} + \frac{\partial(V_y \rho V_z)}{\partial y} + \frac{\partial(V_z \rho V_z)}{\partial z}
$$

$$
= -\frac{\partial p}{\partial z} + \eta \nabla^2 V_z + \frac{1}{3}\eta \frac{\partial}{\partial z}\nabla \cdot \mathbf{V}
$$

$$
+ \left(C_{44}\frac{\partial^2}{\partial x^2} + C_{44}\frac{\partial^2}{\partial y^2} + (C_{33} - C_{13} - C_{44})\frac{\partial^2}{\partial z^2} \right) u_z
$$

$$
+ (C_{13} + C_{44} - B)\frac{\partial}{\partial z}\nabla \cdot \mathbf{u} - (A - B)\frac{1}{\rho_0}\frac{\partial \delta \rho}{\partial z}
$$

where $V \equiv V_z$ and $V_x = V_y = 0$. For incompressible, steady state and omitting nonlinear terms, the above equations are simplified as equations such as

$$
-\frac{\partial p}{\partial z} + \rho_0 B' \frac{\partial^2 u}{\partial z^2} - K_1 \nabla^2 \nabla^2 u - \eta \nabla^2 V = 0
$$

$$
V + \Gamma_u \nabla^2 \nabla^2 u = 0
\tag{16.1.8}
$$

In the case the field variables are independent of variable z and omitting the fluid effect then furthermore we have

$$
\nabla^2 \nabla^2 u = 0
\tag{16.1.9}
$$

16.2   The Kleman-Pershan Solution of Screw Dislocation

If there is screw dislocation with the Burgers vector $(0, 0, b)$, Kleman [3], Pershan [4] solved it under boundary condition

$$\int_\Gamma du = b \tag{16.2.1}$$

and their solution is

$$u = \frac{b\theta}{2\pi} \tag{16.2.2}$$

This solution is not correct, because it causes all stress components to vanish. Although certain researchers, e.g., Pleiner [5], criticized de Kleman-Pershan's solution (16.2.2), he still confirmed that the solution (16.2.2) holds in the region outside the dislocation core. Unfortunately, the idea has been widely accepted. A further discussion on the solution is necessary. The above solution is a solution out of the core of the dislocation. Kralj and Sluckin [6] studied the core structure of a screw dislocation in smectic A liquid crystals, providing very interesting and important results. The core structure naturally influences the solution out of the core. But at present, our attention focuses only on the solution out of the core whatever the core structure is. If we can correctly explore the solution, this may help us to reveal the core structure.

16.3   Common Fundamentals of Discussion

On the solution (16.2.2) there have been many discussions from different perspectives so far, e.g., the magnetism analog [2, 4], the differential geometry [7], the dynamics [8], the structure of dislocation core [6, 9, 10], etc. Although these discussions from different points of view are beneficial, this leads to some difficulties for the readers. To ensure the discussion reaches an agreement, the following common understanding is necessary.

According to the physical facts mentioned above, the mathematical formulation of screw dislocation in smectic A liquid crystals is the boundary value problem of the biharmonic partial differential equation such as

$$\left.\begin{aligned}
&\nabla^2 \nabla^2 u = 0 \\
&(x^2 + y^2)^{1/2} \to \infty : \quad \sigma_{ij} = 0 \\
&\sigma_{zz}(x, 0) = 0 \\
&\int_\Gamma du = b
\end{aligned}\right\} \tag{16.3.1}$$

in which σ_{ij} denotes the elastic stress tensor, and Γ denotes a closed contour enclosing the dislocation core.

If one can solve the boundary value problem, then the solution may be obtained.

16.4 The Simplest and Most Direct Solution and the Additional Boundary Condition

To let readers easily understand the discussion, we suggest to take the simplest, elementary, and straightforward method to solve the boundary value problem (16.3.1), and do not need to use the magnetism analog, or Fourier transforms, or Green function, etc. Some references made the problem complex by using complicated mathematical methods. In contrast, we take an alternative way, in which the analysis is extremely simplified.

Introducing polar coordinate system (r, θ), the biharmonic equation is rewritten as

$$\left(\frac{\partial^2}{\partial r^2} + \frac{1}{r}\frac{\partial}{\partial r} + \frac{1}{r^2}\frac{\partial^2}{\partial \theta^2} \right) \left(\frac{\partial^2}{\partial r^2} + \frac{1}{r}\frac{\partial}{\partial r} + \frac{1}{r^2}\frac{\partial^2}{\partial \theta^2} \right) u(r, \theta) = 0 \qquad (16.4.1)$$

A suitable solution of Eq. (16.4.1) through the variable separation method, i.e., $u(r, \theta) = f(r)\Theta(\theta)$, takes the following form

$$\begin{aligned}
u = \frac{b}{2\pi} \big[& (Dr^2 + Er^2 \ln r + F + G \ln r)\theta \\
& + (D_1 r + E_1 r \ln r)\theta \sin\theta + (F_1 r + G_1 r \ln r)\theta \cos\theta \big]
\end{aligned}$$

where we have neglected the terms which are independent of the solution of the dislocation. In other words, those terms that only cause an increment in angle when going a circuit around the dislocation core are retained. Making use of the boundary conditions in (16.3.1), we find that the parts related to D, E, and G give rise to an increment dependent on r when running around the dislocation core. After removing the terms related to D, E, and G, a suitable solution further takes the following form

$$u = \frac{b}{2\pi}[F + (D_1 r + E_1 r \ln r)\sin\theta + (F_1 r + G_1 r \ln r)\cos\theta]\theta \qquad (16.4.2)$$

in which the unknown constants E_1 and G_1 vanish by considering stress continuity. Consequently, solution (16.4.2) at last becomes

$$u = \frac{b}{2\pi}[F + D_1 r \sin\theta + F_1 r \cos\theta]\theta \qquad (16.4.2b)$$

Furthermore, due to condition of dislocation in (16.4.2b) one can determine

$$F = 1 \text{ and } F_1 = 0 \qquad (16.4.3)$$

However, the unknown constant D_1 still cannot be determined. To determine the value of D_1, let us give the stress field. This can be done by substituting (16.4.2b) into (16.1.7), yields the nonzero components

$$\left. \begin{aligned} \sigma_{zx} = \sigma_{xz} &= \frac{b}{2\pi} \frac{2K_1 D_1 \left(x^2 - y^2\right)}{\left(x^2 + y^2\right)^2} \\ \sigma_{zy} = \sigma_{yz} &= \frac{b}{2\pi} \frac{4K_1 D_1 yx}{\left(x^2 + y^2\right)^2} \end{aligned} \right\} \qquad (16.4.4)$$

In the book on the mathematical theory of solid quasicrystals, e.g., Fan [11], the dislocation solutions are developed, in which the problems of crystals are naturally included (because if the phason field is absent, the elasticity of quasicrystals is reduced to elasticity of crystals). The theory demonstrated that the higher partial differential equations describing dislocations need appropriate additional boundary conditions except for the dislocation condition; otherwise the boundary value problem will not be well-defined. This is valid for boundary value problem (16.3.1) too. It is sufficient to determine the unknown constants F and D_1 with the aid of two conditions in expressions (16.3.1) except the condition at infinity. Because $\sigma_{zz}(x, 0) = 0$ is automatically satisfied, we must now search for an additional condition determining the third constant. We here use the minimization of dislocation energy, i.e.,

$$\frac{\partial U}{\partial D_1} = 0 \qquad (16.4.5)$$

where the energy will be given in the following (i.e., Eqs. (16.5.2)–(16.5.4)) we have

$$D_1 = \frac{-\frac{8}{3}\pi\alpha}{(R_0 + r_0)\left(\frac{\pi}{4}\alpha\beta + \frac{b^2 K_1}{8\pi}\right)\ln\frac{R_0}{r_0} + \frac{\pi}{320}\alpha\gamma(R_0 - r_0)} \qquad (16.4.6)$$

in which

$$\left. \begin{aligned} \alpha &= \left(\frac{b}{2\pi}\right)^4 \rho_0 B' \\ \beta &= 2 + \frac{32\pi^2}{3} \\ \gamma &= 75 - 160\pi^2 + 256\pi^4 \end{aligned} \right\} \qquad (16.4.7)$$

Comparing Eqs. (16.4.2b) and (16.2.2), one can find the solution given by Kleman [3] and Pershan [4] is only one of the terms of the present solution. In other words, the classical solution is the zero-order approximation of the present solution. In particular, the classical solution does not induce any stresses, or the dislocation causes a stress-free state, while according to our solution, the stress field exits, and exhibits a square singularity near the dislocation core. This singularity is also different from the stress field induced by a screw dislocation in conventional crystals. For the latter, the stress field has a r^{-1} singularity, rather than r^{-2} singularity. In addition, this singularity is also different from the square-root singularity near a crack tip in conventional solid.

16.5 Mathematical Mistakes of the Classical Solution

In crystal elasticity (or classical elasticity) the screw dislocation problem is formulated by

$$\left.\begin{array}{l} \nabla^2 u^{(c)} = 0 \\[6pt] (x^2 + y^2)^{1/2} \to \infty : \quad \sigma_{ij}^{(c)} = 0 \\[6pt] \int_\Gamma du^{(c)} = b^{(c)} \end{array}\right\} \tag{16.5.1}$$

and the superscript (c) represents field variables and Burgers vector magnitude of crystal, in which the stresses are

$$\left.\begin{array}{l} \sigma_{zx}^{(c)} = \sigma_{xz}^{(c)} = \mu \dfrac{\partial u^{(c)}}{\partial x} \\[10pt] \sigma_{zy}^{(c)} = \sigma_{yz}^{(c)} = \mu \dfrac{\partial u^{(c)}}{\partial y} \end{array}\right\} \tag{16.5.2}$$

where the μ shear modulus of the crystal.

The solution (16.2.2) is only the solution of boundary value problem (16.5.1). According to the theory of partial differential equations or mathematical theory of elasticity [12, 13], the solution (16.2.2) of boundary value problem (16.5.1) cannot be the solution of boundary value problem (16.3.1) at the same time. The problem cannot be solved by the so-called smallest surface concept, or magnetism analog provided Eqs. (16.1.1) and (16.1.9) (it reduced from (16.1.8)) are invalid. We believe that Eqs. (16.1.1) and (16.1.9) are valid, and the stress field induced by a single straight screw dislocation along the z-axis can be determined by solving Eq. (16.1.9) subjected to boundary conditions given in (16.3.1) and (16.4.5) which is an additional boundary condition. The solution (16.4.2) including (16.4.3) and (16.4.6) is the unique solution of boundary value problem (16.3.1).

16.6   The Physical Mistakes of the Classical Solution

The solution (16.2.2) leads to some physical mistakes too. This can be viewed in the following.

(1)   It leads to a zero stress field. Substituting solution (16.2.2) into Eq. (16.1.7) leads to

$$\sigma_{ij} = 0, \quad i, j = 1, 2, 3 \tag{16.6.1}$$

This is completely wrong physically.

(2)   It leads to wrong energy formulas. The energy induced by the dislocation is one of important aspects of the problem. On the calculation of energy induced by the dislocation, there are many contradictions between de Gennes and Prost [2], Oswald and Pieranski [14], Kleman et al. [7], and Pleiner [8], even if in the monograph [14] there are logic contradiction itself. This shows the difficulty of the problem. According to our understanding, the point of view of Kleman et al. in [7] is correct, though his calculation is not complete, in which there are some mistakes because he used the wrong solution (16.2.2). Adopting the point of view of Kleman et al. [7], the energy consists of three parts: (1) arising from splay, (2) arising from bulk deformation, (3) corresponding to the dislocation core energy, respectively, i.e.,

$$\begin{cases} U_1 = \iint_{A_0} \frac{1}{2} K_1 \left(\nabla^2 u \right)^2 dx dy \\[2ex] U_2 = \iint_{A_0} \frac{1}{2} \rho_0 B' \left[\left(\frac{\partial u}{\partial x} \right)^2 + \left(\frac{\partial u}{\partial y} \right)^2 \right]^2 dx dy \\[2ex] U_3 = \frac{1}{2} \iint_{\Omega} \left(\sigma_{zx} \frac{\partial u}{\partial x} + \sigma_{zy} \frac{\partial u}{\partial y} \right) dx dy \\[2ex] \qquad = \frac{1}{2} \int_{r_0}^{R_0} \int_{0}^{2\pi} r \left(\sigma_{zx} \frac{\partial u}{\partial x} + \sigma_{zy} \frac{\partial u}{\partial y} \right) dr d\theta \end{cases} \tag{16.6.2}$$

$$U = U_1 + U_2 + U_3 \tag{16.6.3}$$

where A_0 represents the integration domain—the total xy-plane, R_0 and r_0 the conventional outer and inner radii in calculating dislocation energy. If substituting solution (16.2.2) into the energy Eq. (16.6.2), one can obtain the wrong results only.

According to Landau-Ginzburg-de Gennes free energy, the first part of Eq. (16.6.2) always vanishes. When substituting solution (16.4.1) including (16.4.2) and (16.4.5) into the first, second, and third parts of (16.6.2), one gets $U_1 = 0$ and

$$U_2 = \frac{\pi}{8}\left(\frac{b}{2\pi}\right)^4 \rho_0 B' \left[\frac{1}{r_0^2} - \frac{1}{R_0^2}\right] + \frac{\pi}{8}\left(\frac{b}{2\pi}\right)^4 \rho_0 B' D_1^2 \left[2 + \frac{32\pi^2}{3}\right] \ln \frac{R_0}{r_0}$$

$$+ \frac{D_1}{240}\frac{\pi}{8}\left(\frac{b}{2\pi}\right)^4 \rho_0 B' \left[\frac{5120}{R_0 + r_0} + 3D_1\left(75 - 160\pi^2 + 256\pi^4\right)\right]\left(R_0^2 - r_0^2\right)$$

$$U_3 = \frac{b^2 K_1 D_1^2}{16\pi} \ln \frac{R_0}{r_0} \tag{16.6.4}$$

The solution (16.2.2) cannot obtain these energy expressions. The solution (16.2.2) does not hold for smectic A liquid crystals, even if under the condition of the continuum model. The invalidity presents not only in the region inside the core of dislocation but also in the region outside the core of dislocation.

16.7 Meaning of the Present Solution

The solution (16.4.2b) connecting (16.4.6) and (16.4.7) overcomes the mistakes of well-known classical solution (16.2.2) mentioned above, and from which we obtain some meaningful and useful results. For example, we can evaluate the dislocation core energy of smectic A liquid crystals, and find that the dislocation energy is correlated to both Young's modulus and splay modulus. Another finding is that the stresses obtained from solution (16.4.3) exhibit singularity near the dislocation core

$$\sigma_{xz}, \sigma_{yz} \sim r^{-2}, \quad r \to 0 \tag{16.7.1}$$

This presents a different singularity as that in solid quasicrystals apart from crystals.

The solution (16.4.2b) given by Fan and Li [15] provides a basis for studying other problems, e.g., crack problem, of smectic A liquid crystals, which will be introduced in the succeeded section.

16.8 Solution of Plastic Crack

Plasticity and crack problems in soft matter are very interesting topics [16–20]. Especially the crack in soft matter is in plastic state in fact, this is a coupling of crack-plasticity. So far there is no plasticity theory of soft matter, there is a lack of condition to study plastic crack. The solutions of dislocation solutions including those given in previous sections are beneficial to study plastic cracks in soft matter.

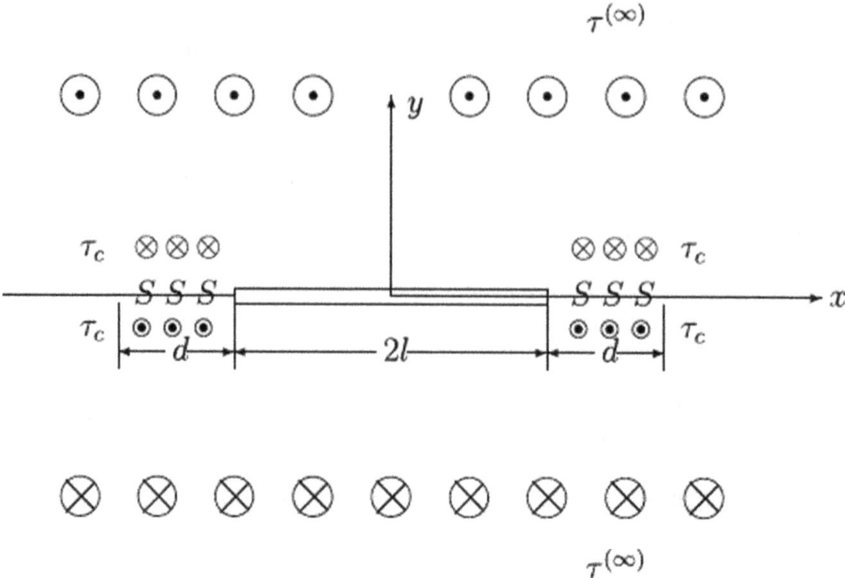

Fig. 16.1 Plastic crack in a soft matter

For simplicity we consider a simple crack model in a soft matter shown (e.g., a smectic A liquid crystal) by Fig. 16.1 the layers are in the.

xy-plane, and a crack dislocation group with length $2l$ along $x-$ axis subjected to uniform shear stress $\sigma_{yz} = \tau^{(\infty)}$ shown in the figure. Because the deformation is assumed to be independent of variable z, the figure depicts any transverse cross section of the body. At the crack tip, there is a screw dislocation pile-up with length d, whose value is temporarily unknown and to be determined, in which the single screw dislocation has Burgers vector $b = (0, 0, b)$. We call the pile-up a sliding dislocation group. Within the zone defined by $y = 0, l \leq |x| \leq l + d$, a counter direction shear stress τ_c is applied, the value of which represents the yield strength of the materials macroscopically. In other words, the dislocation pile-up zone is the plastic zone. The physical meaning of τ_c can be referred to the monograph [2] (p. 499).

The model can be formulated by the following (equivalent) boundary conditions:

$$\begin{cases} (x^2 + y^2)^{1/2} \to \infty : & \sigma_{ij} = 0 \\ y = 0, \quad |x| < l : & \sigma_{yz} = -\tau^{(\infty)} \\ y = 0, \quad l < |x| < l + d : \sigma_{yz} = -\tau^{(\infty)} + \tau_c \end{cases} \tag{16.8.1}$$

The governing equation for the boundary value problem is Eq. (16.1.9). Under boundary conditions (16.8.1) to solve Eq. (16.1.9), but the boundary value problem is not well-conditional, like that of the problem (16.3.1), because the boundary condition is not sufficient to determine solution of governing equation. If we take the dislocation solution (16.4.2b) (connected with (16.4.6) and (16.4.7)), then the problem can be partly solved.

By using dislocation group concept the boundary value problem "(16.1.9) + (16.8.1)" can be transformed to solve the following singular integration equation

$$\int_L \frac{f(\xi)d\xi}{\xi - x} = \frac{\tau(x)}{A} \tag{16.8.2}$$

for a purpose to determine certain key quantities, in which $f(\xi)$ is a dislocation density function unknown, ξ the dislocation source point coordinate, and x the field point coordinate on the real axis, L represents interval $(-(l + d),\, l + d)$, and $\tau(x)$ the shear stress distribution at the region $y = 0,\, |x| \leq l + d$, i.e.,

$$\tau(x) = \begin{cases} -\tau^{(\infty)}, & |x| < l \\ -\tau^{(\infty)} + \tau_c, & l < |x| < l + d \end{cases} \tag{16.8.3}$$

which is given from the physical state of zone, and in (16.8.2) the constant

$$A = \left(\frac{b}{\pi}\right) K_1 D_1 \tag{16.8.4}$$

and D_1 is given by (16.4.5), note that A here is not confused with the same letter in Eq. (16.1.1), but that constant has never been used since then.

In terms of the singular integral equation theory of Muskhelishvili [21] (p. 251), the integral Eqs. (16.8.2) under condition (16.8.3) has the solution [20]

$$\begin{aligned}
f(x) &= -\frac{1}{\pi^2 A}\sqrt{\frac{x + (l + d)}{x - (l + d)}}\int_L \sqrt{\frac{\xi - (l + d)}{\xi + (L + d)}}\,\tau(\xi)\frac{d\xi}{\xi - x} \\
&= -\frac{1}{\pi^2 A}\sqrt{\frac{x + (l + d)}{x - (l + d)}}\left\{i\left[2\tau_c arc\cos\left(\frac{l}{l + d}\right) - \tau^{(\infty)}\pi\right]\right\} \\
&\quad + \frac{\tau_c}{\pi^2 A}\left[arc\cosh\left|\frac{(l + d)^2 - lx}{(l + d)(l - x)}\right| - arc\cos h\left|\frac{(l + d)^2 + lx}{(l + d)(l + x)}\right|\right] \tag{16.8.5}
\end{aligned}$$

(the details of the mathematical calculation are quite lengthy and are omitted here), in which $i = \sqrt{-1}$, and A is defined by Eq. (16.8.4). Because the dislocation density $f(x)$ should be a real function, the factor multiplying the imaginary number i in the first term of right-hand side of formula (16.8.5) must be zero, this leads to

$$2\tau_c \arccos\left(\frac{l}{l+d}\right) - \tau^{(\infty)}\pi = 0$$

i.e.,

$$d = l\left[\sec\left(\frac{\pi\tau^{(\infty)}}{2\tau_c}\right) - 1\right] \tag{16.8.6}$$

This determines the plastic zone size in the matter.

From solution (15.8.5) we evaluate the amount of dislocations $N(x)$ such as

$$N(x) = \int_0^x f(\xi)\mathrm{d}\xi \tag{16.8.7}$$

Substituting (16.8.5) (coupled with (16.8.6)) into (16.8.7) we can get values of $N(l+d)$ and $N(l)$, so the amount of dislocation movement is

$$\delta = b[N(l+d) - N(l)] = \frac{2bl\tau_c}{\pi^2 A}\left(\ln\frac{l+d}{l}\right) = \frac{2\tau_c l}{\pi\rho_0 B'}\ln\sec\left(\frac{\pi\tau^{(\infty)}}{2\tau_c}\right) \tag{16.8.8}$$

This is the crack tip opening (tearing) displacement, which is an important parameter.

We suggest the following rupture criterion

$$\delta = \delta_c \tag{16.8.9}$$

which can be used for determining the thermodynamic stability/instability of the material, δ_c is the critical value of the crack tip sliding displacement, can be measured by experiments, and is a material constant of the liquid crystals. The Eq. (16.8.9) describes a critical state of equilibrium of the plastic crack. When $\delta < \delta_c$, the crack does not propagate but when $\delta > \delta_c$, the crack will propagate. By using this criterion, the limiting value of the applied stress $\tau^{(\infty)}$ or the limit value of the crack size l can be determined.

The above treatment is a macro-description (or the continuum model), but a micro-description (or a micro-mechanism) can be given as follows.

By introducing the de Gennes theory (refer to [2]), the yield stress is

$$\tau_c \sim \frac{\pi\gamma_0^2}{a_0 k_B T \ln(v_0/v_1)} \tag{16.8.10}$$

where

$$\gamma_0 \approx \frac{1}{2}\sqrt{K_1 B}\, a_0^2/\varepsilon, \quad \varepsilon \sim \sqrt{\frac{K_1}{B}}, \quad v_0 = 10^{33}\,\mathrm{s}^{-1}\mathrm{cm}^{-3}, \quad v_1 = 1\,\mathrm{s}^{-1}\mathrm{cm}^{-3}$$

$$(16.8.11)$$

and a_0 represents the thickness of the layer of the smectic, whose value is almost equivalent to the magnitude of a Burgers vector, k_B is the Boltzmann constant, T the absolute temperature, v_0 and v_1 the fluctuation frequencies, respectively.

Substituting expression (16.8.10) into Eqs. (16.8.6) and (16.8.8) respectively one reveals the physical sense of the plastic zone size (or dislocation sliding width) and crack tip sliding displacement (or amount of dislocation movement) in-depth. Owing to the limitation of space a detailed discussion is not given here.

Crack and plasticity are difficult topics in liquid crystals. One of the reasons for this lies in there being lack no theory of plasticity, at least, there is an absence of macroscopic plastic constitutive equation so far. Here we have adopted a phenomenological model to discuss the problem. In this way, we obtain some physical quantities for describing the coupling between rupture and plasticity. In particular, the solution given in this study is exactly satisfying the Peach-Koehler force rule. The methodology developed here is generally effective for other problems in smectics and other classes of liquid crystals. The work is given by Ref [20].

This chapter hints us that one of the distinctions between liquid crystals and quasicrystals generated from liquid crystals lies in the phason elementary excitation. If the elementary excitation is absent, then the latter reduces to the former.

In the meantime, this chapter straightforwardly shows the ill-conditional boundary value problems twice. This indicates, the well-conditionality of initial-boundary value problems of governing equations in hydrodynamics of soft matter has not been proved, i.e., the mathematical solvability of these problems has not been studied.

References

1. Landau, L.D., Lifshitz, E.M.: Theory of Elasticity. Pergamon Press, Oxford (1980)
2. de Gennes, P.D., Prost, J.: The Physics of Liquid Crystals. Clarendon, London (1993)
3. Kléman, M.: Linear theory of dislocations in a smetic A. Journal de Physique **35**, 595–600 (1974)
4. Pershan, P.S.: Dislocation effects in smectic-A liquid crystals. J. Appl. Phys. **45**, 1590–1604 (1974)
5. Pleiner, H.: Structure of the core of a screw dislocation in smectic A liquid crystals. Liq. Cryst. **1**, 197–201 (1986)
6. Kralj, S., Sluckin, T.J.: Landau-de Gennes theory of the core structure of a screw dislocation in smectic A liquid crystals. Liq. Cryst. **18**, 887–902 (1995)
7. Kleman, M., Williams, C.E., Costello, M.J., et al.: Defect structures in isotropic smectic phases revealed by freeze-fracture electron microscopy. Phil. Mag. A **35**, 33–56 (1997)
8. Pleiner, H.: Dynamics of a screw dislocation in smectic A liquid crystals. Phil. Mag. A **54**, 421–439 (1986)
9. Kralj, S., Sluckin, T.J.: Core structure of a screw disclination in smectic A liquid crystals. Phys. Rev E **48**, R3244 (1983)

10. Pleiner, H.: Energetics of screw dislocations in smectic A liquid crystals. Liq. Cryst. **3**, 249–258 (1998)
11. Fan, T.Y.: Mathematical Theory of Elasticity of Quasicrystals and Its Applications. Science Press, Beijing/Springer-Verlag, Heidelberg, 1st edn (2010), 2nd edn (2016)
12. Courant, R., Hilbert, D.: Methods of Mathematical Physics. Interscience Publishers, New York (1953)
13. Muskhelishvili, N.I.: Some Basic Problems of Mathematical Theory of Elasticity, English translation by Radok, J.R.M., Groningen, Noordhoff (1953).
14. Oswald, P., Pieranski, P.: Smectic and Columnar Liquid Crystals. Taylor & Francis, London (2006)
15. Fan, T.Y., Li, X.F.: The stress field and energy of screw dislocation in smectic A liquid crystals and mistakes of the classical solutions. Chin. Phys. B, **23**, 046102 (2014)
16. Bohn, S., Pauchard, L., Couder, Y.: Hierarchical crack pattern as formed by successive domain divisions. I. Temporal and geometrical hierarchy, Phys. Rev. E, **71**, 046214 (2005)
17. Tirumkudulu, M.S.: Cracking in drying latex films. Langmuir **21**, 4938–4948 (2005)
18. Yow, H.N., Goikoetra, M., Goehring, L., Routh, A.F.: Effect of film thickness and particle size on cracking stresses in drying latex films. J. Colloid Interface Sci. **352**, 542–548 (2010)
19. van der Kooij, H.M., Sprakel, J.: Watching paint dry; more exciting than it seems. Soft Matter, **11**, 6353–6359 (2015)
20. Fan, T.Y., Tang, Z.Y.: A model of crack based on dislocations in smectic A liquid crystals. Chin. Phys. B, **23**, 106103(2014)
21. Muskhelishvili, N.I.: Singular Integral Equations, English translation by Radok, J.R.M., Groningen, Noordhoff (1954)

Chapter 17
Conclusion Remarks

The modification and supplementary contents in the new edition have been introduced in the text.

To some extent, the dynamics of soft-matter quasicrystals is a continuation, extension, and development of hydrodynamics of solid quasicrystals by Lubensky and his group at the University of Pennsylvania, the USA. In the meantime, the symmetry group theory by Hu and his group in Wuhan University, China presents its importance as well. The facts show the theory of solid quasicrystals is fundamental for the present study.

In this book, we preliminarily solve the initial- or boundary- or initial, and boundary-value problems of the final governing equations of the generalized dynamics of soft-matter quasicrystals. The solutions verify the theory at first and give some applications in the meantime. The constructing of the theory and the solving of the initial- or boundary- or initial, and boundary-value problems of the final governing equations of the generalized dynamics are still at a preliminary level because there is a lack of experimental data and experimental verification. It is noted that however a part of computational results have been compared with those of the classical fluid dynamics or solid quasicrystals.

More comprehensive, strict, and reliable verifications using experiments, in particular, are required to be further carried out.

The photonic band-gap for both solid and soft-matter quasicrystals is very interesting, which presents an important application in quantum electronics and integrated photonics. The related discussion is presented in Chap. 14. The applications in thermodynamic stability of soft-matter quasicrystals and in general soft matter are also discussed, for example, in Chaps. 13, 15, and 16 respectively.

Some important topics (e.g., the correlation between Frank-Kasper phase and soft-matter quasicrystals), which we have been concerned about, are not given detailed discussion. In addition, the most interesting event--observation of tenfold symmetry soft-matter quasicrystals is not introduced in detail because the experimental results have not been openly reported yet.

© The Author(s), under exclusive license to Springer Nature Singapore Pte Ltd. 2022

T.-Y. Fan et al., *Generalized Dynamics of Soft-Matter Quasicrystals*, Springer Series in Materials Science 260, https://doi.org/10.1007/978-981-16-6628-5_17

Correction to: Introduction on Elasticity and Hydrodynamics of Solid Quasicrystals

Correction to:
Chapter 3 in: T.-Y. Fan et al., *Generalized Dynamics of Soft-Matter Quasicrystals*, Springer Series in Materials Science 260, https://doi.org/10.1007/978-981-16-6628-5_3

The original version of the book was inadvertently published with incorrect equation in Chapter 3. The book and the chapter have been updated with the changes.

The updated version of this chapter can be found at
https://doi.org/10.1007/978-981-16-6628-5_3

Printed in the USA
CPSIA information can be obtained
at www.ICGtesting.com
LVHW011351110923
757743LV00005B/247